PLANETARY AND LUNAR COORDINATES

FOR THE YEARS

2001 — 2020

HM NAUTICAL ALMANAC OFFICE

Rutherford Appleton Laboratory

LONDON: THE STATIONERY OFFICE

ISBN 0 11 887312 1

First published 2000.

Typeset using TEX

Printed in the United Kingdom for The Stationery Office
TJ002075 019585 541875 08/00

PREFACE

This volume of *Planetary and Lunar Coordinates for the years 2001–2020* follows the same general arrangement as the previous volume for the years 1984–2000. The coordinates are based on the planetary and lunar ephemerides given in DE245/LE245 respectively generated using numerical integration by the Jet Propulsion Laboratory, California. DE245 provides improved positions for the outer planets over those given in DE200 used in the previous edition. The phenomena section has been expanded to give more comprehensive graphical information on the visibility of both solar and lunar eclipses. The coordinates of the planets and the Moon have been moved from the paper edition and put on a CD-ROM in both ASCII format and Adobe Acrobat portable document format files.

The primary aim of this publication is the provision of low-precision data for planning purposes in advance of the annual publication of *The Astronomical Almanac* and of the booklet *Astronomical Phenomena* which contains extracts from it. It is hoped that it will be found useful for many different applications, including the computation of the orbits of comets and minor planets, which was the main purpose of the earlier volumes.

This volume has been prepared at the Rutherford Appleton Laboratory, under the direction of the head of HM Nautical Almanac Office, P.T. Wallace. The staff concerned were S.A. Bell, C.Y. Hohenkerk and D.B. Taylor.

RICHARD HOLDAWAY
Director

Space Science and Technology Department
Rutherford Appleton Laboratory
Chilton, Didcot
OX11 0QX
England

2000 June

Acknowledgements

We are grateful to the Astronomical Applications Department of the US Naval Observatory for their continuing support in the production of this volume. We would especially like to thank Dr Alan Fiala and Mr John Bangert for the use of their computer code for calculating the circumstances of solar and lunar eclipses.

USEFUL WORLD WIDE WEB ADDRESSES ON THE INTERNET

Please refer to the relevant World Wide Web address for further details about the publications and services provided by the following organisations.

- H.M. Nautical Almanac Office at http://www.nao.rl.ac.uk/nao/
- U.S. Naval Observatory at http://aa.usno.navy.mil/AA/
- The Stationery Office (UK) at http://www.itsofficial.net/
- U.S. Government Printing Office at http://www.access.gpo.gov/
- Willmann-Bell Inc. at http://www.willbell.com/
- University Science Books at http://www.uscibooks.com/
- Earth and Sky at http://www.earthandsky.co.uk/

CONTENTS

RELATED PUBLICATIONS

The Astronomical Almanac contains ephemerides of the Sun, Moon, planets and their natural satellites, as well as data on eclipses and other astronomical phenomena.

Astronomical Phenomena† contains data on the principal astronomical phenomena of the Sun, Moon and planets (including eclipses), the times of rising and setting of the Sun and Moon at latitudes between S 55° and N 66°, and data on the calendar.

The Nautical Almanac contains ephemerides at an interval of one hour and auxiliary astronomical data for marine navigation.

Sight Reduction Tables for Air Navigation (AP3270 (NP 303)), 3 volumes. Volume 1, selected stars for epoch 2000·0, containing the altitude to 1′ and true azimuth to 1° for the seven stars most suitable for navigation, for the complete range of latitudes and hour angles of Aries. Volumes 2 and 3 contain values of the altitude to 1′ and azimuth to 1° for integral degrees of declination from N 29° to S 29°, for the complete range of latitudes and for all hour angles at which the zenith distance is less than 95° providing for sights of the Sun, Moon and planets.

All the above publications are prepared jointly by H. M. Nautical Almanac Office, and the Nautical Almanac Office of the United States Naval Observatory, and are published jointly by TSO and the United States Government Printing Office.

†*Astronomical Phenomena* is no longer published by TSO. However, it is available in the UK from Earth and Sky, Field View Astronomy Centre, East Barsham, North Norfolk NR21 0AR, telephone +44 (0) 1328 820083 and their Internet address is www.earthandsky.co.uk.

NavPac and Compact Data 2001–2005. The positions for Sun, Moon, navigational planets and bright stars (to navigational accuracy or better) are mainly represented in the compact form of polynomial coefficients for use with small programmable calculators or personal computers. This volume contains algorithms required by navigators and astronomers and includes a CD-ROM with the software package NavPac for an IBM PC or compatible, together with the ASCII files containing the printed ephemerides.

Sight Reduction Tables for Marine Navigation (NP 401), 6 volumes. This series is designed to effect all solutions of the navigational triangle, given two sides and the included angle to find the third side and an adjacent angle; the tables are arranged to facilitate rapid position finding and are intended for use with *The Nautical Almanac*, Explanatory material and auxiliary tables are included in all volumes.

The Star Almanac for Land Surveyors contains the Greenwich hour angle of Aries and the position of the Sun, tabulated for every six hours, and represented by monthly polynomial coefficients. Positions of all stars brighter than magnitude 4·0 are tabulated monthly to a precision of $0^{s}.1$ in right ascension and 1″ in declination. Coefficients representing the data are also available, to those who have purchased the book, in ASCII files and may be downloaded from our web page www.nao.rl.ac.uk/online. This publication is available from TSO and from UNIPUB, 4611/F Assembly Drive, Lanham, MD 20706-4391, U.S.A.

Further details about the publications and services provided by H. M. Nautical Almanac Office may be found via the Internet, on the World Wide Web (WWW).

EXPLANATION

1. INTRODUCTION

This Explanation is limited to giving brief descriptions of the tabulations in this volume, and indications of how they may be used in the derivation of related data. Additional background information and more precise data for the year concerned are given in *The Astronomical Almanac*, while the earlier volume *Planetary Coordinates, 1960–1980* contains a detailed account of the methods for the integration of the orbits of comets and minor planets. Details of our related publications are given opposite.

All the times in this volume may be regarded as being in universal time (UT), or mean solar time on the meridian of Greenwich, since the difference between UT and the dynamical time scale of the fundamental ephemerides is expected to be about only 1 minute during the period 2001–2020. The argument is always given for midnight (0^h) at the beginning of the calendar day on the Greenwich meridian.

All the coordinates in this volume are referred to the mean equinox and equator, or ecliptic, of the standard epoch of J2000·0. For many purposes the effects of precession can be ignored, but formulae and auxiliary tables are given so that they may be reduced to the mean equinox and equator (or ecliptic) of date. The effects of nutation and aberration are small to the precision of the tabulation; for the inner planets the combined effect may occasionally reach about $0^{h}.001$ in right ascension and $0^{\circ}.01$ in declination.; for the outer planets the maxima are smaller.

The following notes may be found helpful by those who are not familiar with the technical terms used in this explanation, but otherwise further guidance should be sought from, for example, text-books on astronomy.

The *celestial equator* is defined to lie in the plane that is normal to the axis of rotation of the Earth. *Right ascension* is measured (from 0^h to 24^h) around this axis in the right-handed sense (eastwards) with respect to the north pole. *Declination* is measured (from 0° to $\pm 90^\circ$) from the equator, northern declinations being positive, and southern declinations being negative. The *ecliptic* is defined to be the plane of the orbit of the Earth around the Sun. *Celestial longitude* is measured (from 0° to 360°) around the normal to the ecliptic in the right-handed sense (eastwards) with respect to the pole of the ecliptic lying north of the celestial equator. *Celestial latitude* is measured (from 0° to $\pm 90^\circ$) from the ecliptic, with the same sign convention as for declination. The *equinox*, which is the zero point of both right ascension and longitude, is defined to be in the direction of the ascending node of the ecliptic on the equator (i.e. the direction at which the declination of a point moving in a positive direction on the ecliptic changes from negative to positive). The adjective "Celestial" is usually omitted in the rest of the volume.

Owing to the effects of *precession* and *nutation*, the equator and ecliptic, and hence the equinox and poles, are continuously in motion, and so the current celestial coordinates of a "fixed" direction change continuously. The motion of the equator is primarily due to the action of the Sun and Moon, and so is usually referred to as the luni-solar precession and nutation, while the motion of the ecliptic is primarily due to the perturbing action of the planets and so is known as planetary precession.

The terms *heliocentric* and *geocentric* are used to indicate whether the origin of the coordinate system is assumed to be at the centre of the Sun or the centre of the Earth. To the precision of the coordinates in this volume the parallactic effect due to displacement from the centre of the Earth of an observer on the surface of the Earth can be ignored for the outer planets.

2. ECLIPSES AND PHENOMENA (Pages 1–133)

Eclipses. The purpose of this section is to show the visibility of both solar and lunar eclipses over the next twenty years. The first page lists the eclipses in chronological order and by type, followed by visibility maps and timing information. More thorough predictions can be found in *The Astronomical Almanac* and *Astronomical Phenomena* for the relevant years.

All of the maps provided in this section adopt an orthographic projection. In the case of the solar eclipse diagrams, the central longitude and latitude are given below the map. The assumed value of Delta T (terrestrial time minus universal time) is given. All timing information is given in terms of universal time (UT).

Solar Eclipses. The upper panel of the solar eclipse pages shows the visibility of the eclipse. Each diagram shows at least two types of shaded area. The region shown in light gray is the part of the Earth from which the whole eclipse can be observed. From the stippled areas, only part of the eclipse can be seen. Where both northern and southern penumbral limits exist, these two stippled areas form two separate loops. If only one penumbral limit exists these loops are connected. The western stippled area indicates the region where the Sun rises during the eclipse and the eastern stippled area where the Sun sets during the eclipse. The situation is somewhat more complicated for an eclipse such as that of 2006 September 22 where the eastern stippled area forms a distorted figure of eight. Here the Sun sets during the eclipse in the upper half of the figure of eight and rises during the eclipse in the lower half.

From the innermost half of these stippled areas more than half of the eclipse can be observed. From the outermost areas less than half of the eclipse can be seen. The first contact of the Earth with the penumbral shadow is denoted by (⊕). Similarly, the last contact point is denoted by (⊙). These two points indicate the direction of motion of the shadow across the surface of the Earth.

Where the eclipse is either annular or total, a thin dark gray strip indicates the path of annularity or totality respectively. Tick marks are plotted within this region every 15 minutes past the hour. In the case of a partial eclipse, the point of greatest eclipse is denoted by (⊗). A summary of the circumstances of the solar eclipse is given in the lower panel of each page.

Lunar Eclipses. The upper panel of the lunar eclipse pages show the position of the Moon relative to the umbral and penumbral shadows cast by the Earth. The number of Moon symbols depends on the type of eclipse. The ecliptic has been plotted for reference purposes and the direction of motion of the Moon is shown by the arrows.

For a penumbral eclipse, the times of first contact of the Moon with the penumbral shadow (P1), mid-eclipse and last contact with the penumbral shadow (P4) are given.

For a partial eclipse, the times of first contact of the Moon with the umbral shadow (U1) and last contact with the umbral shadow (U4) are given in addition to the information provided for a penumbral eclipse.

For a total eclipse, the times when the Moon enters totality (U2) and leaves totality (U3) are given in addition to the information provided for a partial eclipse.

The lower panel of the lunar eclipse pages shows the visibility of the Moon at different key stages throughout the eclipse. The Moon is above the horizon for all locations on each of the hemispheres of the Earth displayed and is directly overhead at the point denoted by ($\oplus$). Timing information is also given below each map.

Phases of the Moon. The times of the principal phases of the Moon are tabulated in UT to the nearest minute on pages 94–98 for the years 2001–2020. The times are for the instants when the excess of the apparent longitude of the Moon over the apparent longitude of the Sun is 0° (new moon), 90° (first quarter), 180° (full moon) and 270° (last quarter). The dates and times when the Moon is nearest to (perigee) and furthest from (apogee) the Earth are also given.

Seasons. The times of the beginning of the astronomical seasons are tabulated in UT on page 99 for the years 2001–2020. They refer to the instants when the apparent longitude of the Sun is 0° (ascending equinox), 90° (northern solstice), 180° (descending equinox) and 270° (southern solstice). The dates when the Earth is nearest to (perihelion) and furthest from (aphelion) the Sun are also given.

Special phenomena. The table at the foot of page 101 gives the dates on which the Earth passes through the ring-plane of Saturn; the short periods around these dates are the most suitable for the detection of close satellites. The table also includes dates on which the planets Mercury and Venus may be seen in transit across the disc of the Sun. As in the case of a solar eclipse the times and appearance of the transit depend on the position of the observer on the Earth. Further details are given in *The Astronomical Almanac*.

Planetary phenomena. The times in UT of the occurrence of the principal phenomena of the major planets are given to the nearest hour on pages 99–101 for the years 2001–2020. They are given to facilitate the planning of future observations or other studies. They have been computed on the same basis as the corresponding values in the current editions of *The Astronomical Almanac*.

More detailed information relating to the observability of the major planets is given on pages 102–133. The time of Greenwich transit, the elongation from the Sun and the apparent visual magnitude are given for every tenth day for Mercury, Venus, Mars, Jupiter and Saturn, and for every fortieth day for Uranus, Neptune and Pluto. The elongations refer to the apparent angular separation of the centres of the Sun and the planet, and do not in general pass through 0° or 180° as they change from west to east or east to west.

3. COORDINATES OF THE PLANETS (Pages *134–309*)

Coordinate systems. The coordinates of the planets Mercury, Venus, Mars, Jupiter, Saturn, Uranus, Neptune and Pluto are tabulated for the following four coordinate systems;

(a) geocentric equatorial spherical coordinates, i.e. right ascension, α, declination, δ, and light-time, τ;
(b) heliocentric equatorial rectangular coordinates, x, y, z;
(c) geocentric ecliptic spherical coordinates, i.e. longitude, λ, latitude, β, and distance, ρ;
(d) heliocentric ecliptic spherical coordinates, i.e. longitude, l, latitude, b, and distance, r.

The geocentric coordinates of the Sun are tabulated with the geocentric coordinates of Venus and Mars, while the heliocentric coordinates of the barycentre (centre of mass) of the Earth–Moon system are tabulated with the heliocentric coordinates of Venus and Mars.

All of these coordinates are referred to the equinox and equator, or ecliptic, of the standard epoch J2000·0. This implies, for example, that the right ascensions and declinations of the planets will be directly comparable with those of the stars given in new star atlases and catalogues. Corrections for aberration have not been applied in forming these coordinates. Formulae and tables for the conversion of the coordinates to other systems are discussed on pages A9-A12; precise values of the apparent spherical coordinates are given in the annual volumes of *The Astronomical Almanac*.

Form of tabulations. The coordinates of Mercury are tabulated separately on pages *134–164*, since they require an interval of 5 days; all four sets of coordinates are given on each page. For the other planets, the coordinates of only one system are given at each opening as indicated in the list of contents on page v; the coordinates of Venus and Mars, are given, with those of the Sun or Earth-Moon on the left-hand page, while the coordinates of the outer planets are given on the right-hand page; the interval of tabulation is 10 days except for Uranus, Neptune and Pluto, for which an interval of 40 days is suitable. Techniques for the interpolation of these coordinates are discussed on page A13.

The units and precision of the tabulations are, in general, the same as in previous volumes. In particular the heliocentric equatorial rectangular coordinates are tabulated in astronomical units of distance (au) to the third-decimal place for Mercury, to the fourth decimal place for Uranus, Neptune and Pluto, and to the fifth decimal place for the other planets. This precision should be sufficient for most purposes, including the computation of the orbits of comets. The angular coordinates are give to 0°.001 for declination, longitude and latitude and to 0h.000 1 for right ascension, except for Mercury for which one less figure is given.

The unit of tabulation of the light-time, τ, is the microday, μd. This unit has been chosen in preference to the second, s, since it can be used directly in the calculation of the corrections for aberration. The light-time in seconds is approximately one-tenth of the tabulated value. The equivalent distance in astronomical units is given with the tabulations of the geocentric ecliptic spherical coordinates. Light-time and distance in the other units may be calculated from the data on page A1.

The time-argument is given as a Julian date on all pages, and on most pages the equivalent Gregorian calendar date is also given. The tabular values were calculated for 0^h dynamical time, but for most purposes can be considered to be for 0^h UT, since the difference (ΔT) is expected to be about 0d.000 7. A table of ΔT is given on page A8.

Basis of the tabulations. The tabulated coordinates of the planets were derived from fundamental ephemerides computed at the Jet Propulsion Laboratory, Pasadena, California. These ephemerides were obtained as a result of a series of numerical integrations of the equations of motion of the Sun, Moon and planets; the values of the numerical constants and starting values of the coordinates were chosen to give a best fit to a very extensive set of observational data; relativistic effects were taken into account.

Coordinates of the barycentre S_4. The heliocentric equatorial rectangular coordinates of the barycentre S_4 are tabulated on pages *230–245* at an interval of 10 days. The barycentre S_4 is the centre of mass of the Sun and the four inner planets and its coordinates (x_b, y_b, z_b) were calculated from formulae of the type

$$M_4 x_b = m_2 x_2 + m_3 x_3 + m_4 x_4$$

where M_4 is the sum of the masses of the Sun and the four inner planets (in units of the mass of the Sun), and $m_n x_n$ are the mass and x-coordinate of the n-th planet (Venus = 2, etc); the orbital motion of Mercury (n = 1) is ignored. The coordinates are tabulated in units of 10^{10} au for consistency with the previous volume. The maximum displacement of S_4 from the centre of the Sun is so small that the tabulated heliocentric coordinates of the inner planets can also be regarded as being referred to an origin at S_4. The use of these coordinates is explained briefly in the next paragraph.

Indirect attractions. The quantities X, Y, Z tabulated on pages *230–245* with the coordinates of the barycentre S_4 are known as the indirect attractions of the planets on the Sun and may be used to simplify the numerical integration of the equations of motion of, say, a comet. They have been calculated by summing quantities of the form

$$X_n = -k^2 m_n x_n / r_n^3$$

where X_n is the indirect attraction of the n-th planet on the Sun, k is the Gaussian constant of gravitation, and m_n, x_n, r_n are the mass, x-coordinate, and distance from the Sun of the n-th planet. Values are tabulated for the summations for the two most massive planets, Jupiter and Saturn, and for the four planets, Jupiter, Saturn, Uranus and Neptune ($n = 5$ to 8); the unit of tabulation is 10^{-9} au/d^2.

These indirect attractions may be regarded as being multiplied by the factor w^2, where w is the interval in days, so that the unit of tabulation is effectively 10^{-7} au, as in the previous volume.

The indirect attractions of the inner planets are not tabulated in this volume since they do not appear in the equations for the accelerations of the coordinates ($\bar{x}$, $\bar{y}$, $\bar{z}$) of the comet with respect to the barycentre S_4. The perturbations by Pluto are generally negligible and so its indirect attractions are not tabulated.

Use of the tabulations for numerical integration. The tabulations of the rectangular coordinates of the planets and barycentre S_4 and of the indirect attractions are primarily intended for use in the numerical integration of the orbits of comets, etc, using an interval of 10 days and a working unit of 10^{-7} au for the coordinates of the comets. The earlier volume, *Planetary Co-ordinates, 1960–1980* contains an extensive collection of formulae and numerical examples for several different methods of integration and a detailed illustration of the use of the Cowell and Encke methods for which the tables were more particularly designed. The tabulations in *Planetary and Lunar Coordinates, 1980–1984*, Planetary and Lunar Coordinates, 1984-2000 and this volume are intended for use with these methods, although it is recognised that the details of the techniques require revision in view of the general replacement of mechanical and electromechanical calculators with electronic calculators. The widespread availability of powerful electronic computers capable of integrating the basic equations rigorously and with greater precision has meant, however, that this volume is not likely to be used extensively for orbit computations and so *Planetary Co-ordinates, 1960–1980* should be consulted for further details.

4. COORDINATES OF THE MOON (Pages *310–462*)

Spherical coordinates. Equatorial and ecliptic coordinates for equinox J2000·0 of the Moon are given to low precision at an interval of 1 day on pages *310–462*. Right ascension (α) is given to $0^{\rm h}\!\!.001$, ($= 0\!\!'.9$ at the equator), while declination (δ), longitude (λ) and latitude (β) are given to $0\!\!^\circ.01$ ($= 0\!\!'.6$), but the coordinates cannot be easily interpolated to this precision. The light-time (τ) is given in units of nanodays (1 nd $= 10^{-9}$ d, or about $0\!\!^{\rm s}.000\,1$), while the distance (ρ) is given in units of 10^{-7} au. Formulae and constants for the computation of horizontal parallax and semi-diameter are given with other data for the Moon on pages A1–A2.

Rectangular coordinates. Equatorial rectangular coordinates of the Moon are also given on pages *310–462*; they are referred to the standard equinox of J2000·0 and are given in units of 10^{-7} au. Conversion factors for other units are given on page A1.

The origin of these coordinates is the centre of the Earth, not the centre of mass of the Earth-Moon system. The coordinates of the centres of the Moon and of the Earth referred to Earth-Moon barycentre may be formed by multiplying the tabulated coordinates by

$$+81{\cdot}30/82{\cdot}30 = +0{\cdot}987\,849 \text{ for the Moon}$$
$$\text{and by } - 1{\cdot}00/82{\cdot}30 = -0{\cdot}012\,151 \text{ for the Earth.}$$

These modified coordinates may be added directly to the heliocentric rectangular coordinates of the Earth-Moon Barycentre given on pages *198–229* in order to obtain the heliocentric coordinates of the centres of the Moon and of the Earth.

Elongation and Transit. The last two columns on the pages for the coordinates of the Moon give, respectively, the elongation of the Moon from the Sun at 0^h and the time of transit of the Moon. The elongation is given to 1°. The quantity given is the angular distance between the centres of the Sun and Moon and so does not normally pass through 0° at new moon nor reach 180° at full moon.

The column headed transit may be taken to be the universal time of transit over the Greenwich meridian. An estimate of the universal time of transit over another meridian may be obtained by linear interpolation between the tabular values using an interpolation factor λ, where λ is the longitude of the place west of Greenwich expressed as a fraction of a revolution (24^h or 360°). The value so obtained may be converted to standard time, and is unlikely to be in error by more than a minute. During each lunar month there is one day on which no transit occurs, and there is a discontinuity of 24^h in the tabular values. In all cases the actual difference in universal time between the tabular values is about 25^h.

Basis of the coordinates. The coordinates for the Moon have been derived from the fundamental ephemerides described on page x.

5. AUXILIARY DATA (Pages A1–A13)

Constants related to units. For the computation of motions in the solar system it is customary to use the Gaussian system of astronomical units of distance (length), mass and time. The astronomical unit of mass is the mass of the Sun, the astronomical unit of time is the day (d), and the astronomical unit of distance (au) is such that the Gaussian gravitational constant (k) has the exact value 0·017 202 098 95 in these units. The conversion factors that relate these units to those of the International System of Units (SI) are related to other astronomical quantities, and their values are included in a self-consistent system of astronomical constants. In the system that has been adopted by the International Astronomical Union (IAU) in 1976 the unit of time is a day of 86 400 SI-seconds. Most of the values given on page A1 are taken from the 1976 system, but a few other quantities have been added for the convenience of the users of this volume. A full statement of the IAU (1976) system of astronomical constants is given in *The Astronomical Almanac*.

Masses of the planets and the Moon. The values given in the table for the reciprocal masses of the planetary systems are those used in the computation of the ephemerides in this volume; the values for the Earth, Moon, Jupiter, Saturn and Uranus differ slightly from the values in the IAU (1976) system of astronomical constants, while the value for Pluto is smaller by a factor of about 50.

Other data for the planets and the Moon. A selection of other data on the orbits and physical properties of the planets and the Moon are given on pages A1–A2. The sidereal and synodic periods are mean values; the sidereal period (in tropical years) refers to the heliocentric motion of the planet with respect to the stars, while the synodic period refers to the relative apparent geocentric motions of the planet and the Sun. Two values are given for the apparent angular semi-diameter; the first is the adopted value corresponding to a distance of 1 au while the second corresponds to the value when the planet is in a favourable position for observing. In the case of the inferior planets the value is for a typical distance when the planet is at maximum elongation from the Sun: Venus may be observed easily over a wide range of elongations and the apparent semi-diameter varies between, say 5″ and 35″ during a synodic period. The values for the superior planets refer to the mean distance $(a - 1)$ from the Earth at opposition.

The values for the radii of the planets are in units of 10^6 m and refer to the equators of the planets; they are not always consistent with the adopted values of the angular semi-diameters. The values for the radii and mean densities are based on recent determinations. The rotation periods refer to the solid surface or to the equatorial region, as appropriate. The column headed obliquity refers to the inclination of the equator of the planet to the orbit of the planet.

Similar considerations apply to the data for the Moon.

Orbital elements of the planets. The heliocentric motions of the planets over the period 2001–2020 are represented by four sets of orbital elements given on pages A2–A5. The precision of representation varies from planet to planet. It corresponds to about 0·0001 au or 0°01 for the inner planets, 0·001 au or 0°01 for Mars and Jupiter, and 0·02 au or 0°05 for Saturn, Uranus and Pluto. The precision for Neptune is rather worse, being about 0·06 au or 0°1.

Formulae for the computation of heliocentric rectangular coordinates, and hence of geocentric spherical coordinates from orbital elements are given on pages A6-A7.

Data related to time-scales. A limited selection of data related to time-scales is given on pages A7–A8.

The expression for Greenwich sidereal time in terms of universal time will give a precision of better than 1^s over the period 2001–2020. The maximum difference between mean and apparent sidereal time is $1^s\!.1$ and so can be ignored. The table of local sidereal time at 0^h local mean time gives values to $0^h\!.1$ on the first day of each month; it is intended only to provide a rough guide to the orientation of the celestial sphere. Sufficient space has been left to permit the manuscript addition of the corresponding values for local clock time.

A brief table of the difference ΔT between ephemeris time or dynamical time and universal time is given to indicate how this quantity has varied since 1900 and what values have been adopted in preparing UT ephemerides in this volume. However for the eclipse predicitions a more recent prediction has been used which is given on each map. Observed values and new estimates are given in *The Astronomical Almanac*.

The tabulated values of the equation of time are the average values of the differences between apparent and mean solar time for the beginning and middle of each month. The actual values may differ from these average values by up to $0^m\!.2$.

The dates of the principal standard epochs are also given. it should be noted that in the IAU (1976) system of astronomical constants the standard Julian epoch of J2000·0 and the Julian century are used with both the precessional constants and the orbital elements. The Besselian epoch and the tropical century are no longer used.

Precessional constants. The tabulations of precessional constants in this volume are much less extensive than those in the first three volumes since it is anticipated that most users will prefer to evaluate them directly from simple series rather than interpolate in a printed table. The precision of the constants and formulae corresponds to that of the quantities tabulated in this volume.

In the formulae given on pages A9–A10, the subscript zero always refers to a quantity for the standard epoch J2000·0, while an unsubscripted quantity is for the mean equinox of date.

Nutation. The principal terms in the expression for the nutations in longitude and obliquity are given on page A12 so that the corresponding corrections to geocentric coordinates for mean equinox of date may be computed. The total contribution of the neglected terms is always less that $0^{\circ}\!.0002$.

Aberration. Approximate formulae are given on page A12 for the effects of aberration on the positions of stars and planets.

Interpolation formulae. A brief selection of interpolation formulae is given on page A13 for occasional use with the tabulations in this volume. More extensive collections of formulae and tables are given in *Interpolation and Allied Tables* and *Subtabulation*.

The choice of formula will be affected by the precision required, the number of interpolations to be made, and the calculating facilities available. Differences should be formed if the appropriate order of formula is not known in advance. Providing that the differences are available, interpolation formulae that use differences are usually more convenient than interpolation formulae that use tabular values of the function.

6. THE CD-ROM

'Coordinates of the Planets' and 'Coordinates of the Moon', printed in the previous edition, are not given in the book, but are distributed as Adobe Acrobat portable document format files (pdf). The page numbers of these pages are given in italics, and their filenames with extension of pdf are given in the Contents.

The CD-ROM also contains computer readable ASCII files of these planetary and lunar coordinates. The ASCII files have no column headings, gaps or pagination and Julian dates are given in full. A read.me file giving the structure of these files is provided together with the freely available Adobe Acrobat reader.

LUNAR ECLIPSES

				Page
2001	Jan.	9	T	2
	July	5	P	4
	Dec.	30	Pen	6
2002	May	26	Pen	7
	June	24	Pen	9
	Nov.	† 19	Pen	10
2003	May	16	T	12
	Nov.	† 8	T	14
2004	May	4	T	17
	Oct.	28	T	19
2005	Apr.	24	Pen	21
	Oct.	17	P	23
2006	Mar.	† 14	Pen	24
	Sep.	7	P	26
2007	Mar.	† 3	T	28
	Aug.	28	T	30
2008	Feb.	21	T	33
	Aug.	16	P	35
2009	Feb.	9	Pen	37
	Jul.	7	Pen	38
	Aug.	† 5	Pen	40
	Dec.	31	P	41
2010	June	26	P	43
	Dec.	21	T	45
2011	June	15	T	48
	Dec.	10	T	51
2012	June	4	P	53
	Nov.	28	Pen	55
2013	Apr.	25	P	56
	May	25	Pen	58
	Oct.	† 18	Pen	59
2014	Apr.	15	T	61
	Oct.	8	T	63
2015	Apr.	4	T	66
	Sep.	28	T	68
2016	Mar.	23	Pen	70
	Aug.	18	Pen	71
	Sep.	16	Pen	73
2017	Feb.	† 10	Pen	74
	Aug.	7	P	76
2018	Jan.	31	T	78
	Jul.	27	T	81
2019	Jan.	21	T	84
	Jul.	† 16	P	86
2020	Jan.	10	Pen	88
	June	5	Pen	89
	July	5	Pen	91
	Nov.	30	Pen	92

Total

2001	Jan.	9
2003	May	16
2003	Nov.	† 8
2004	May	4
2004	Oct.	28
2007	Mar.	† 3
2007	Aug.	28
2008	Feb.	21
2010	Dec.	21
2011	June	15
2011	Dec.	10
2014	Apr.	15
2014	Oct.	8
2015	Apr.	4
2015	Sep.	28
2018	Jan.	31
2018	Jul.	27
2019	Jan.	21

Partial

2001	July	5
2005	Oct.	17
2006	Sep.	7
2008	Aug.	16
2009	Dec.	31
2010	June	26
2012	June	4
2013	Apr.	25
2017	Aug.	7
2019	Jul.	† 16

Penumbral

2001	Dec.	30
2002	May	26
2002	June	24
2002	Nov.	† 19
2005	Apr.	24
2006	Mar.	† 14
2009	Feb.	9
2009	Jul.	7
2009	Aug.	† 5
2012	Nov.	28
2013	May	25
2013	Oct.	† 18
2016	Mar.	23
2016	Aug.	18
2016	Sep.	16
2017	Feb.	† 10
2020	Jan.	10
2020	June	5
2020	July	5
2020	Nov.	30

SOLAR ECLIPSES

				Page
2001	June	21	T	3
	Dec.	14	A	5
2002	June	† 10	A	8
	Dec.	4	T	11
2003	May	31	A	13
	Nov.	† 23	T	15
2004	Apr.	19	P	16
	Oct.	14	P	18
2005	Apr.	8	A/T	20
	Oct.	3	A	22
2006	Mar.	29	T	25
	Sep.	22	A	27
2007	Mar.	19	P	29
	Sep.	11	P	31
2008	Feb.	7	A	32
	Aug.	1	T	34
2009	Jan.	26	A	36
	July	† 21	T	39
2010	Jan.	15	A	42
	July	11	T	44
2011	Jan.	4	P	46
	June	1	P	47
	July	1	P	49
	Nov.	25	P	50
2012	May	† 20	A	52
	Nov.	† 13	T	54
2013	May	† 9	A	57
	Nov.	3	A/T	60
2014	Apr.	29	A	62
	Oct.	23	P	64
2015	Mar.	20	T	65
	Sep.	13	P	67
2016	Mar.	† 8	T	69
	Sep.	1	A	72
2017	Feb.	26	A	75
	Aug.	21	T	77
2018	Feb.	15	P	79
	July	13	P	80
	Aug.	11	P	82
2019	Jan.	† 5	P	83
	July	2	T	85
	Dec.	26	A	87
2020	June	21	A	90
	Dec.	14	T	93

Total

2001	June	21
2002	Dec.	4
2003	Nov.	† 23
2006	Mar.	29
2008	Aug.	1
2009	July	† 21
2010	July	11
2012	Nov.	† 13
2015	Mar.	20
2016	Mar.	† 8
2017	Aug.	21
2019	July	2
2020	Dec.	14

Annular-Total

2005	Apr.	8
2013	Nov.	3

Annular

2001	Dec.	14
2002	June	† 10
2003	May	31
2005	Oct.	3
2006	Sep.	22
2008	Feb.	7
2009	Jan.	26
2010	Jan.	15
2012	May	† 20
2013	May	† 9
2014	Apr.	29
2016	Sep.	1
2017	Feb.	26
2019	Dec.	26
2020	June	21

Partial

2004	Apr.	19
2004	Oct.	14
2007	Mar.	19
2007	Sep.	11
2011	Jan.	4
2011	June	1
2011	July	1
2011	Nov.	25
2014	Oct.	23
2015	Sep.	13
2018	Feb.	15
2018	July	13
2018	Aug.	11
2019	Jan.	† 5

T = Total, A = Annular, A/T = Annular-Total, P = Partial, Pen = Penumbral, † Ends on the next day

Total Eclipse of the Moon 2001 January 09

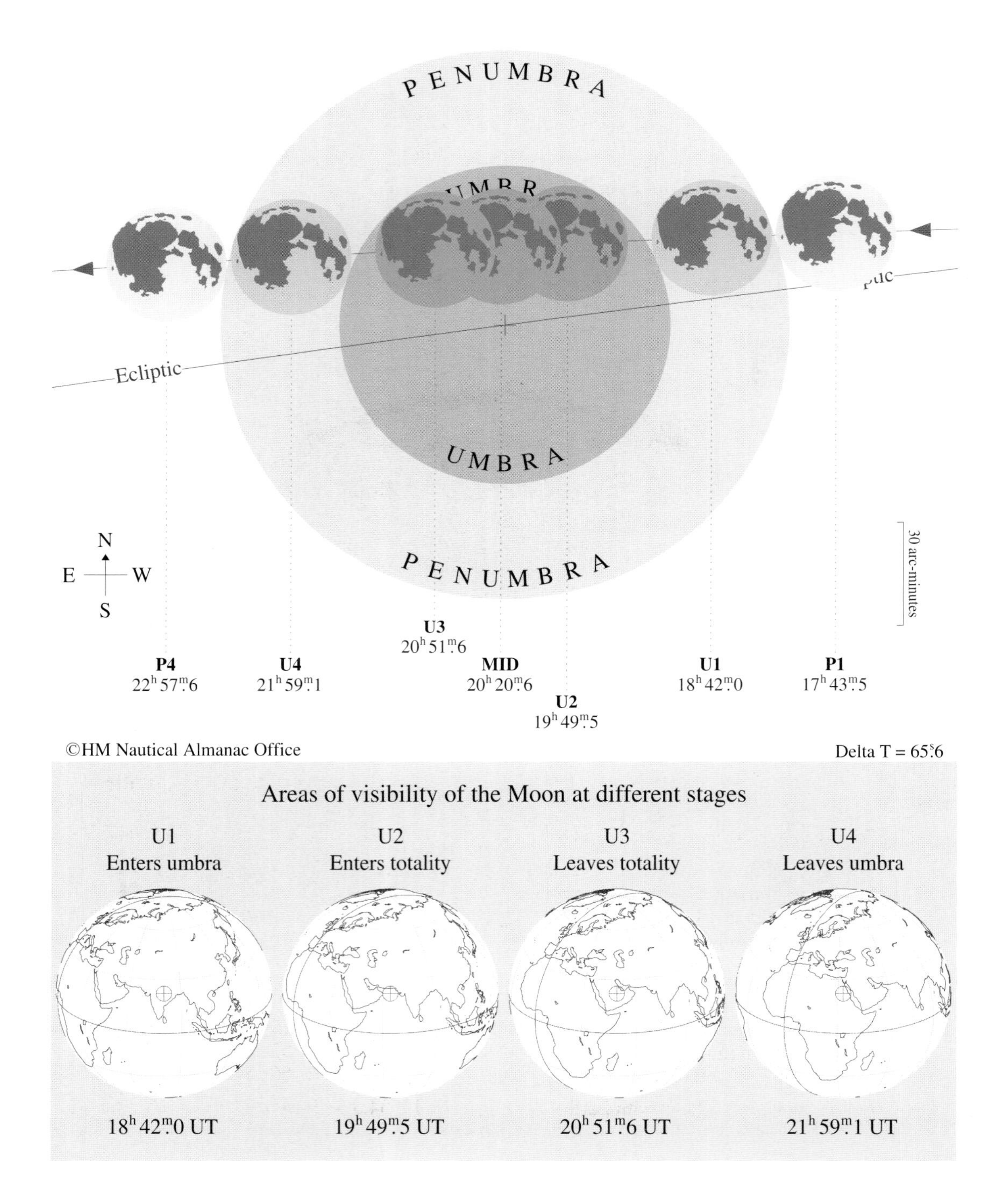

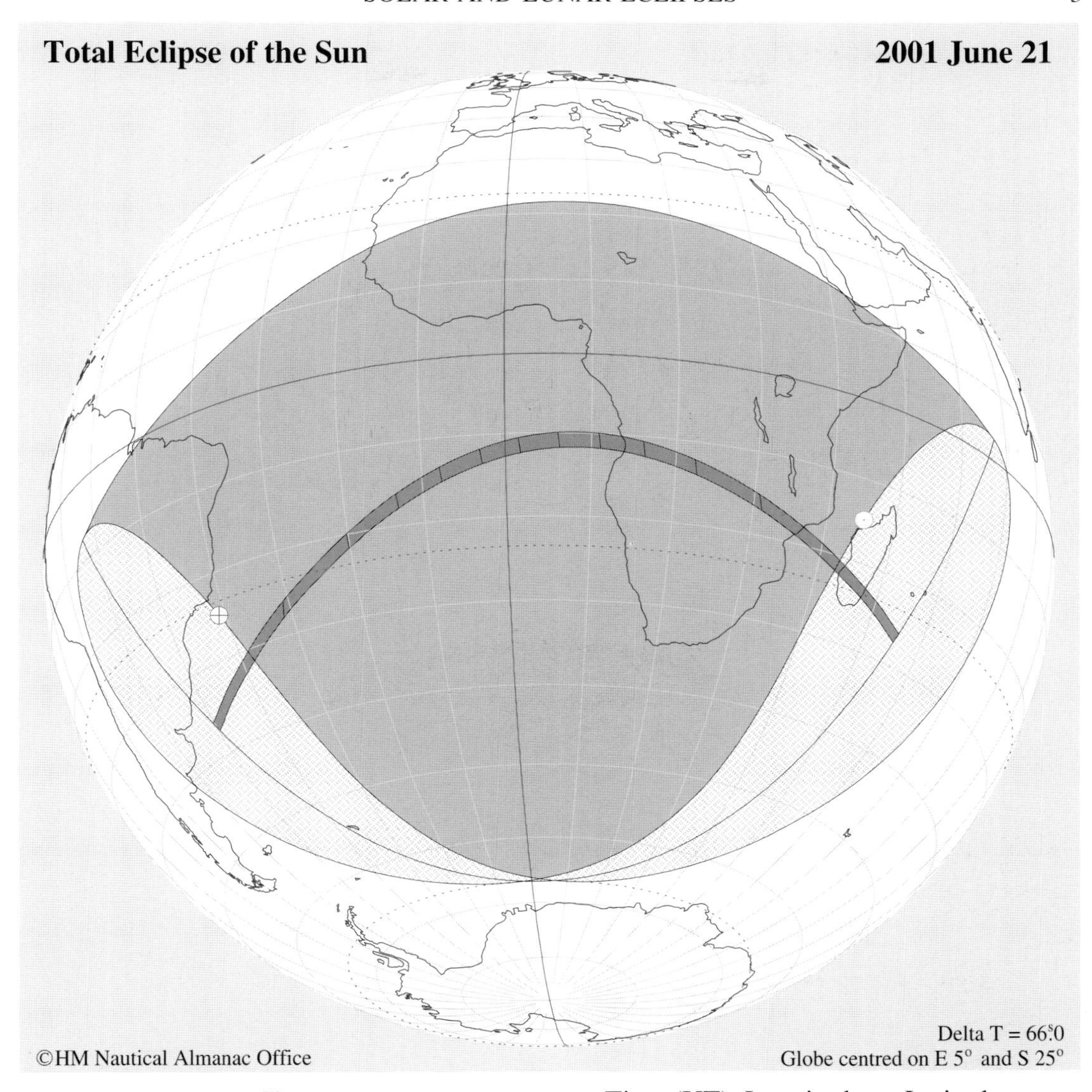

Circumstances	Time (UT)	Longitude	Latitude
	h m	° ′	° ′
⊕ Eclipse begins; first contact with Earth	9 33.0	W 41 05.7	S 25 04.8
Beginning of northern limit of penumbra	10 28.4	W 63 26.2	S 7 22.0
Beginning of northern limit of umbra	10 36.6	W 50 17.3	S 36 04.3
Beginning of centre line; central eclipse begins	10 37.1	W 50 01.4	S 36 37.5
Beginning of southern limit of umbra	10 37.6	W 49 45.3	S 37 10.9
Central eclipse at local apparent noon	11 57.8	E 0 59.6	S 11 35.7
End of southern limit of umbra	13 29.8	E 55 02.4	S 27 20.8
End of centre line; central eclipse ends	13 30.3	E 55 14.3	S 26 45.3
End of northern limit of umbra	13 30.8	E 55 26.1	S 26 10.1
End of northern limit of penumbra	13 38.9	E 66 57.5	N 2 53.0
⊙ Eclipse ends; last contact with Earth	14 34.3	E 45 12.7	S 14 57.5

Partial Eclipse of the Moon 2001 July 05

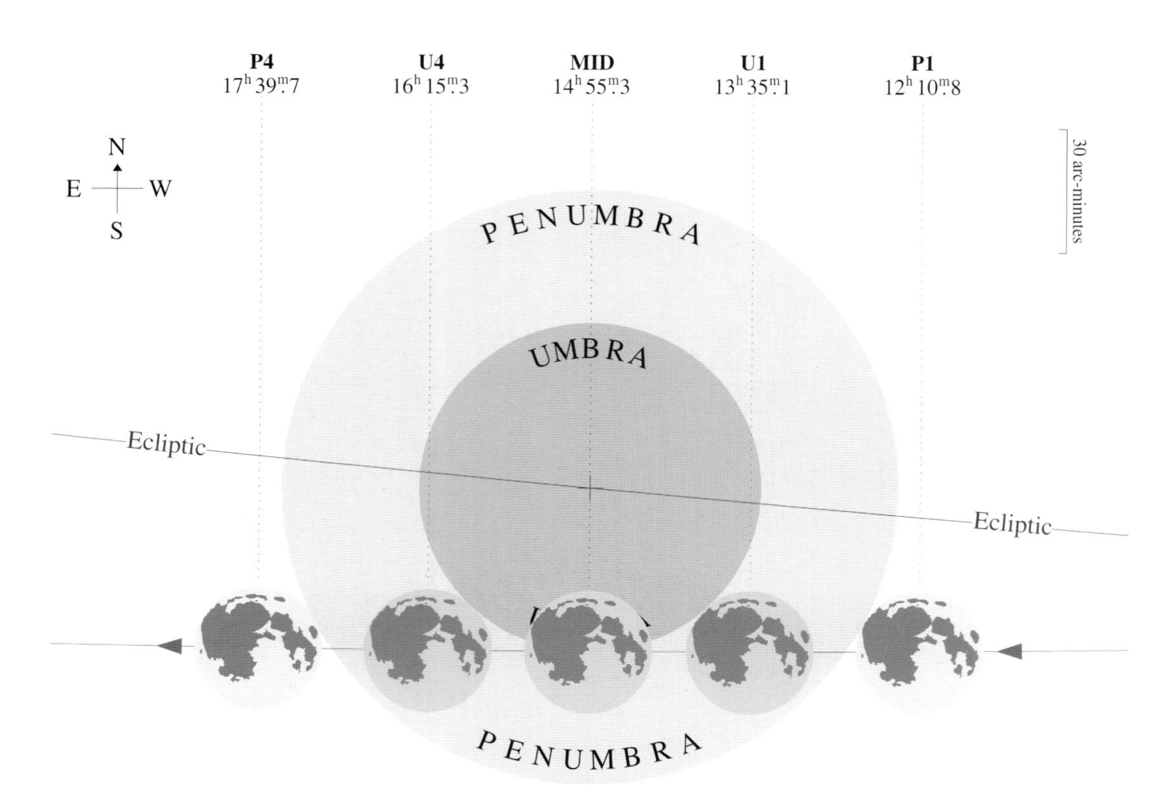

Delta T = $66^{s}.0$

Areas of visibility of the Moon at different stages

U1	MID	U4
Moon enters umbra	Middle of eclipse	Moon leaves umbra
$13^{h}35^{m}.1$ UT	$14^{h}55^{m}.3$ UT	$16^{h}15^{m}.3$ UT

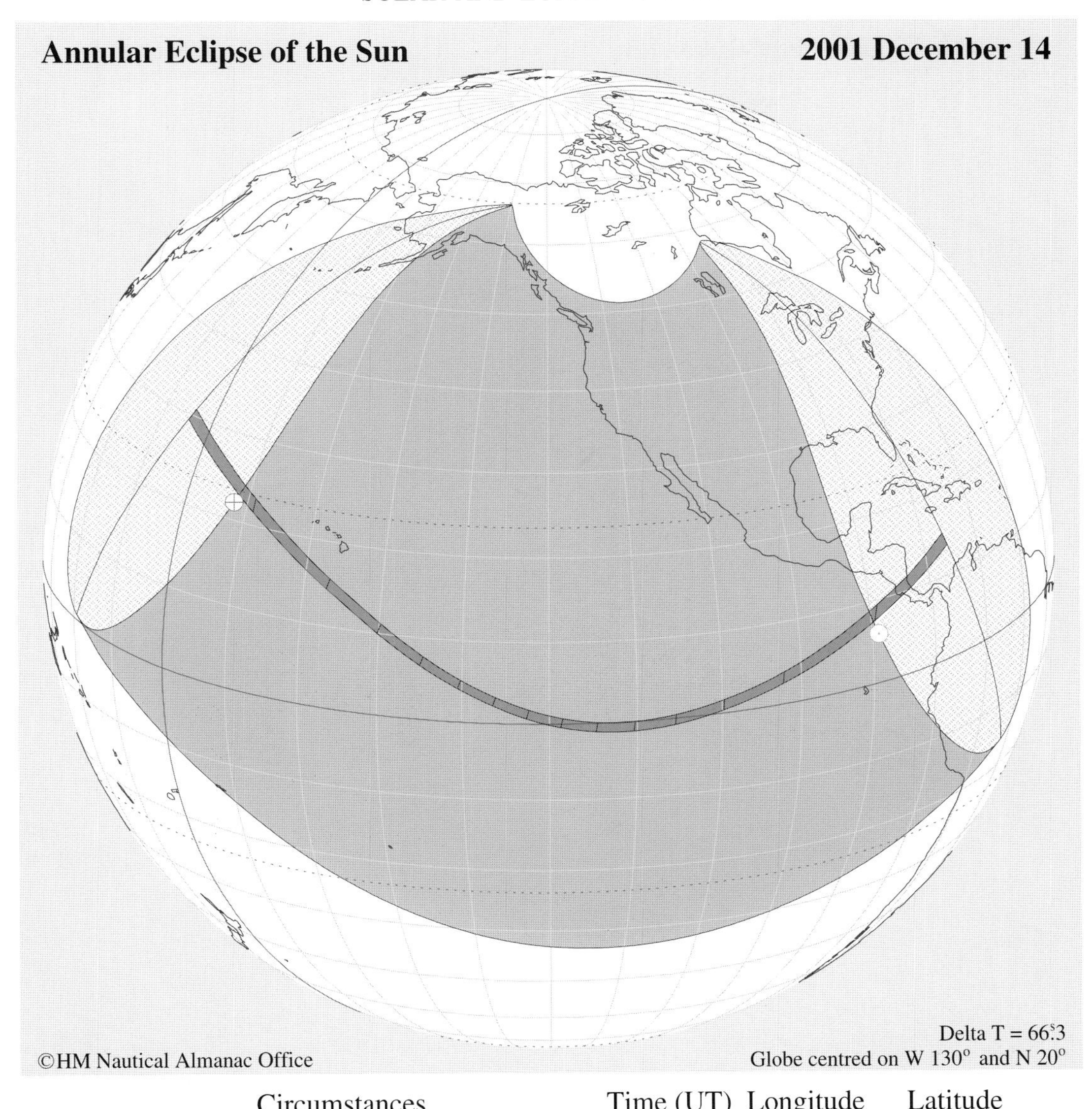

Circumstances	Time (UT)	Longitude	Latitude
	h m	° ′	° ′
⊕ Eclipse begins; first contact with Earth	18 03.3	W172 05.7	N22 02.7
Beginning of southern limit of umbra	19 09.3	E175 22.6	N29 22.8
Beginning of centre line; central eclipse begins	19 09.7	E175 42.2	N30 05.2
Beginning of northern limit of umbra	19 10.1	E176 01.9	N30 47.8
Beginning of southern limit of penumbra	19 12.3	E160 53.7	N 0 36.0
Beginning of northern limit of penumbra	20 21.1	W139 43.8	N66 11.5
Central eclipse at local apparent noon	20 44.8	W132 29.4	N 1 11.3
End of northern limit of penumbra	21 22.9	W 95 16.7	N57 55.2
End of southern limit of penumbra	22 31.5	W 62 16.6	S15 32.3
End of northern limit of umbra	22 33.8	W 76 19.1	N14 58.1
End of centre line; central eclipse ends	22 34.2	W 76 04.0	N14 12.9
End of southern limit of umbra	22 34.5	W 75 48.9	N13 28.0
⊙ Eclipse ends; last contact with Earth	23 40.7	W 89 01.3	N 6 00.0

Penumbral Eclipse of the Moon — 2001 December 30

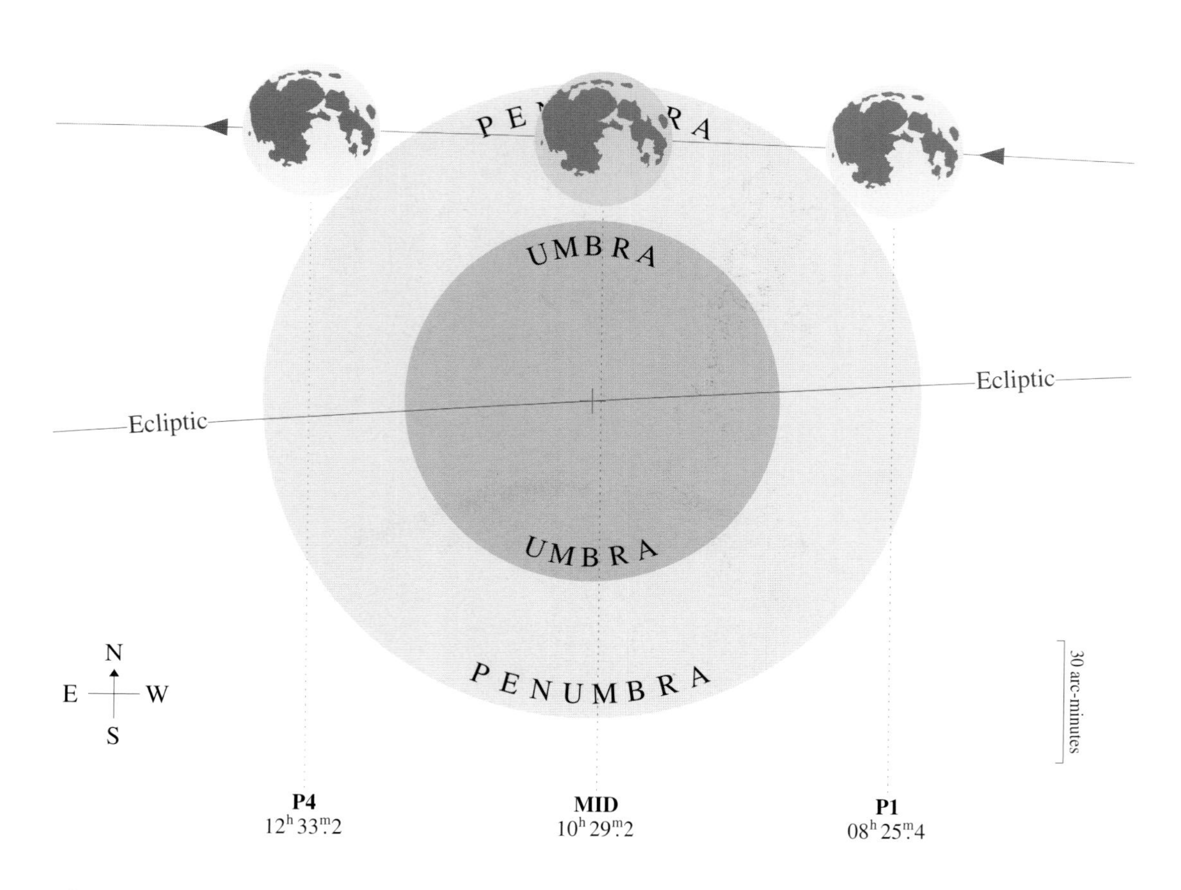

Delta T = $66^{s}\!.4$

Areas of visibility of the Moon at different stages

P1	MID	P4
Moon enters penumbra	Middle of eclipse	Moon leaves penumbra
$08^{h}\,25^{m}\!.4$ UT	$10^{h}\,29^{m}\!.2$ UT	$12^{h}\,33^{m}\!.2$ UT

Penumbral Eclipse of the Moon 2002 May 26

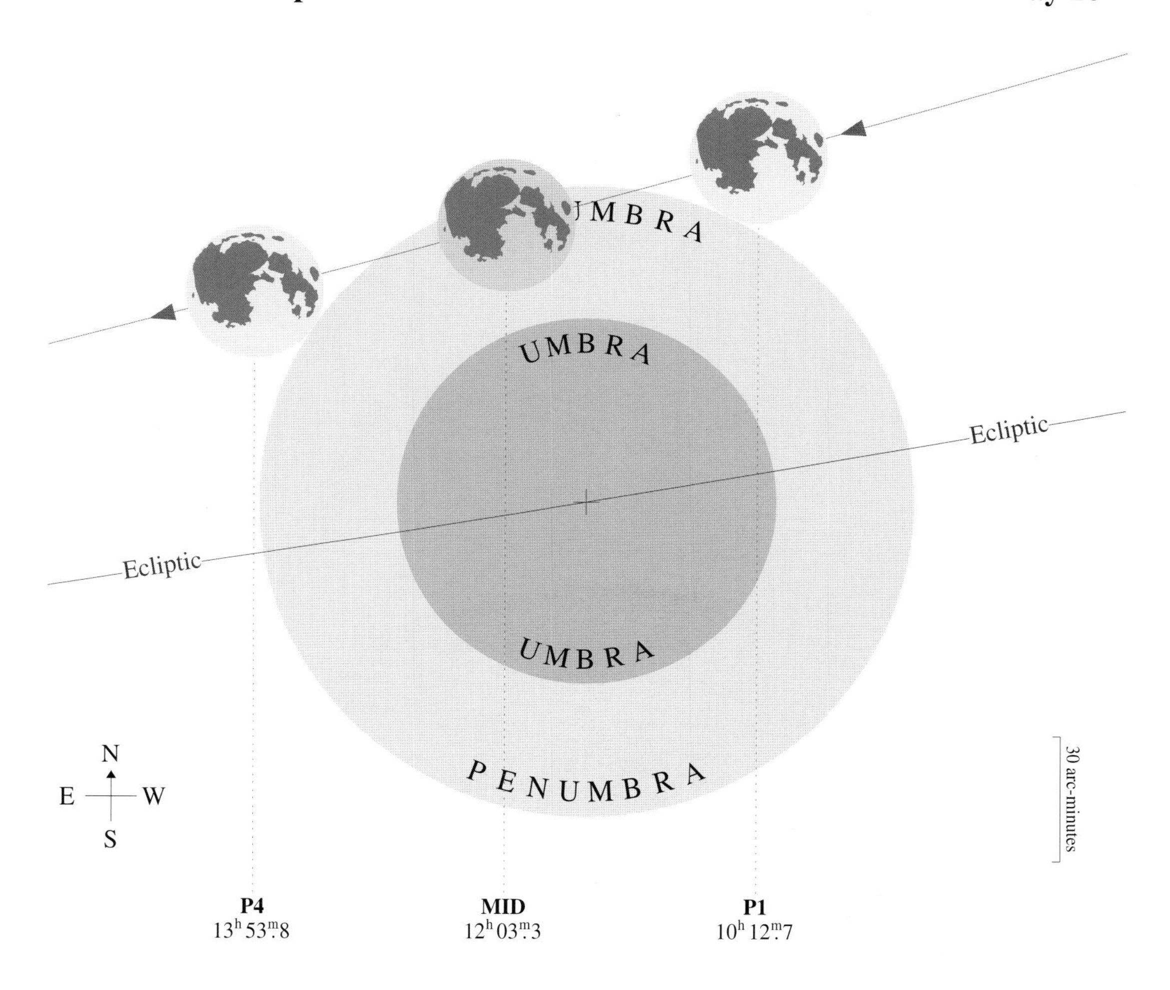

Delta T = 66s.7

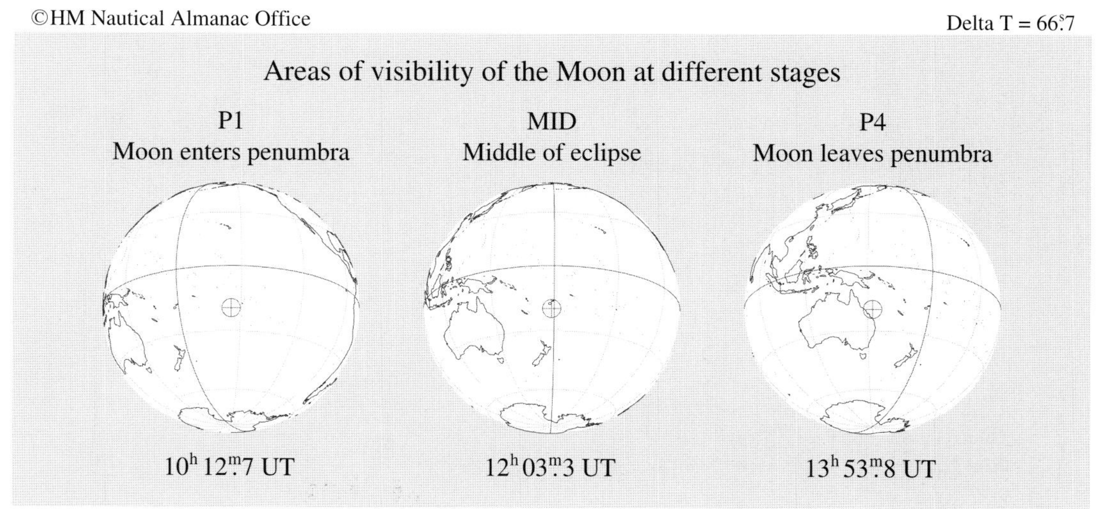

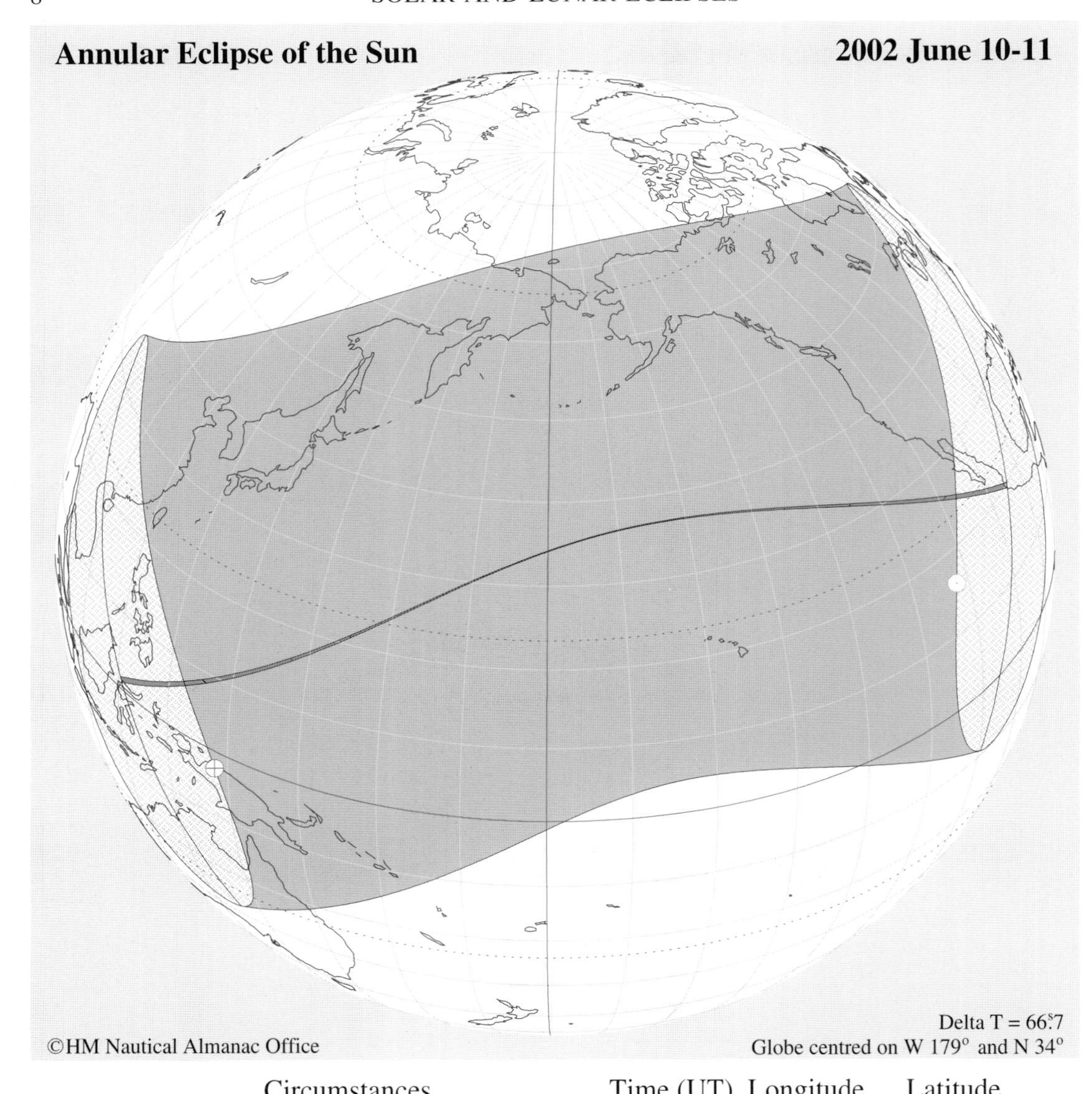

	Circumstances	Time (UT)		Longitude		Latitude	
		h	m	°	′	°	′
⊕	Eclipse begins; first contact with Earth	20	51.8	E 137	59.2	S 2	30.0
	Beginning of southern limit of penumbra	21	47.2	E 135	55.3	S 27	36.7
	Beginning of southern limit of umbra	21	54.2	E 120	52.4	N 1	02.0
	Beginning of centre line; central eclipse begins	21	54.4	E 120	41.5	N 1	19.5
	Beginning of northern limit of umbra	21	54.7	E 120	30.5	N 1	37.0
	Beginning of northern limit of penumbra	22	36.8	E 93	19.6	N 35	01.0
	Central eclipse at local apparent noon	23	48.2	W 177	10.8	N 34	55.3
	End of northern limit of penumbra	0	51.5	W 70	16.2	N 51	48.0
	End of northern limit of umbra	1	33.8	W 104	37.6	N 20	04.1
	End of centre line; central eclipse ends	1	34.0	W 104	48.6	N 19	48.1
	End of southern limit of umbra	1	34.2	W 104	59.5	N 19	32.2
	End of southern limit of penumbra	1	41.3	W 119	26.4	S 9	15.1
○	Eclipse ends; last contact with Earth	2	36.6	W 122	15.2	N 16	01.0

Penumbral Eclipse of the Moon 2002 June 24

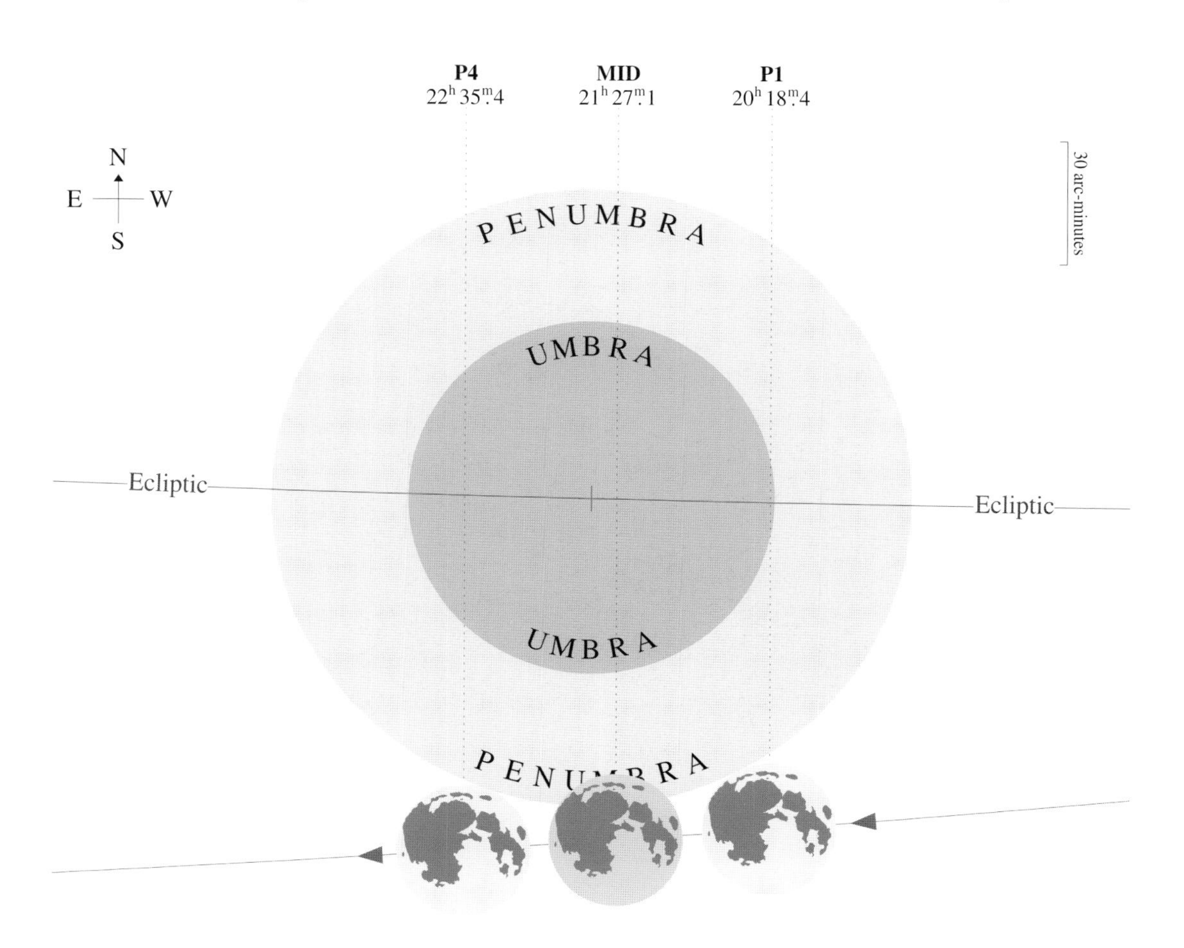

Delta T = 66.s8

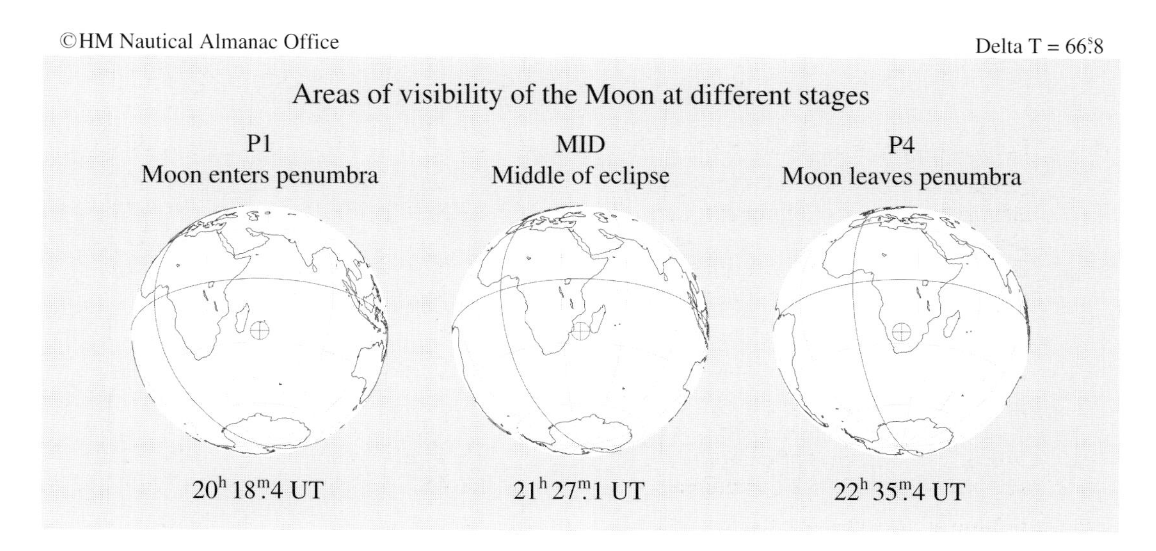

Penumbral Eclipse of the Moon — 2002 November 19-20

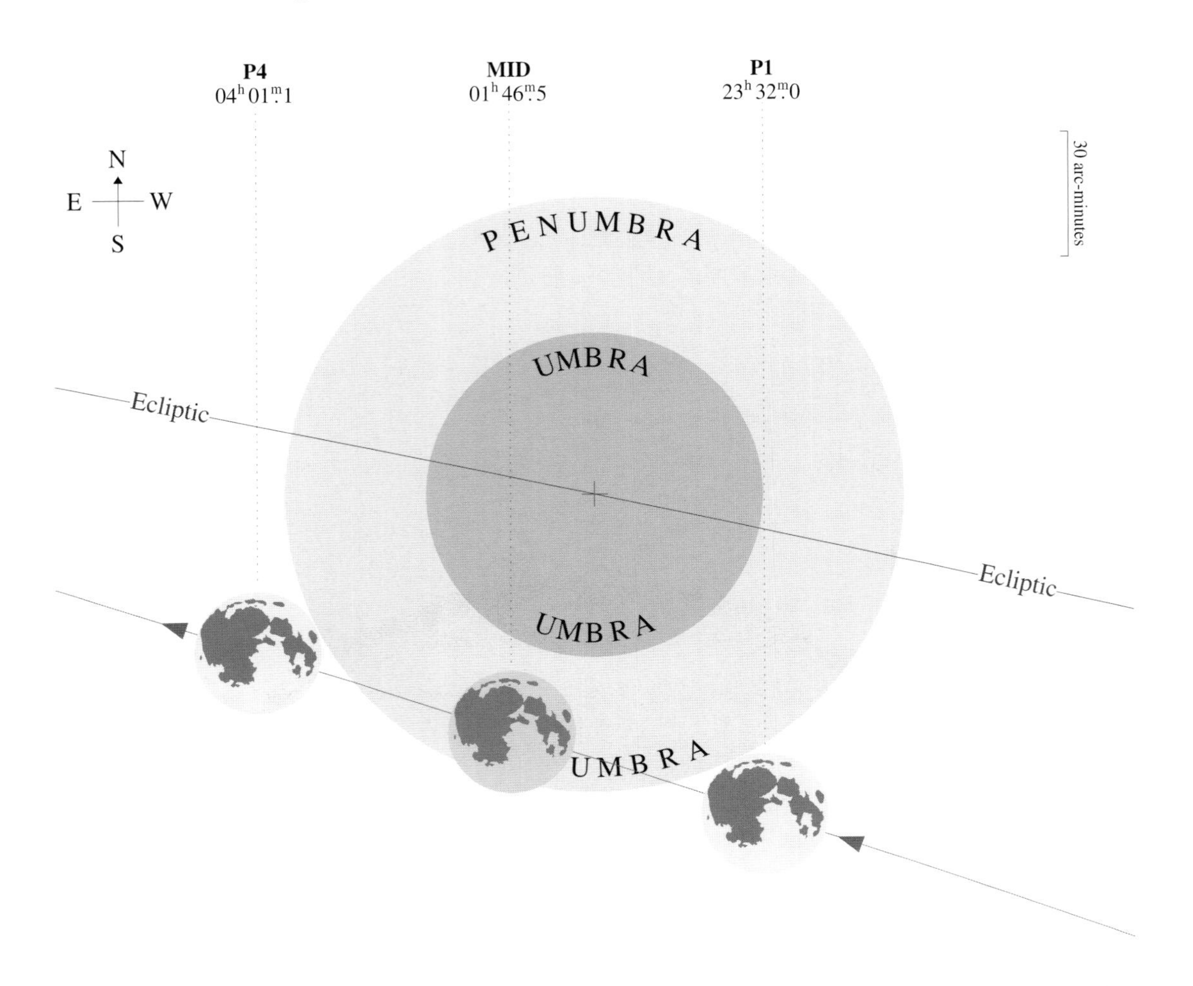

Delta T = $67^{s}.1$

Areas of visibility of the Moon at different stages

P1	MID	P4
Moon enters penumbra	Middle of eclipse	Moon leaves penumbra
$23^{h}32^{m}.0$ UT	$01^{h}46^{m}.5$ UT	$04^{h}01^{m}.1$ UT

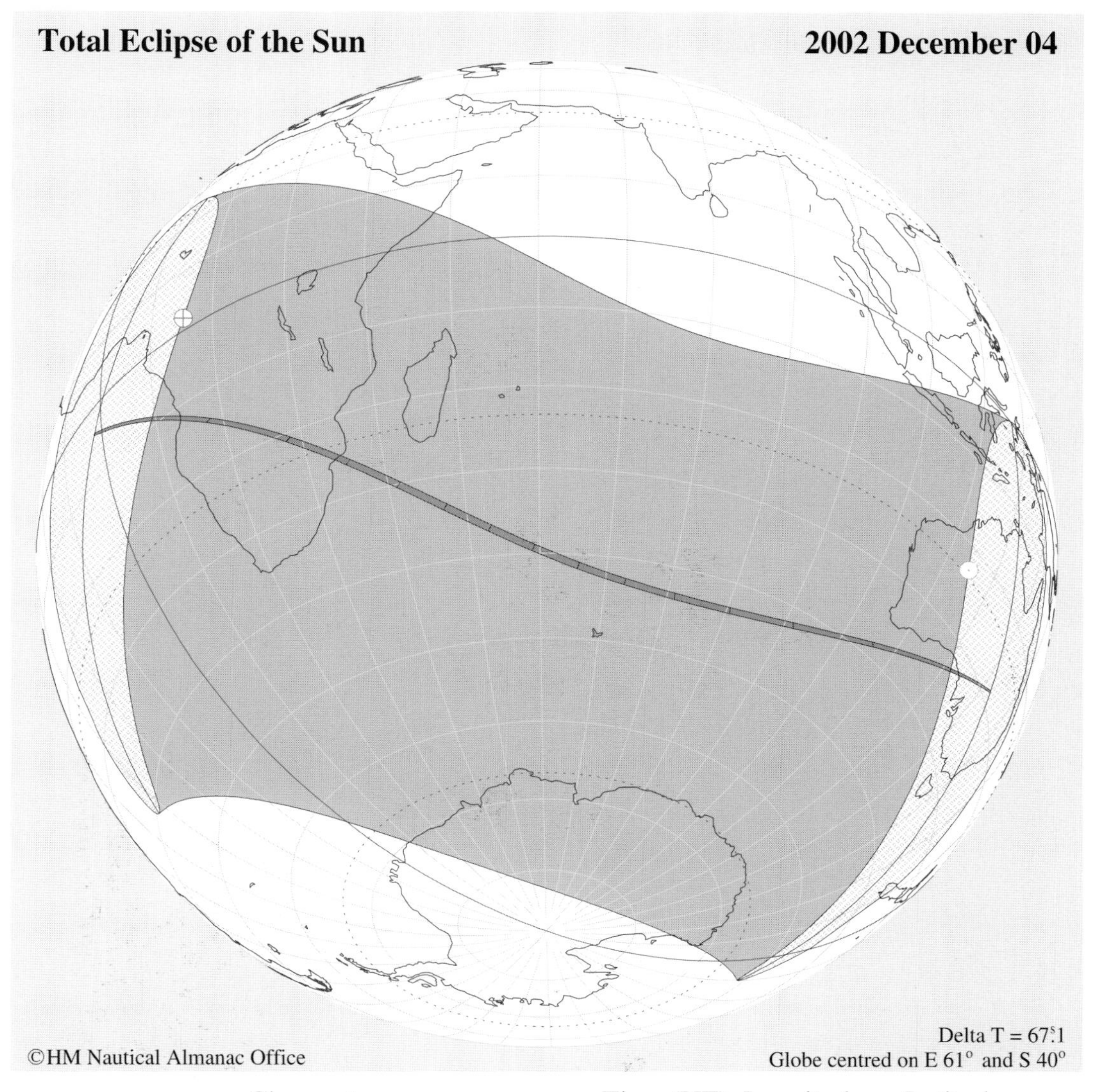

	Circumstances	Time (UT)	Longitude	Latitude
		h m	° ′	° ′
⊕	Eclipse begins; first contact with Earth	4 51.3	E 15 28.6	N 1 56.9
	Beginning of northern limit of penumbra	5 38.5	E 13 48.1	N 24 51.2
	Beginning of northern limit of umbra	5 50.4	W 1 38.2	S 3 47.9
	Beginning of centre line; central eclipse begins	5 50.5	W 1 43.3	S 3 55.9
	Beginning of southern limit of umbra	5 50.6	W 1 48.4	S 4 04.0
	Beginning of southern limit of penumbra	6 39.4	W 32 46.1	S 40 30.3
	Central eclipse at local apparent noon	7 38.7	E 62 50.9	S 40 31.7
	End of southern limit of penumbra	8 22.5	W 169 58.6	S 61 14.7
	End of southern limit of umbra	9 11.6	E 142 31.4	S 28 37.4
	End of centre line; central eclipse ends	9 11.7	E 142 26.1	S 28 30.6
	End of northern limit of umbra	9 11.8	E 142 20.8	S 28 23.9
	End of northern limit of penumbra	9 23.9	E 126 27.8	N 0 12.0
○	Eclipse ends; last contact with Earth	10 11.0	E 124 39.5	S 22 44.8

Total Eclipse of the Moon 2003 May 16

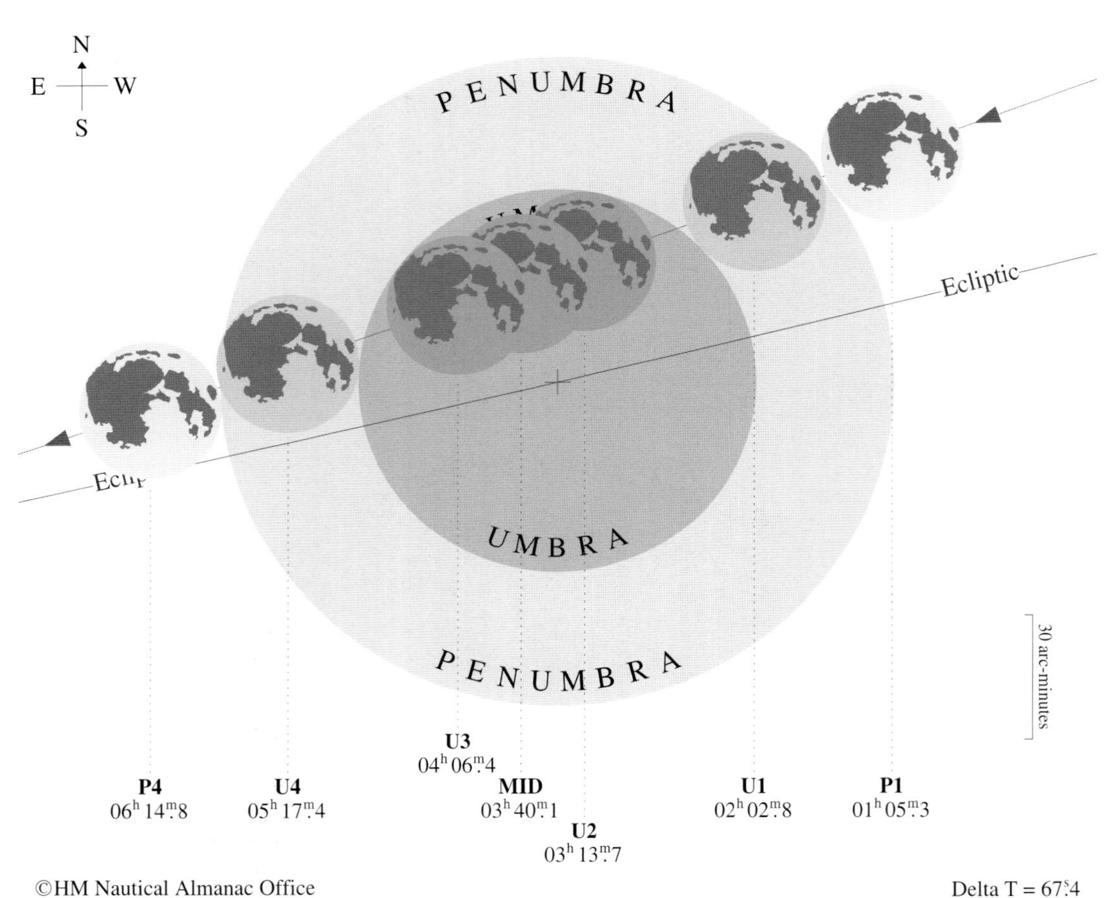

Delta T = $67^{s}.4$

Areas of visibility of the Moon at different stages

U1 Enters umbra	U2 Enters totality	U3 Leaves totality	U4 Leaves umbra
$02^{h}\,02^{m}.8$ UT	$03^{h}\,13^{m}.7$ UT	$04^{h}\,06^{m}.4$ UT	$05^{h}\,17^{m}.4$ UT

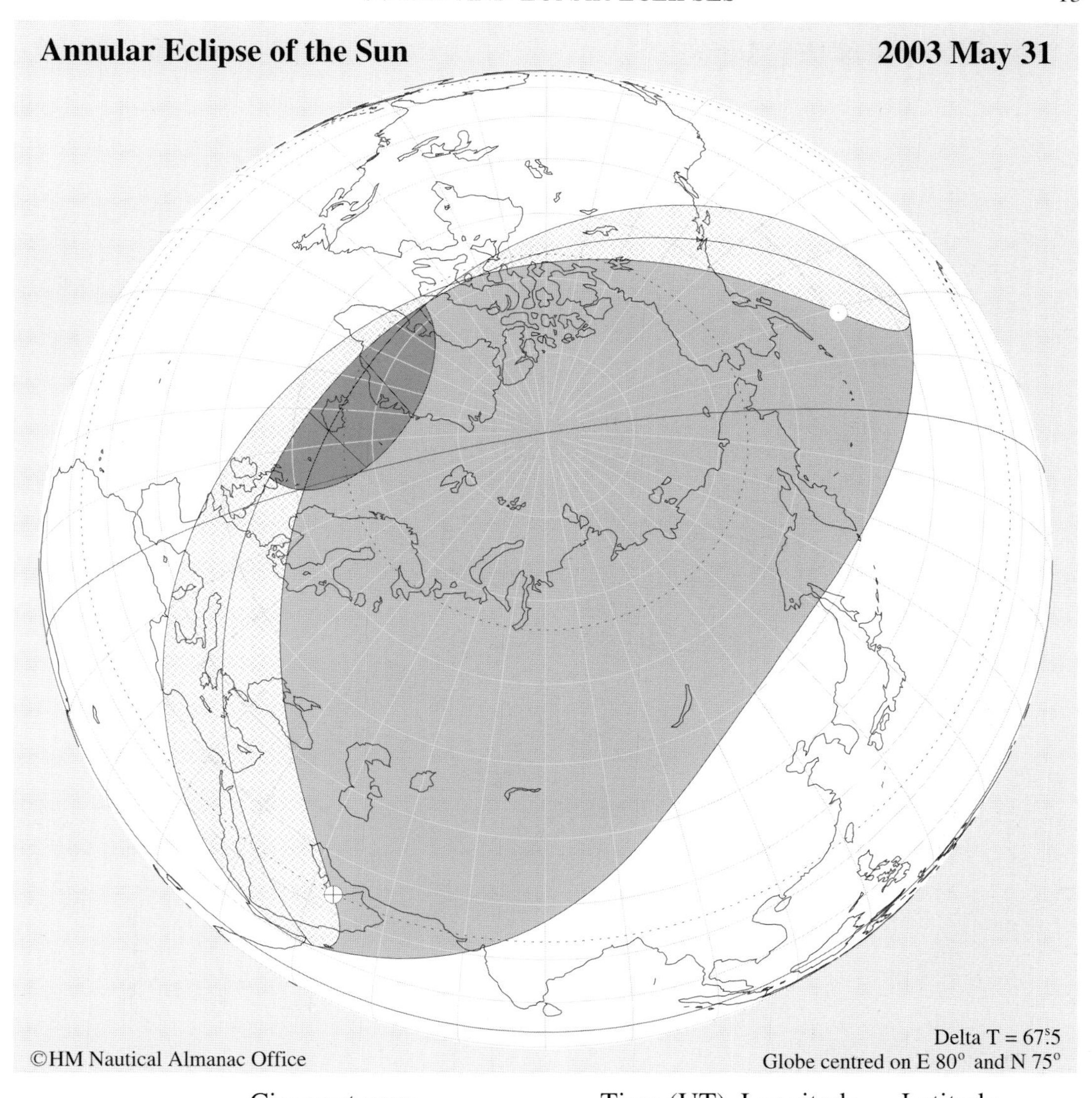

	Circumstances	Time (UT)	Longitude	Latitude
		h m	° ′	° ′
⊕	Eclipse begins; first contact with Earth	1 46.2	E 52 52.4	N 23 21.9
	Beginning of southern limit of penumbra	2 11.8	E 52 01.6	N 10 51.2
	Beginning of southern limit of umbra	3 44.9	W 4 38.6	N 56 48.8
	Beginning of centre line; central eclipse begins	4 01.8	W 21 00.2	N 62 21.2
	End of centre line; central eclipse ends	4 14.3	W 35 56.4	N 65 31.7
	End of southern limit of umbra	4 31.2	W 60 11.6	N 67 56.8
	End of southern limit of penumbra	6 04.4	W 164 03.7	N 37 05.3
○	Eclipse ends; last contact with Earth	6 30.0	W 160 55.9	N 48 42.4

Total Eclipse of the Moon 2003 November 08-09

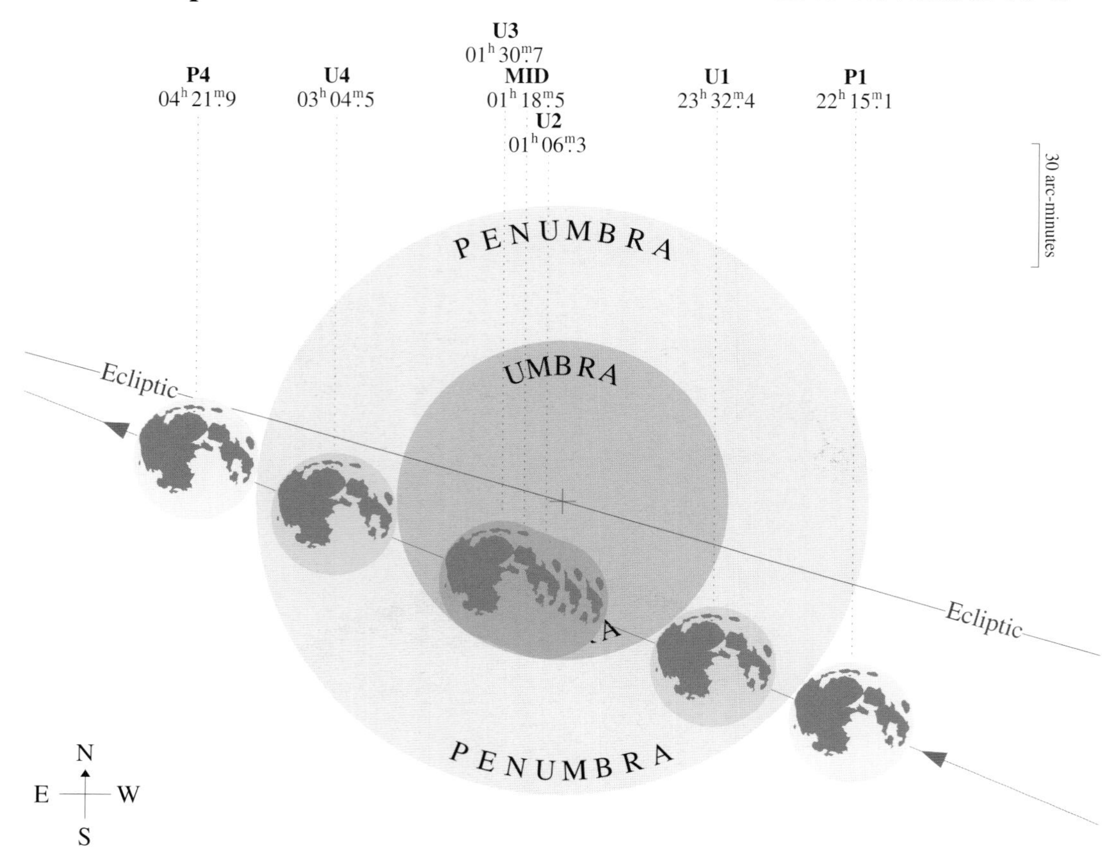

 Delta T = 67s.8

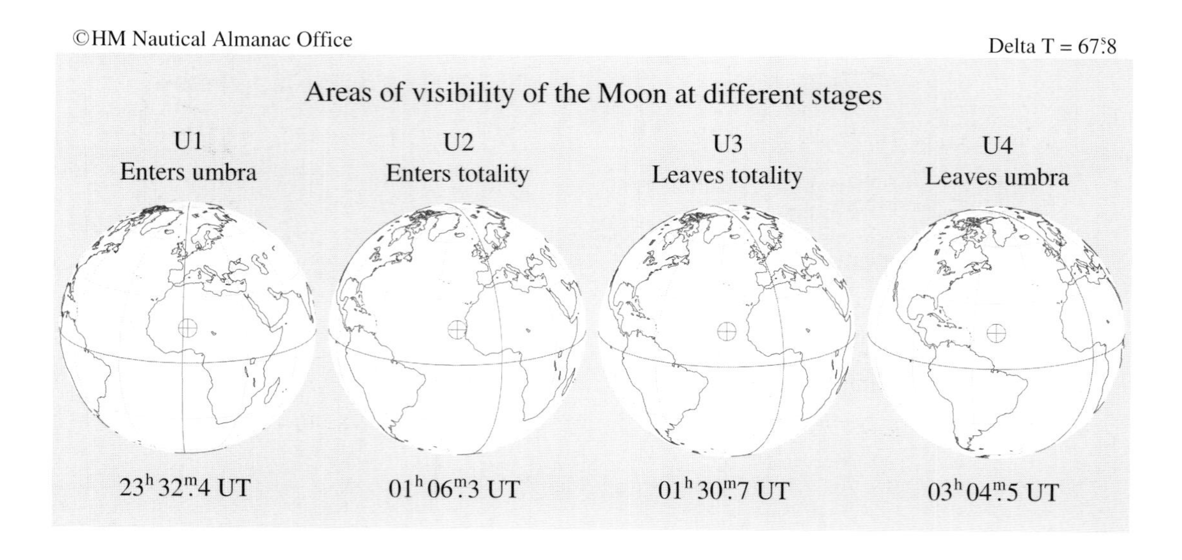

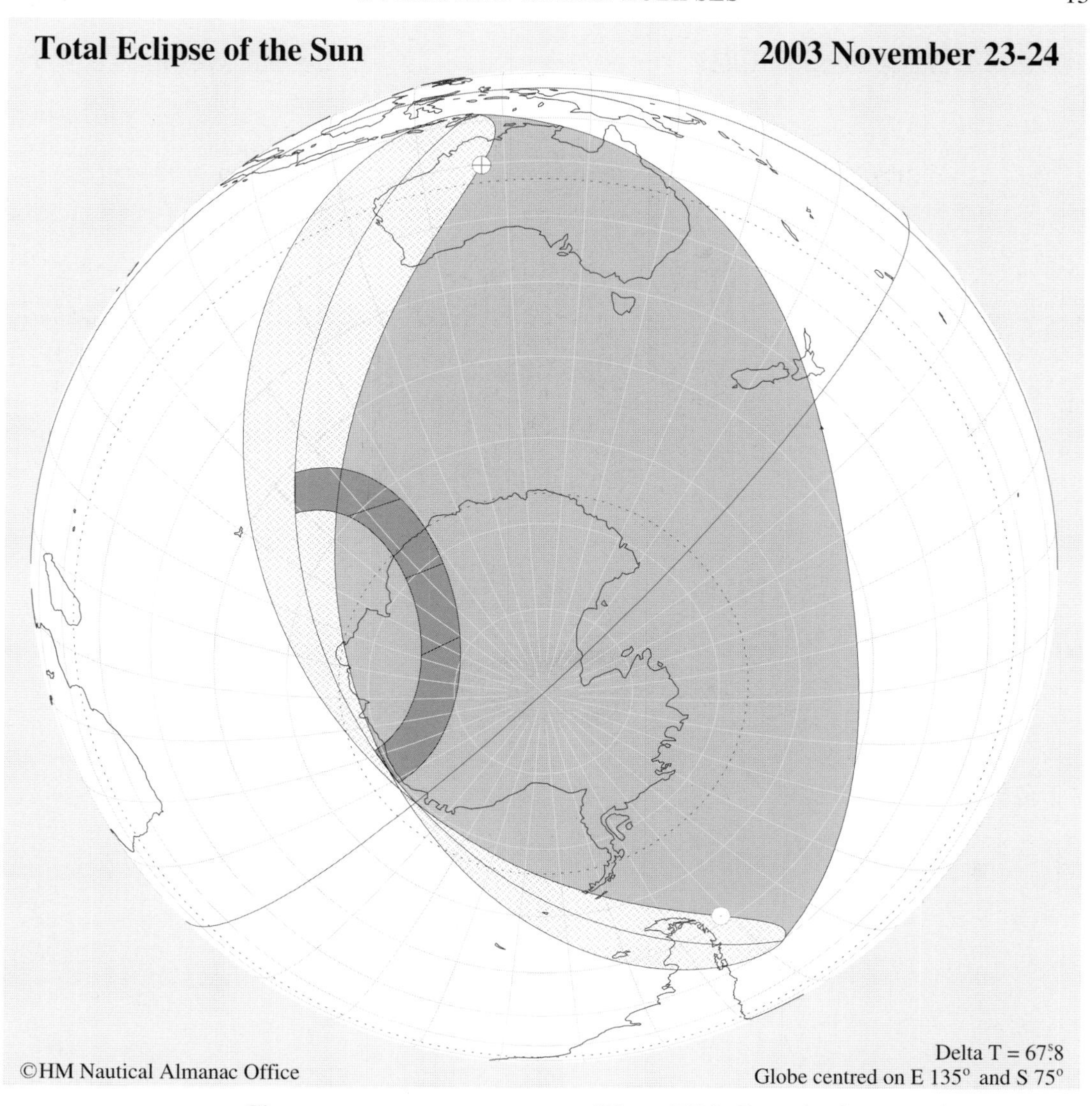

Circumstances	Time (UT)	Longitude	Latitude
	h m	° ′	° ′
⊕ Eclipse begins; first contact with Earth	20 46.0	E 127 15.5	S 20 09.1
Beginning of northern limit of penumbra	21 09.0	E 126 28.2	S 7 42.0
Beginning of northern limit of umbra	22 19.3	E 84 32.3	S 50 54.2
Beginning of centre line; central eclipse begins	22 22.6	E 81 59.7	S 52 28.0
Beginning of southern limit of umbra	22 26.4	E 78 55.6	S 54 13.3
End of southern limit of umbra	23 11.7	E 21 50.8	S 69 05.4
End of centre line; central eclipse ends	23 15.5	E 14 52.6	S 69 26.8
End of northern limit of umbra	23 18.8	E 8 40.0	S 69 35.0
End of northern limit of penumbra	0 29.2	W 82 38.3	S 39 47.6
○ Eclipse ends; last contact with Earth	0 52.2	W 78 46.4	S 51 16.5

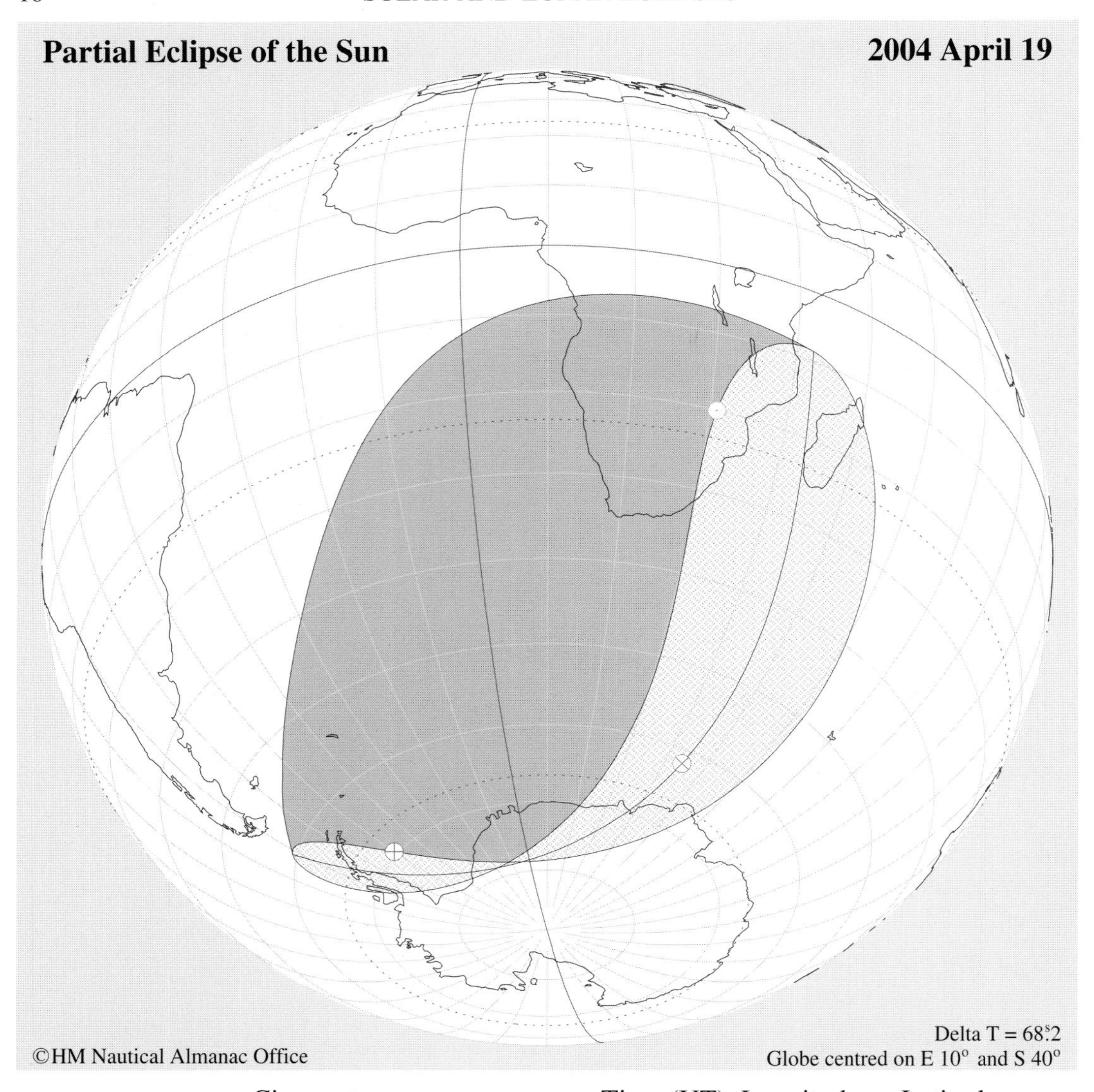

Circumstances	Time (UT)	Longitude	Latitude
	h m	° ′	° ′
⊕ Eclipse begins; first contact with Earth	11 29.8	W 49 58.4	S 69 34.3
Beginning of northern limit of penumbra	12 04.9	W 71 32.1	S 59 24.6
⊗ Greatest eclipse (magnitude = 0.7369)	13 33.9	E 44 13.6	S 61 43.5
End of northern limit of penumbra	15 03.4	E 42 08.4	S 8 38.6
⊙ Eclipse ends; last contact with Earth	15 38.5	E 30 53.3	S 20 01.5

Total Eclipse of the Moon — 2004 May 04

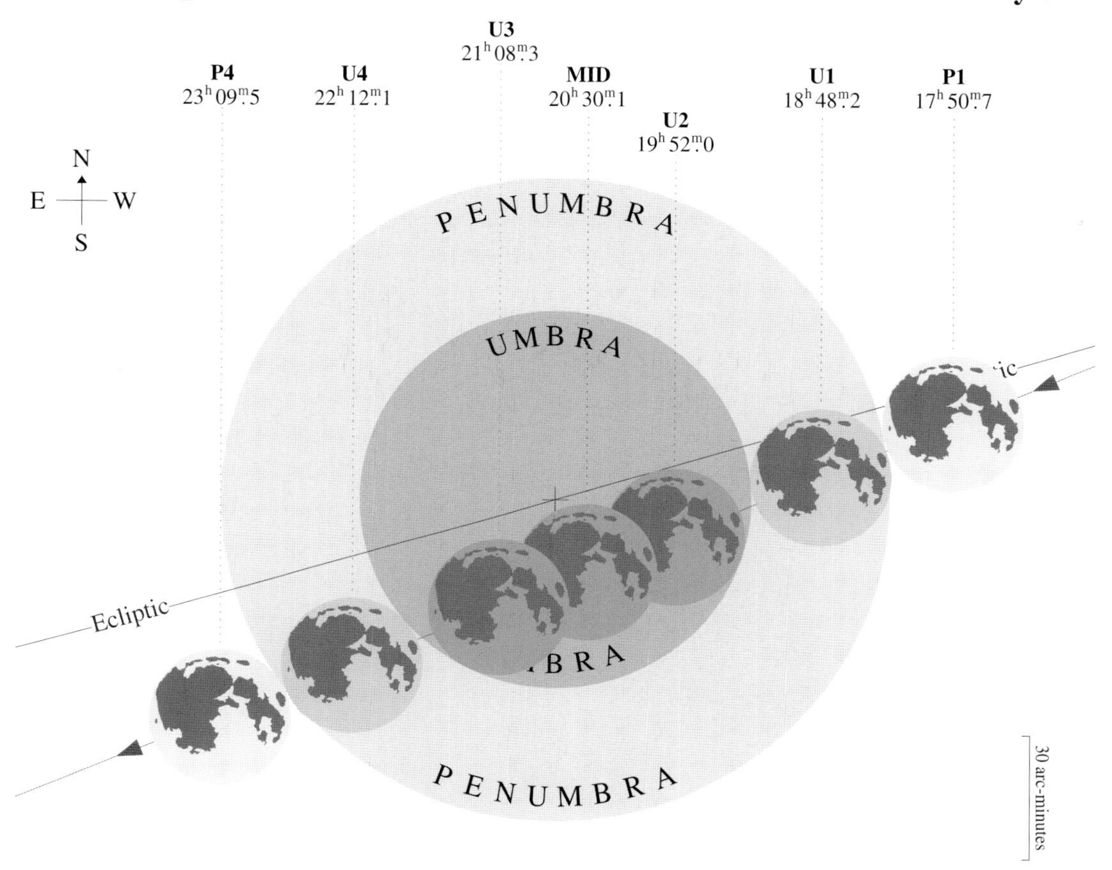

Delta T = 68s.2

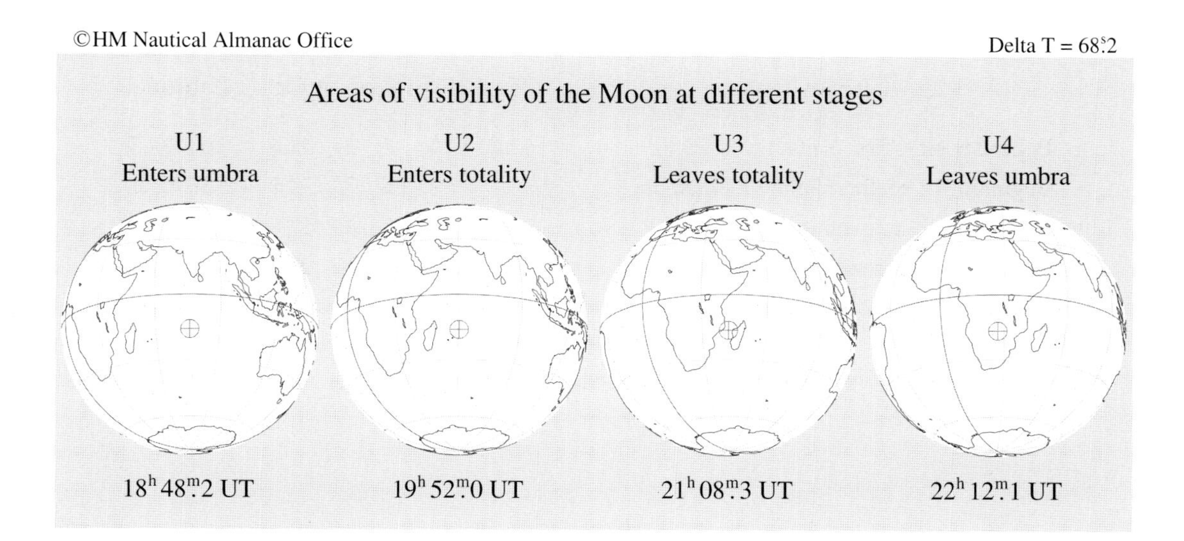

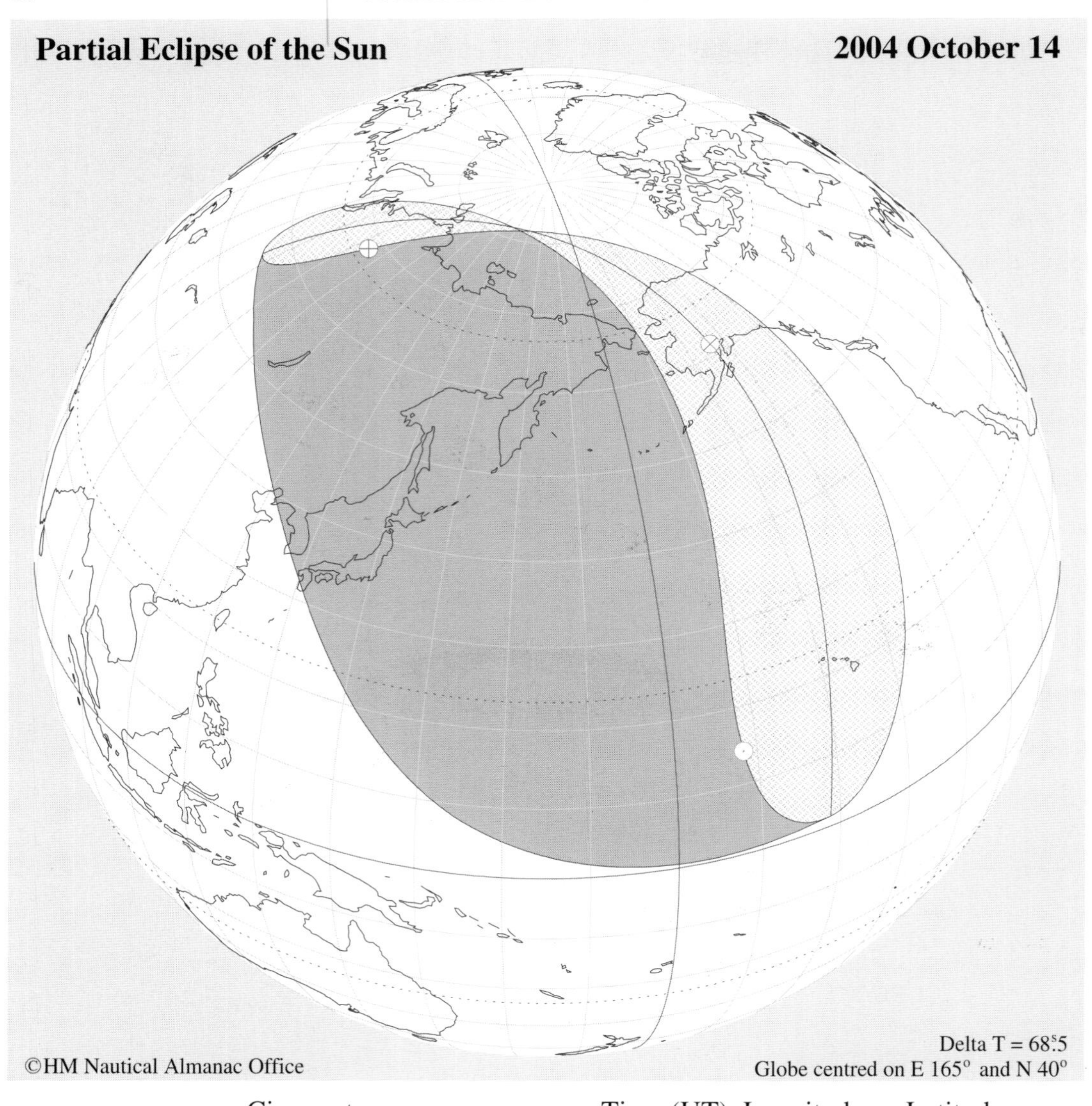

Circumstances	Time (UT)	Longitude	Latitude
	h m	° ′	° ′
⊕ Eclipse begins; first contact with Earth	0 54.5	E 94 04.4	N 68 15.3
Beginning of southern limit of penumbra	1 28.2	E 76 53.1	N 56 09.4
⊗ Greatest eclipse (magnitude = 0.9287)	2 59.2	W 153 42.2	N 61 23.8
End of southern limit of penumbra	4 30.5	W 161 21.0	N 1 32.4
⊙ Eclipse ends; last contact with Earth	5 04.2	W 171 40.5	N 14 15.1

Total Eclipse of the Moon 2004 October 28

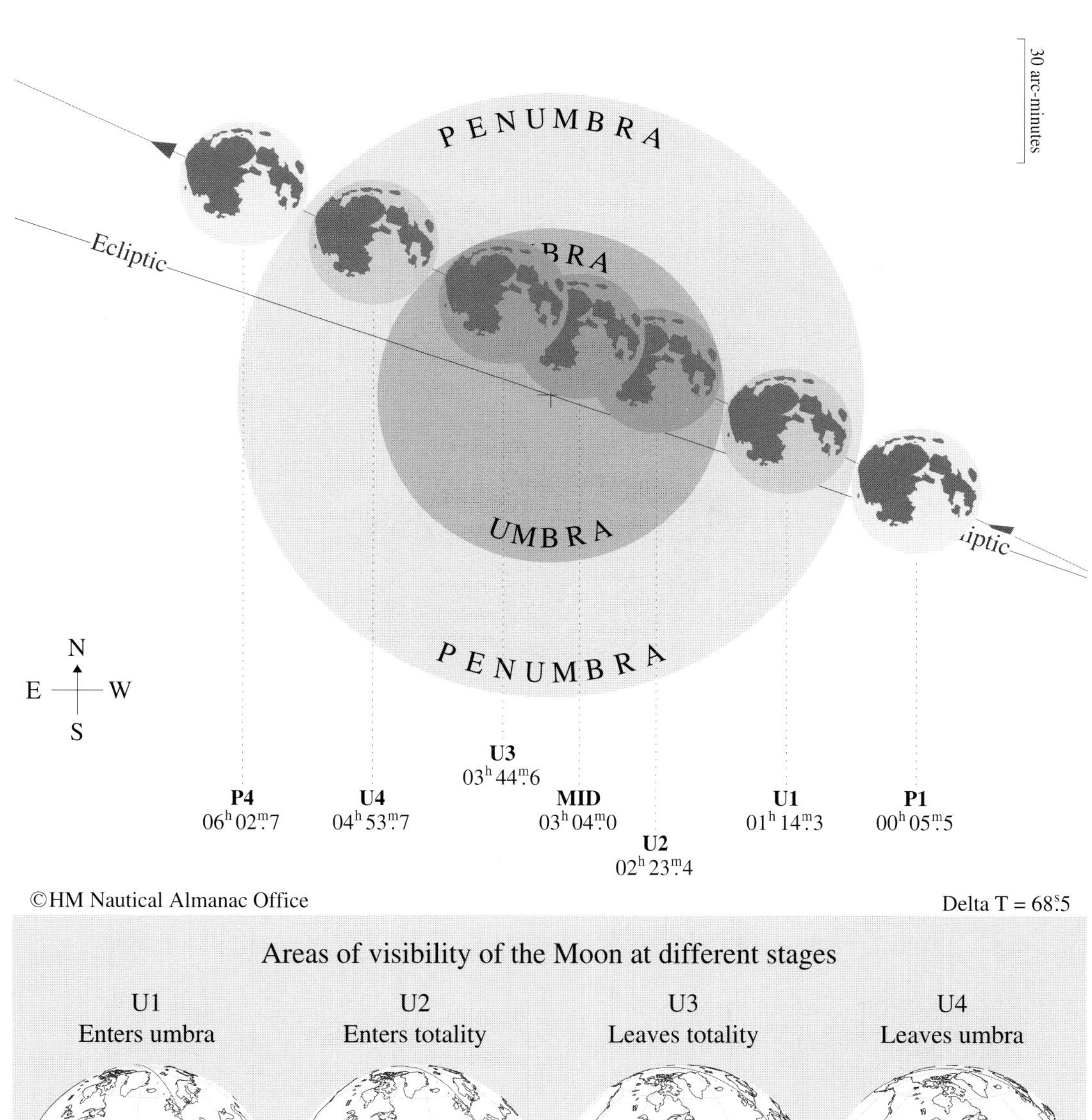

Delta T = $68^{s}.5$

Areas of visibility of the Moon at different stages

U1	U2	U3	U4
Enters umbra	Enters totality	Leaves totality	Leaves umbra
$01^{h}\,14^{m}.3$ UT	$02^{h}\,23^{m}.4$ UT	$03^{h}\,44^{m}.6$ UT	$04^{h}\,53^{m}.7$ UT

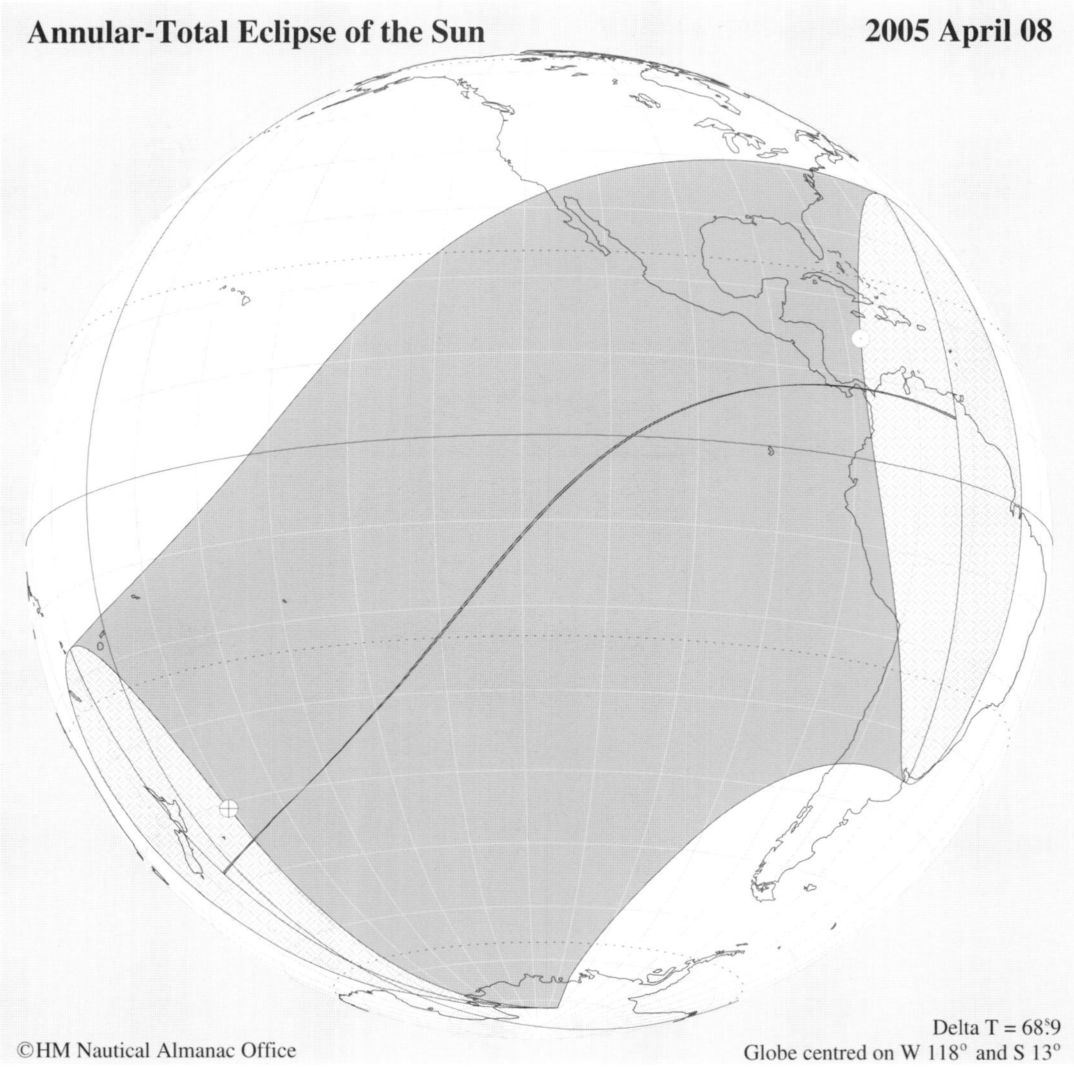

Circumstances	Time (UT)	Longitude	Latitude
	h m	° ′	° ′
⊕ Eclipse begins; first contact with Earth	17 51.2	W170 55.0	S 40 41.5
Beginning of northern limit of penumbra	18 52.3	E 169 34.3	S 16 21.8
Beginning of northern limit of umbra	18 53.5	E 175 22.1	S 47 50.4
Beginning of centre line; central eclipse begins	18 53.6	E 175 23.1	S 47 57.2
Beginning of southern limit of umbra	18 53.6	E 175 24.1	S 48 04.1
Beginning of southern limit of penumbra	19 46.3	W101 49.4	S 82 17.7
Central eclipse at local apparent noon	20 15.6	W123 27.8	S 15 47.3
End of southern limit of penumbra	21 25.8	W 56 25.6	S 35 37.8
End of southern limit of umbra	22 17.9	W 63 04.3	N 7 26.7
End of centre line; central eclipse ends	22 18.0	W 63 04.4	N 7 35.1
End of northern limit of umbra	22 18.1	W 63 04.4	N 7 43.4
End of northern limit of penumbra	22 19.0	W 58 07.7	N 39 17.0
○ Eclipse ends; last contact with Earth	23 20.4	W 77 39.5	N 14 55.5

Penumbral Eclipse of the Moon — 2005 April 24

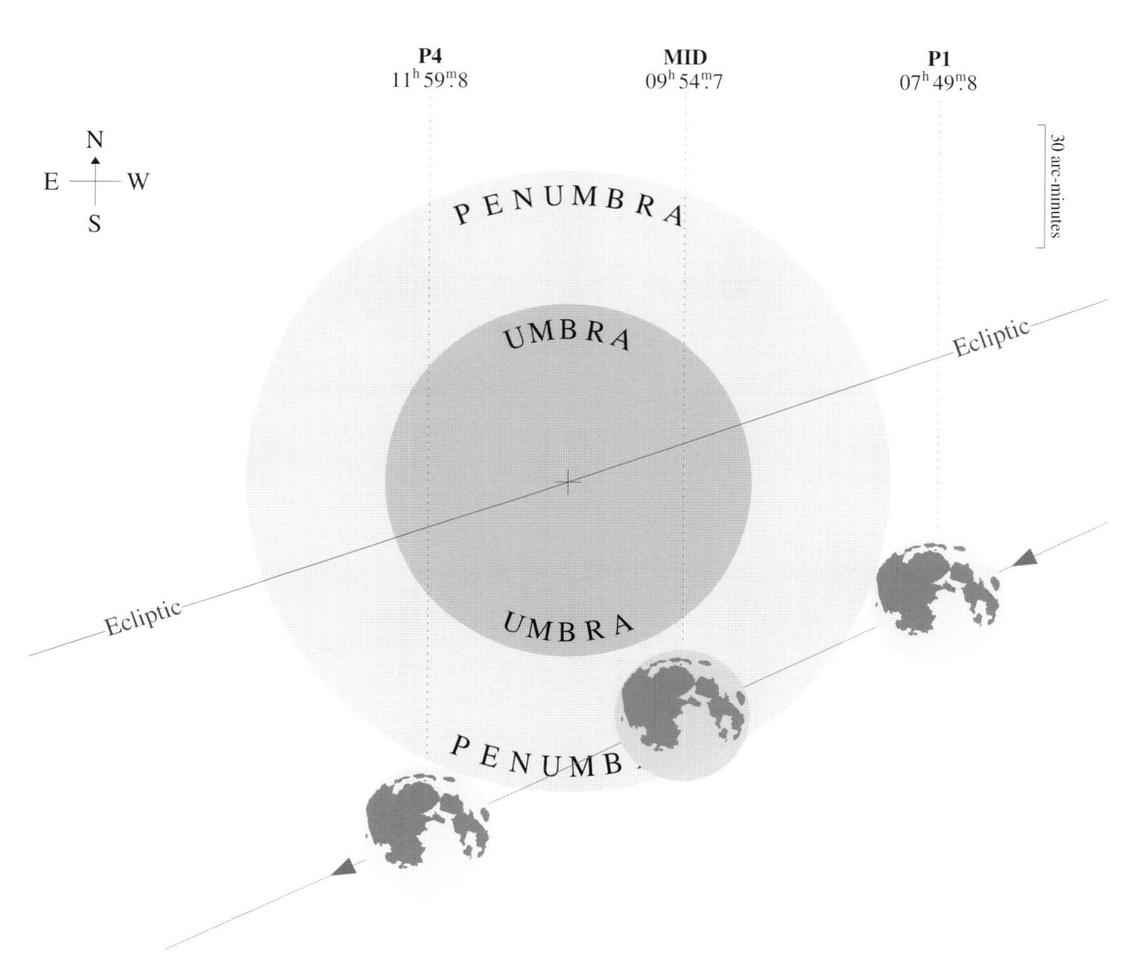

©HM Nautical Almanac Office

Delta T = 68s.9

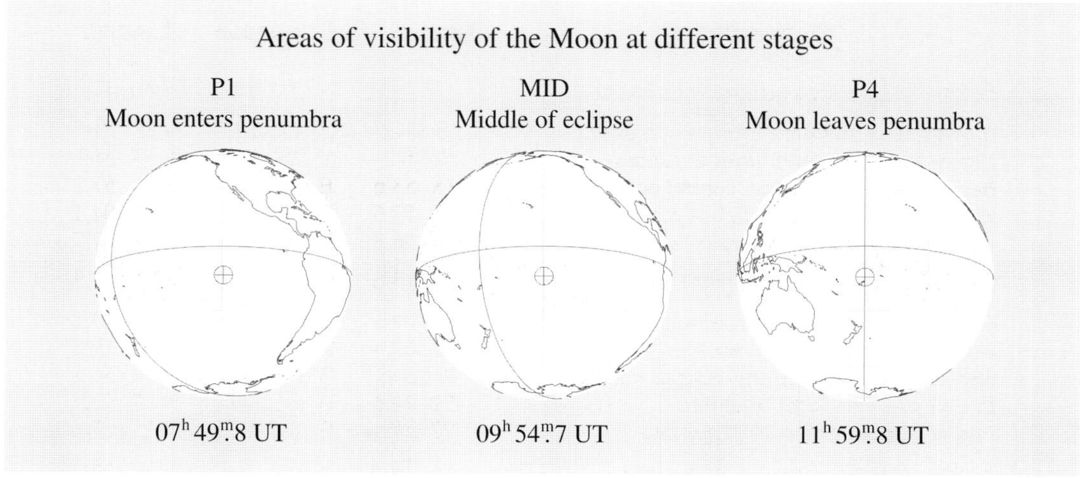

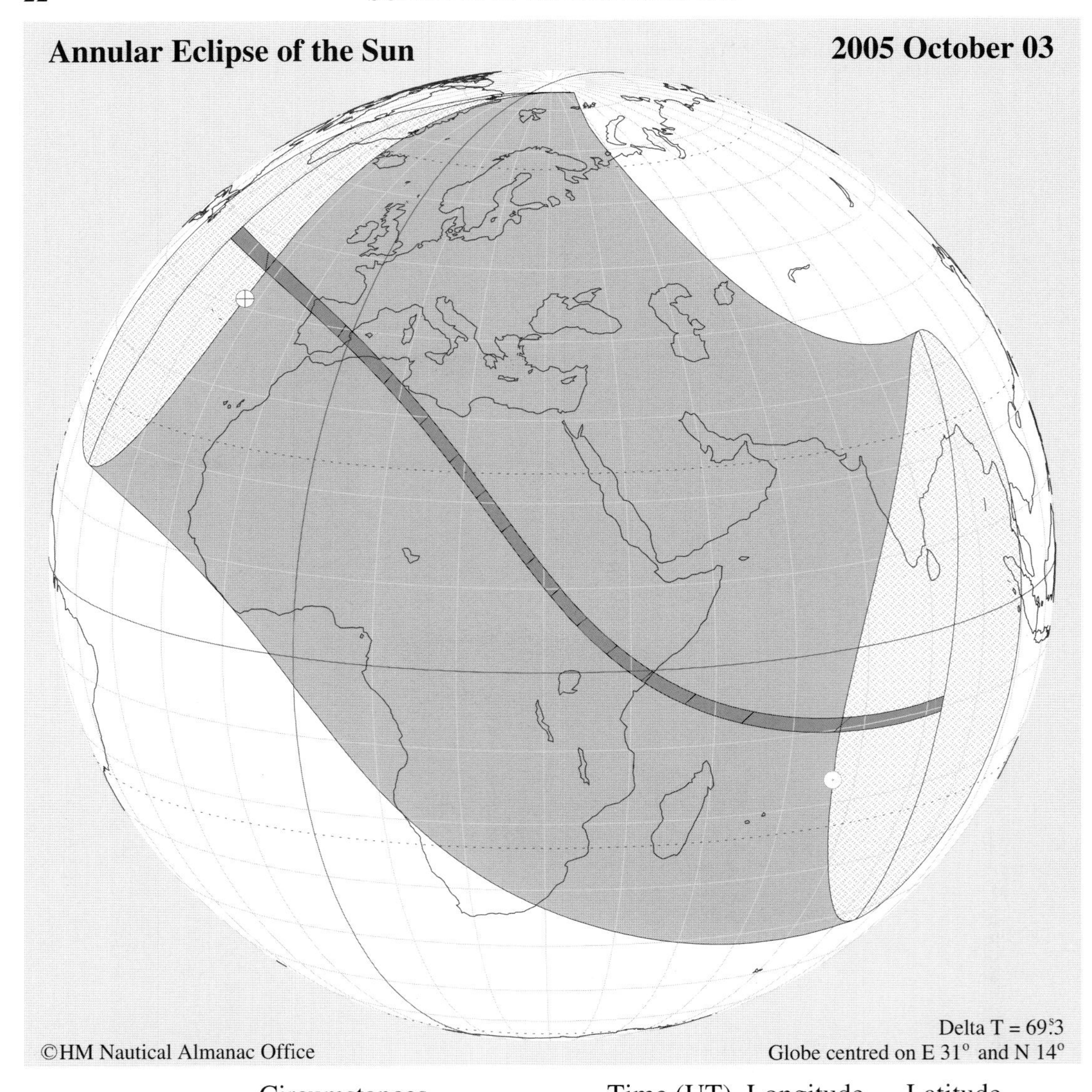

	Circumstances	Time (UT)		Longitude			Latitude	
		h	m		°	′	°	′
⊕	Eclipse begins; first contact with Earth	7	35.5	W	23	04.4	N 41	07.7
	Beginning of southern limit of penumbra	8	41.5	W	41	59.1	N 15	28.8
	Beginning of southern limit of umbra	8	42.3	W	38	55.8	N 47	13.2
	Beginning of centre line; central eclipse begins	8	42.9	W	38	55.3	N 48	12.5
	Beginning of northern limit of umbra	8	43.5	W	38	54.9	N 49	12.2
	Beginning of northern limit of penumbra	9	39.5	E	63	52.1	N 85	13.7
	Central eclipse at local apparent noon	10	10.6	E	24	35.9	N 18	12.6
	End of northern limit of penumbra	11	24.5	E	93	17.8	N 34	37.3
	End of northern limit of umbra	12	19.9	E	82	53.2	S 8	35.6
	End of centre line; central eclipse ends	12	20.5	E	82	48.5	S 9	34.4
	End of southern limit of umbra	12	21.1	E	82	44.4	S 10	32.9
	End of southern limit of penumbra	12	21.5	E	85	36.9	S 42	18.7
⊙	Eclipse ends; last contact with Earth	13	27.8	E	66	31.7	S 16	40.9

Partial Eclipse of the Moon 2005 October 17

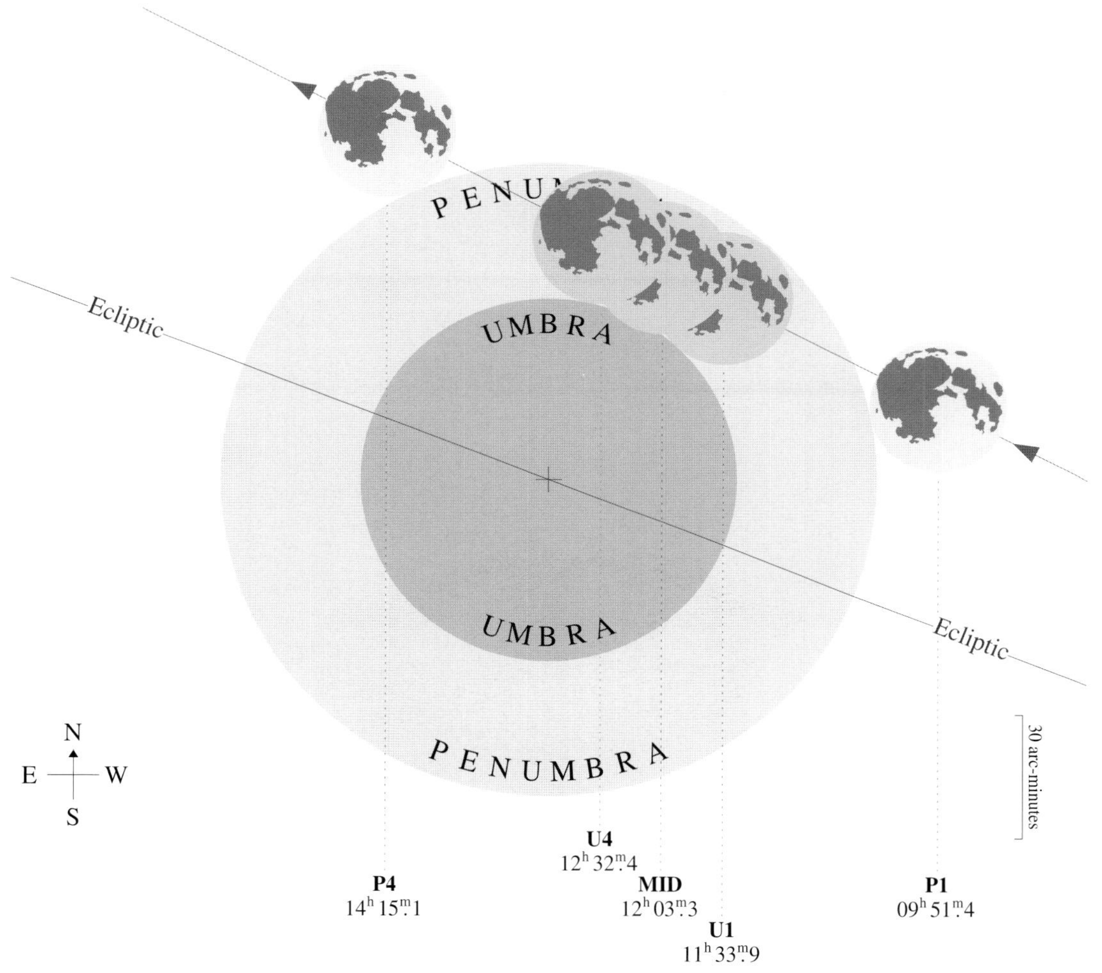

Delta T = 69s.3

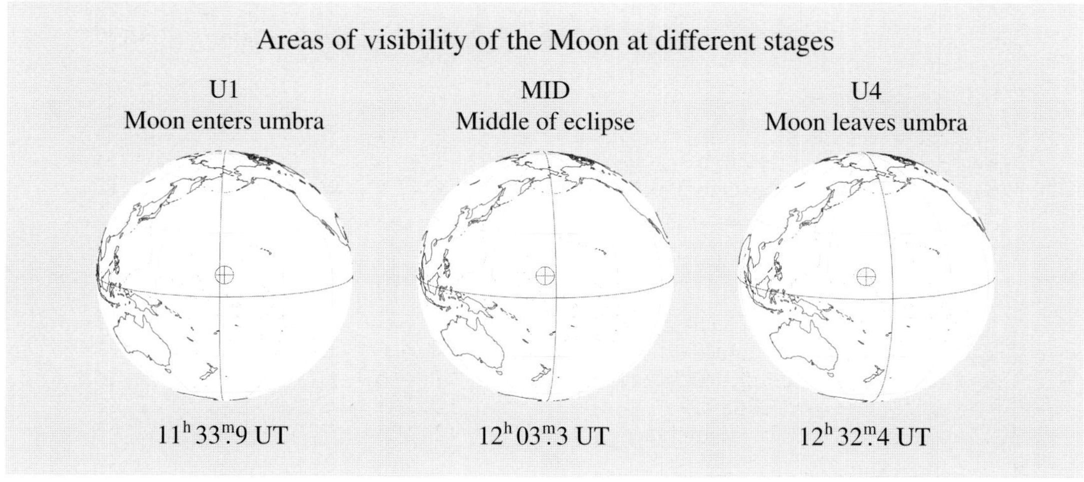

Penumbral Eclipse of the Moon

2006 March 14-15

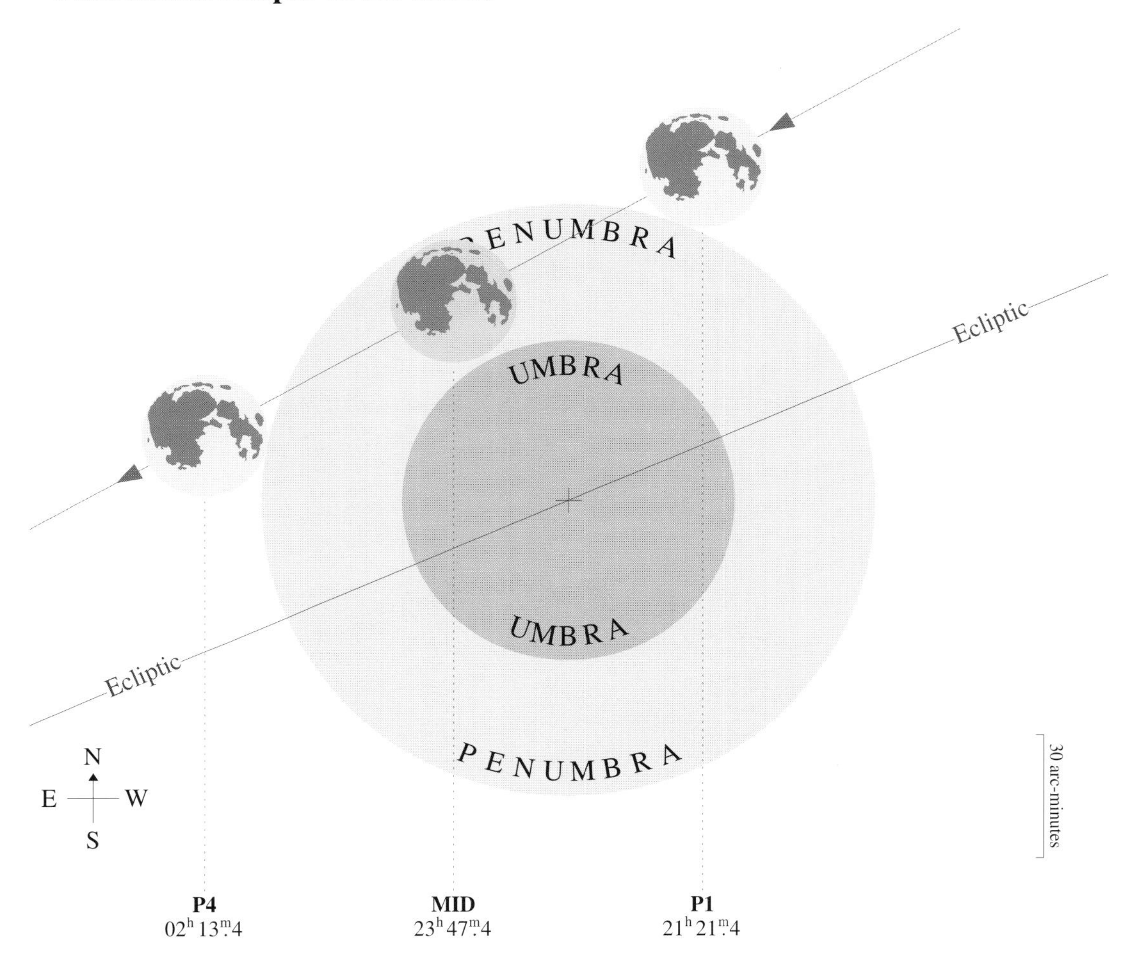

Delta T = $69^s.6$

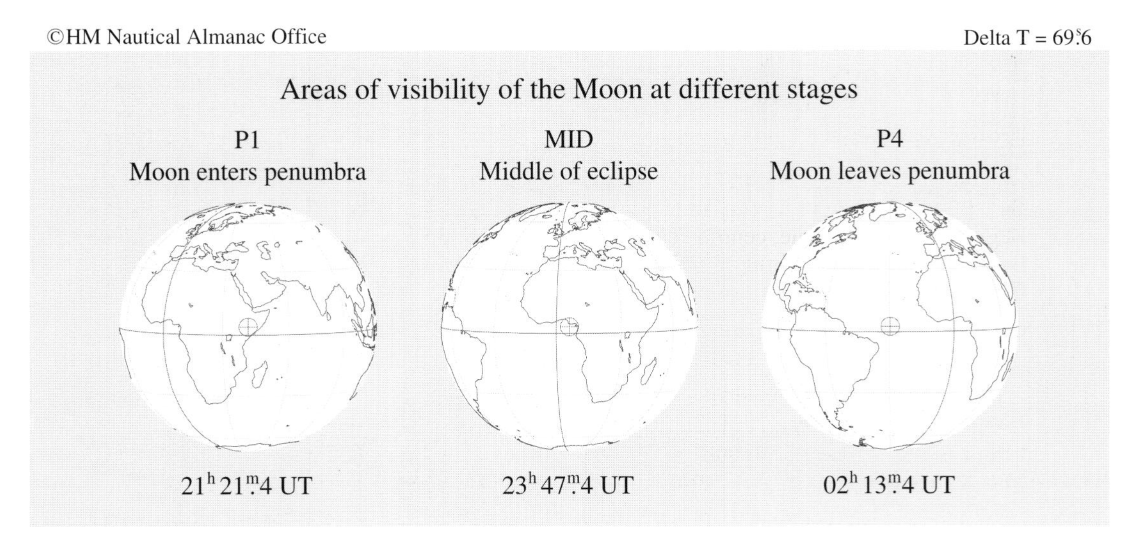

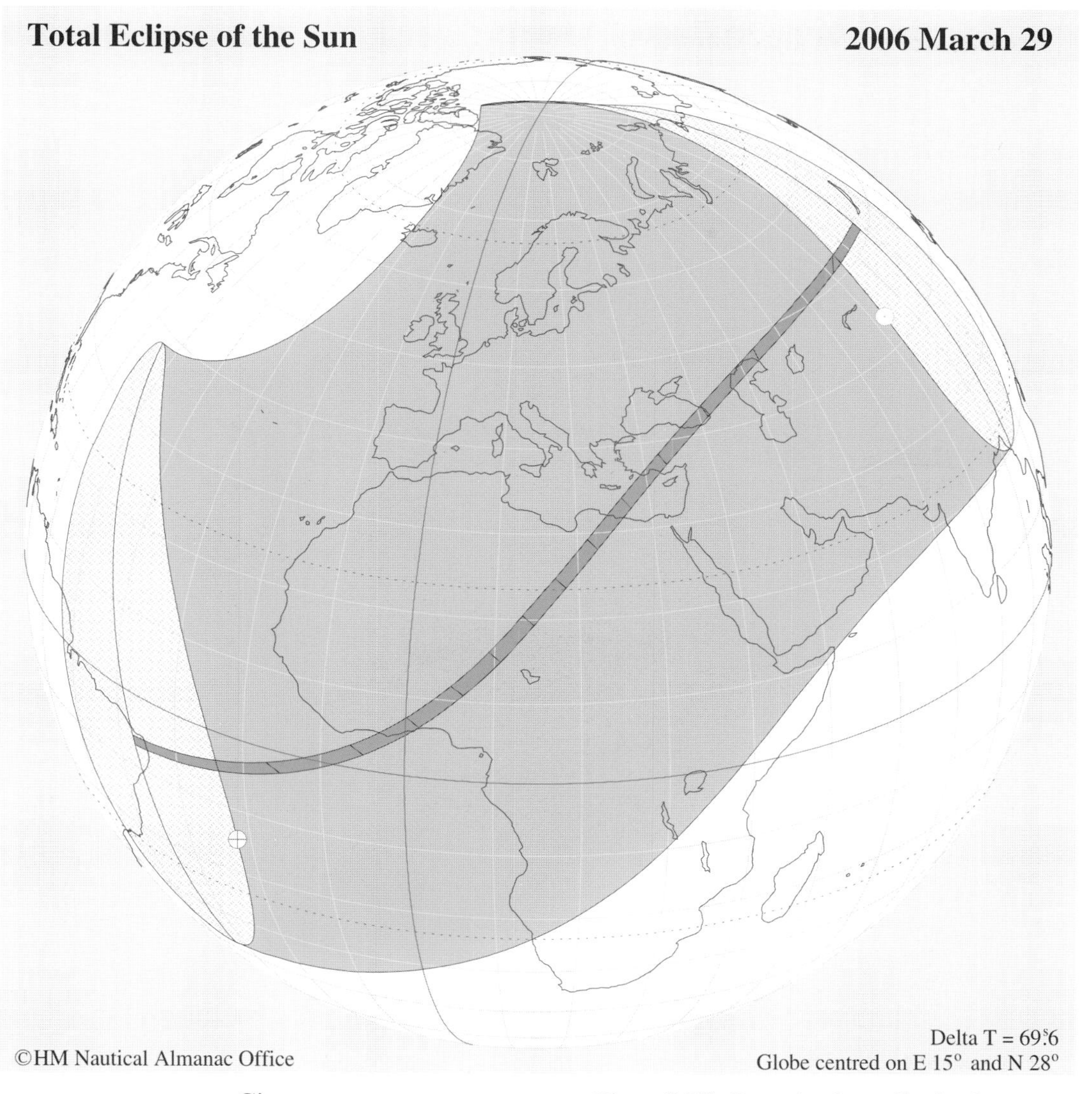

	Circumstances	Time (UT)	Longitude	Latitude
		h m	° ′	° ′
⊕	Eclipse begins; first contact with Earth	7 36.7	W 22 05.5	S 14 27.7
	Beginning of southern limit of penumbra	8 27.1	W 32 57.3	S 37 45.6
	Beginning of southern limit of umbra	8 34.9	W 37 06.5	S 6 53.5
	Beginning of centre line; central eclipse begins	8 35.3	W 37 14.9	S 6 18.3
	Beginning of northern limit of umbra	8 35.8	W 37 23.6	S 5 42.9
	Beginning of northern limit of penumbra	9 31.6	W 54 24.2	N 38 31.2
	Central eclipse at local apparent noon	10 33.2	E 22 54.4	N 29 37.2
	End of northern limit of penumbra	10 50.1	W 99 19.2	N 82 45.6
	End of northern limit of umbra	11 46.4	E 99 01.0	N 52 08.1
	End of centre line; central eclipse ends	11 46.8	E 98 49.0	N 51 33.7
	End of southern limit of umbra	11 47.3	E 98 37.5	N 50 59.4
	End of southern limit of penumbra	11 55.4	E 93 36.6	N 20 09.7
○	Eclipse ends; last contact with Earth	12 45.6	E 83 03.8	N 43 26.3

Partial Eclipse of the Moon 2006 September 07

P4
21h 00m.2
U4
19h 37m.6
MID
18h 51m.2
U1
18h 05m.0
P1
16h 42m.3
N
E
W
S
30 arc-minutes
PENUMBRA
UMBRA
Ecliptic
Ecliptic

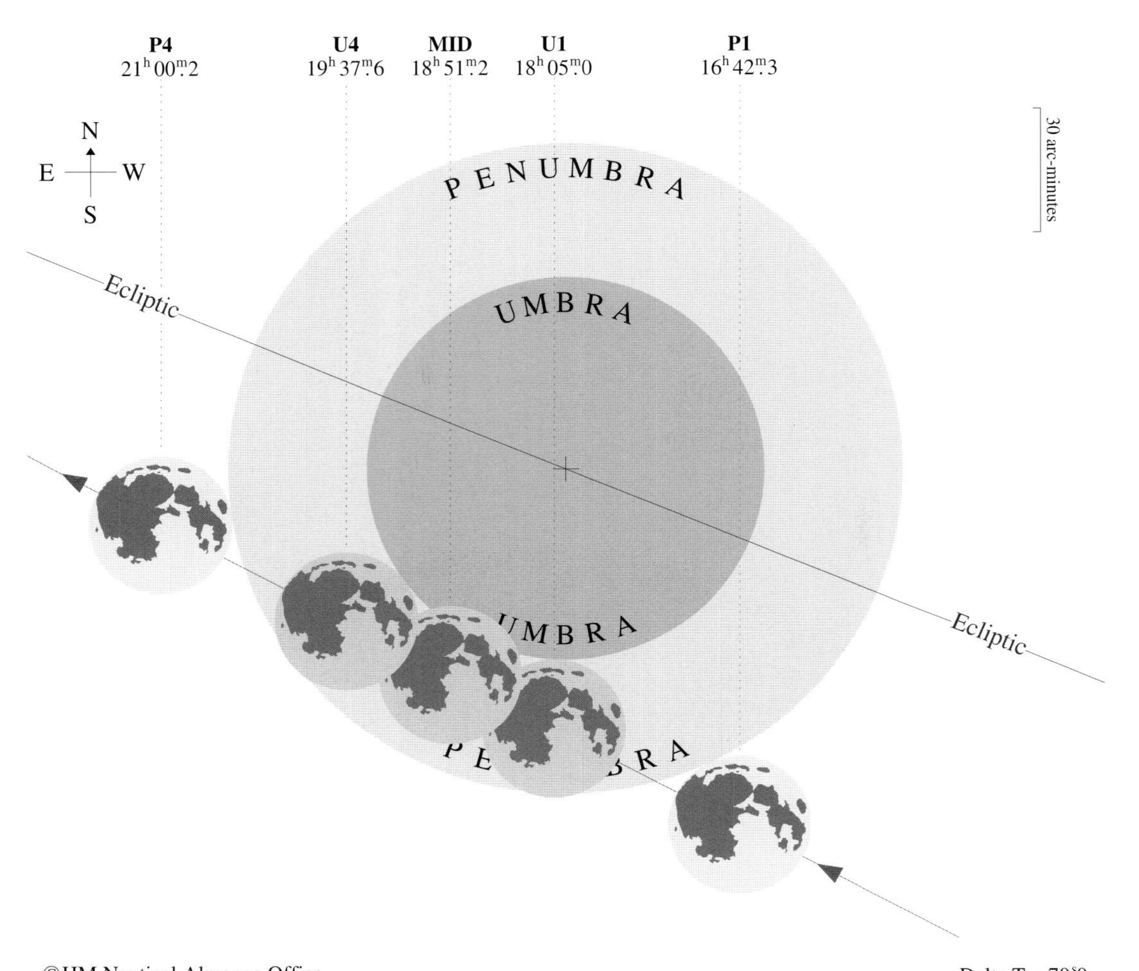

Delta T = 70.s0

Areas of visibility of the Moon at different stages
U1
Moon enters umbra
MID
Middle of eclipse
U4
Moon leaves umbra
18h 05m.0 UT
18h 51m.2 UT
19h 37m.6 UT

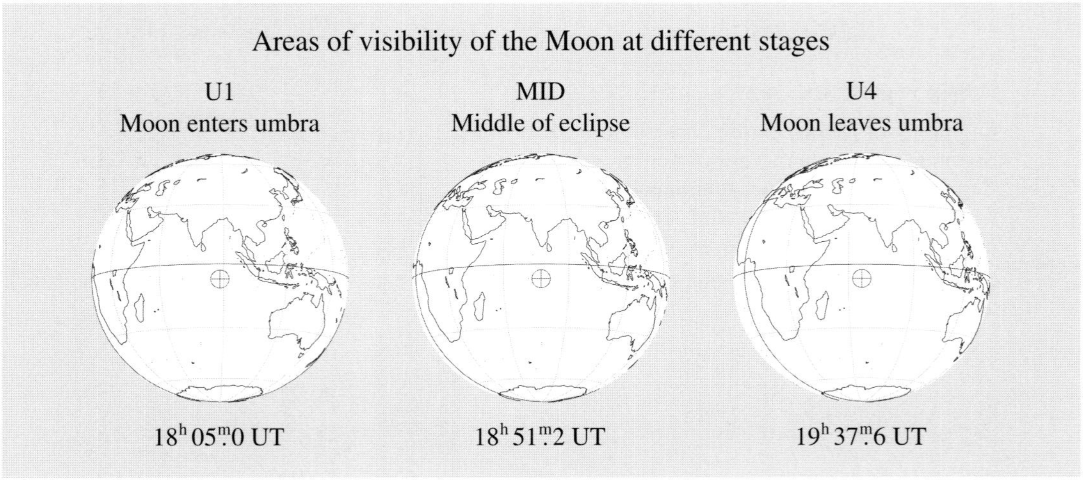

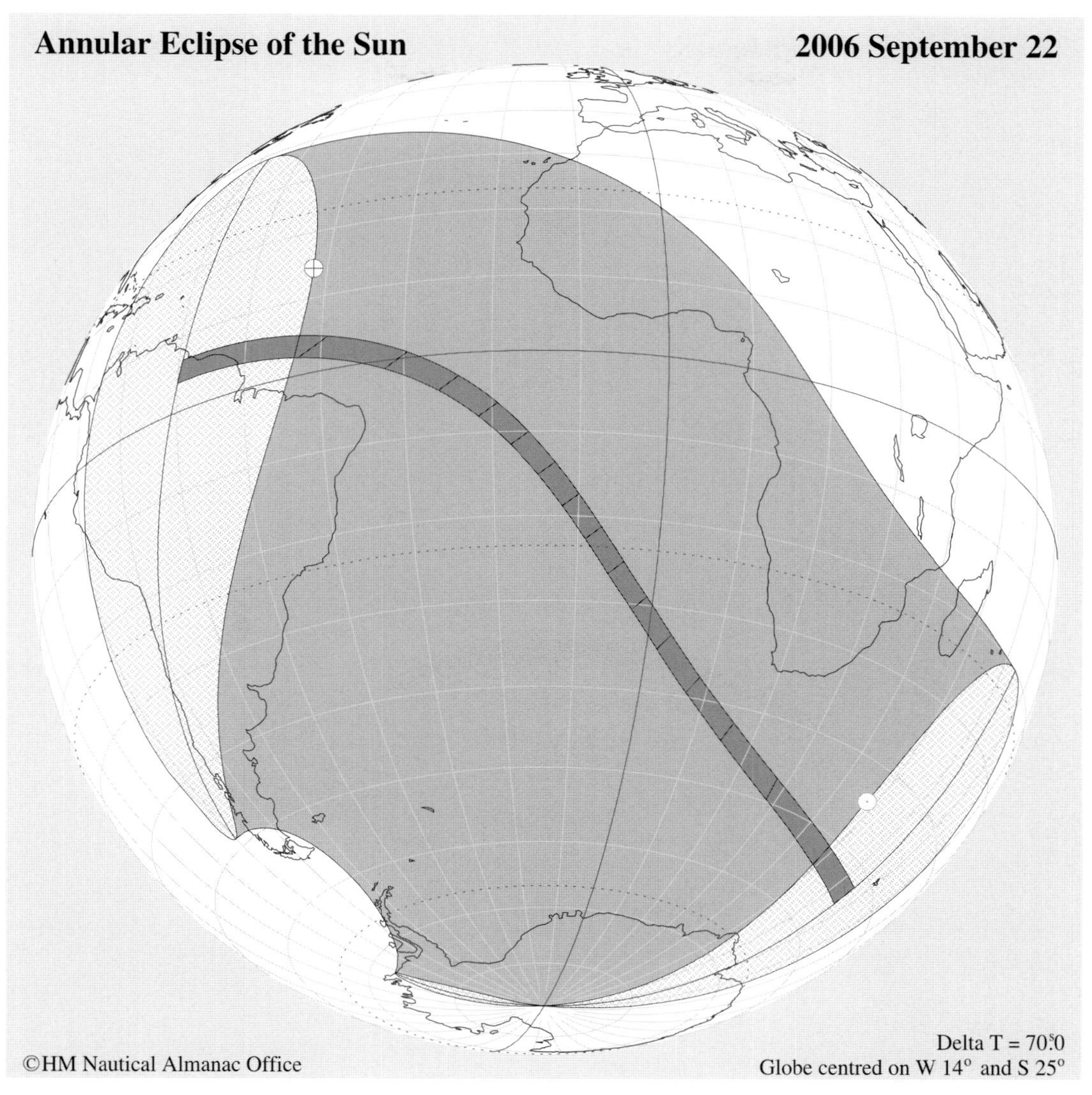

	Circumstances	Time (UT)		Longitude			Latitude	
		h	m		°	′	°	′
⊕	Eclipse begins; first contact with Earth	8	39.8	W	41	50.2	N 14	17.1
	Beginning of northern limit of penumbra	9	43.0	W	57	47.0	N 38	47.1
	Beginning of northern limit of umbra	9	50.3	W	59	24.1	N 6	41.8
	Beginning of centre line; central eclipse begins	9	51.4	W	59	41.2	N 5	15.5
	Beginning of southern limit of umbra	9	52.7	W	59	59.4	N 3	48.3
	Beginning of southern limit of penumbra	11	15.0	W	80	14.2	S 48	50.7
	End of southern limit of penumbra	12	04.4	W	92	04.1	S 72	45.0
	Central eclipse at local apparent noon	12	07.1	W	3	34.9	S 27	39.5
	End of southern limit of umbra	13	27.2	E	66	02.6	S 54	51.1
	End of centre line; central eclipse ends	13	28.5	E	65	44.8	S 53	24.2
	End of northern limit of umbra	13	29.6	E	65	28.0	S 51	58.3
	End of northern limit of penumbra	13	37.4	E	63	45.2	S 19	55.4
⊙	Eclipse ends; last contact with Earth	14	40.1	E	47	55.5	S 44	24.1

Total Eclipse of the Moon 2007 March 03-04

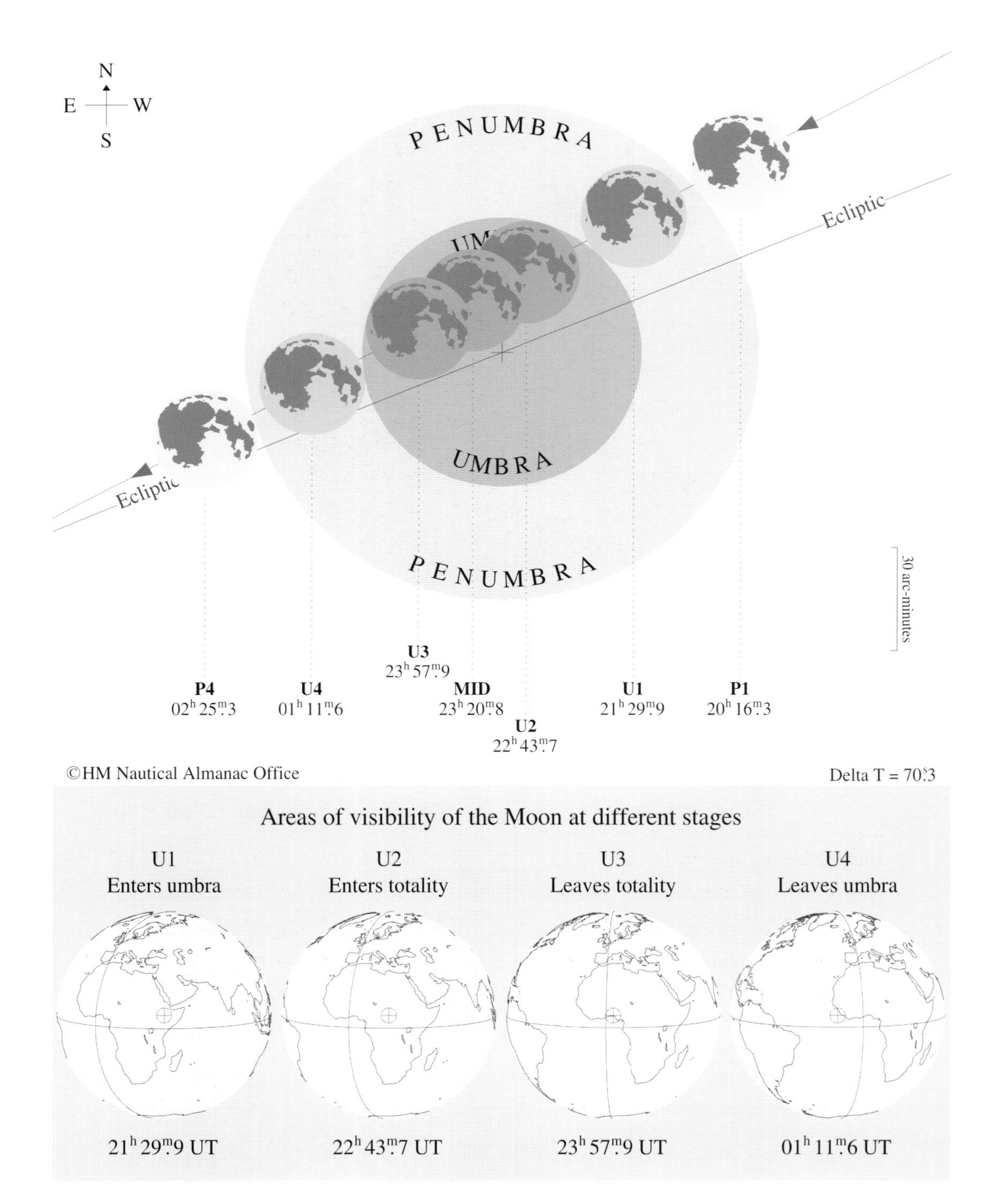

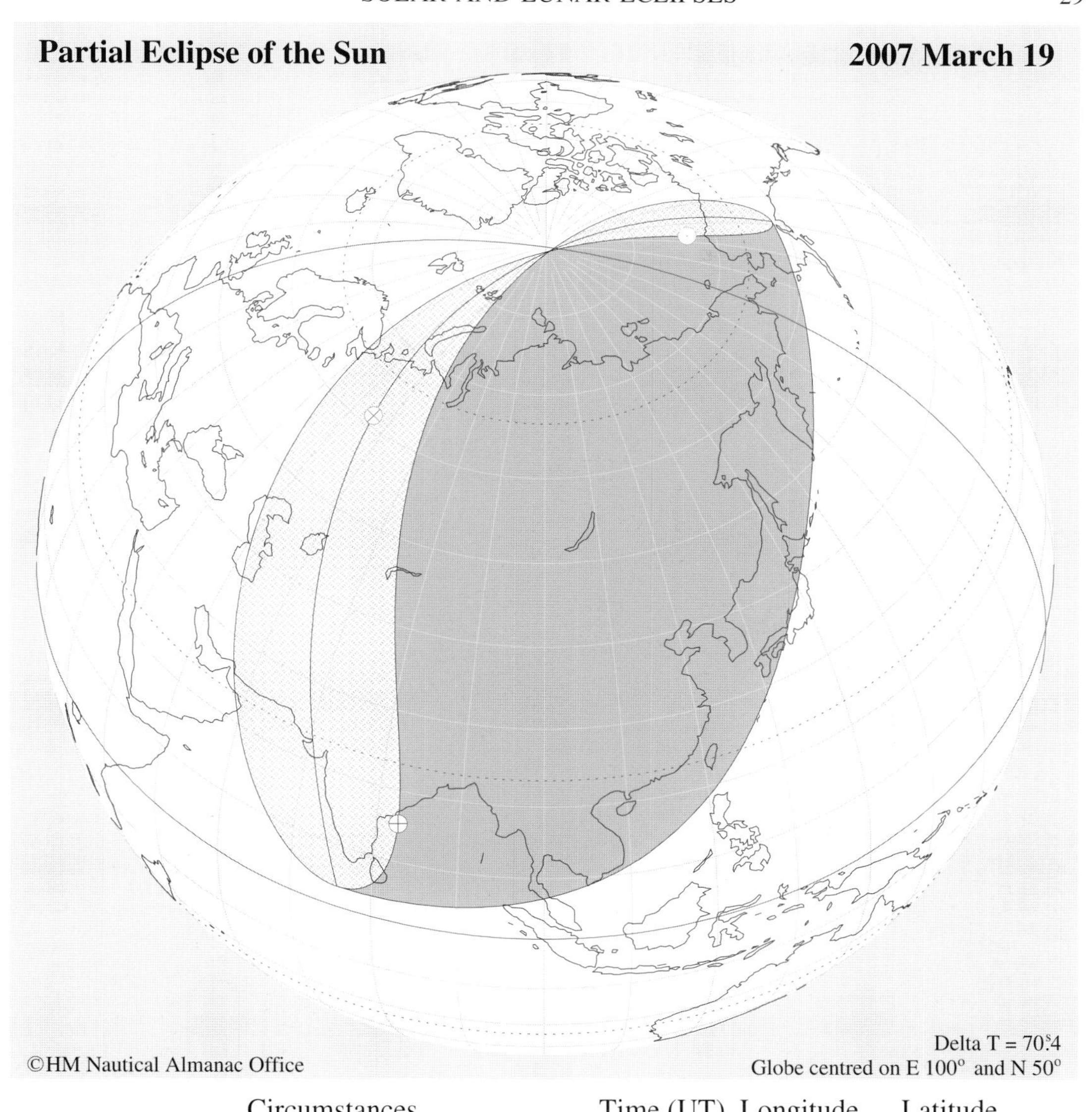

Circumstances	Time (UT)	Longitude	Latitude
	h m	° ′	° ′
⊕ Eclipse begins; first contact with Earth	0 38.2	E 82 39.8	N 15 26.2
Beginning of southern limit of penumbra	1 04.8	E 75 51.4	N 3 30.7
⊗ Greatest eclipse (magnitude = 0.8761)	2 31.8	E 55 25.8	N 61 12.7
End of southern limit of penumbra	3 58.4	W 148 57.1	N 61 37.7
⊙ Eclipse ends; last contact with Earth	4 24.9	W 156 39.3	N 73 25.5

Total Eclipse of the Moon

2007 August 28

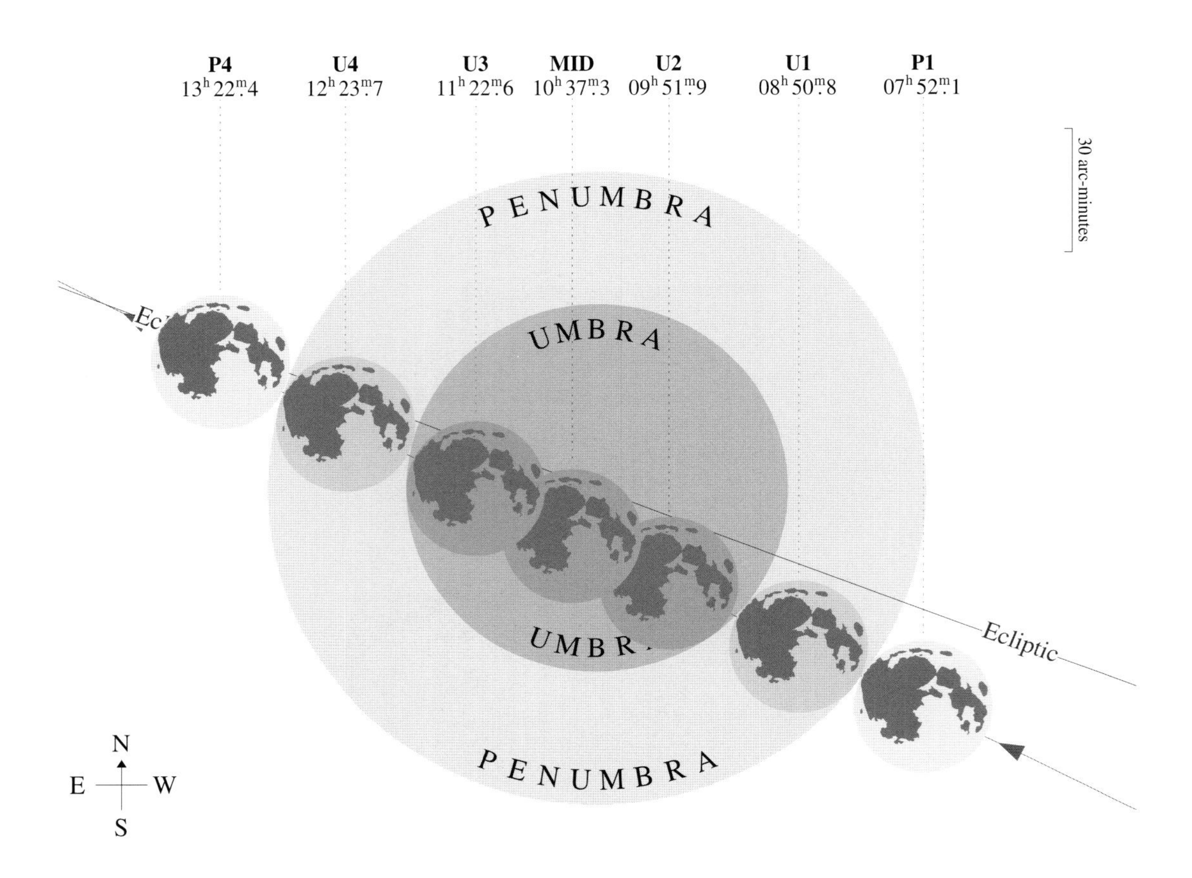

Delta T = $70^s.7$

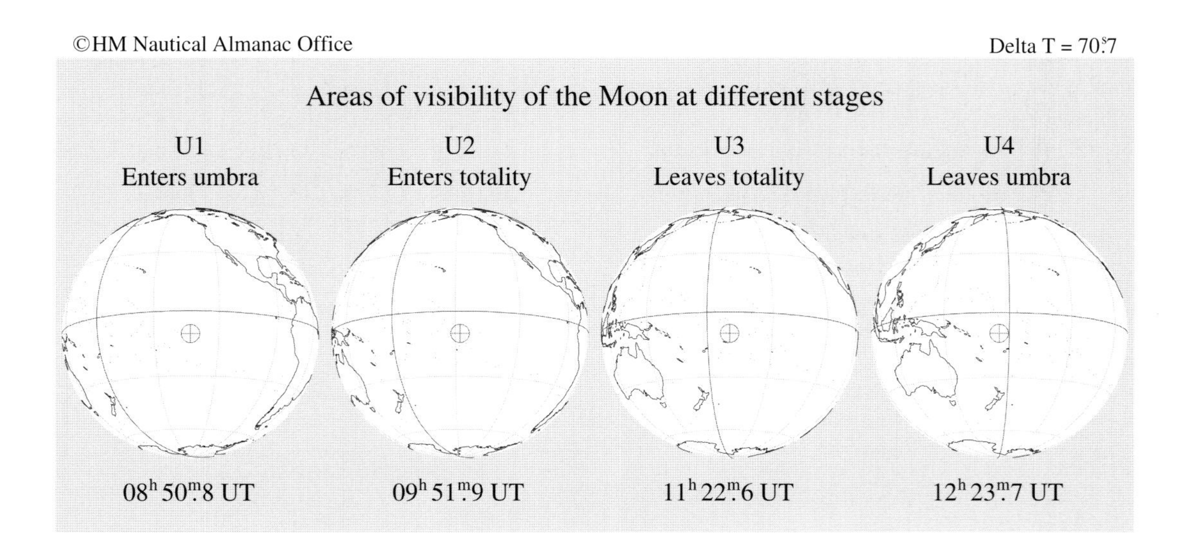

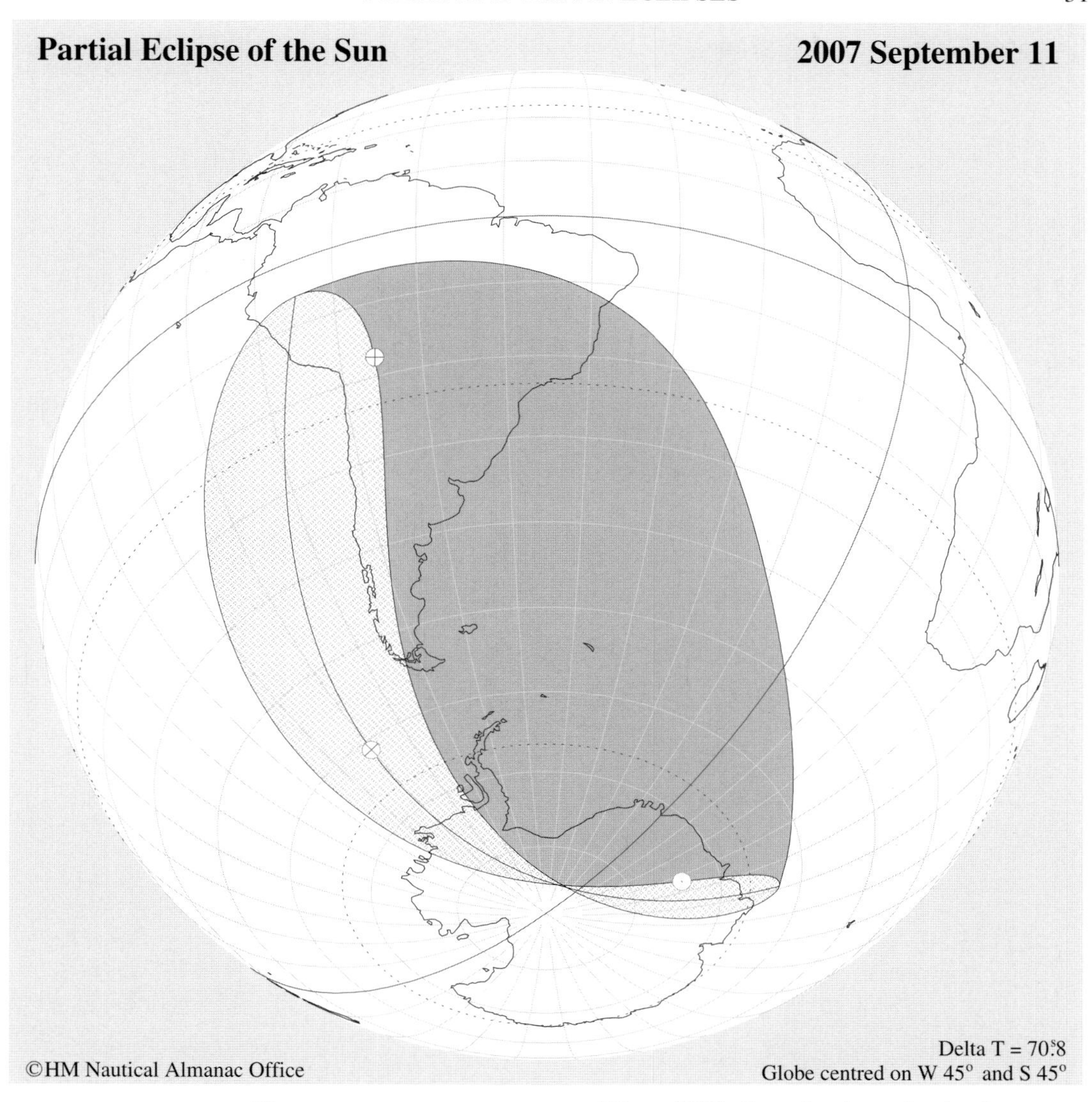

Circumstances	Time (UT)	Longitude	Latitude
	h m	° ′	° ′
⊕ Eclipse begins; first contact with Earth	10 25.6	W 65 44.8	S 17 29.7
Beginning of northern limit of penumbra	10 57.5	W 74 43.4	S 5 40.0
⊗ Greatest eclipse (magnitude = 0.7509)	12 31.2	W 90 14.0	S 61 09.8
End of northern limit of penumbra	14 04.6	E 49 04.9	S 62 51.0
⊙ Eclipse ends; last contact with Earth	14 36.4	E 33 36.0	S 74 17.8

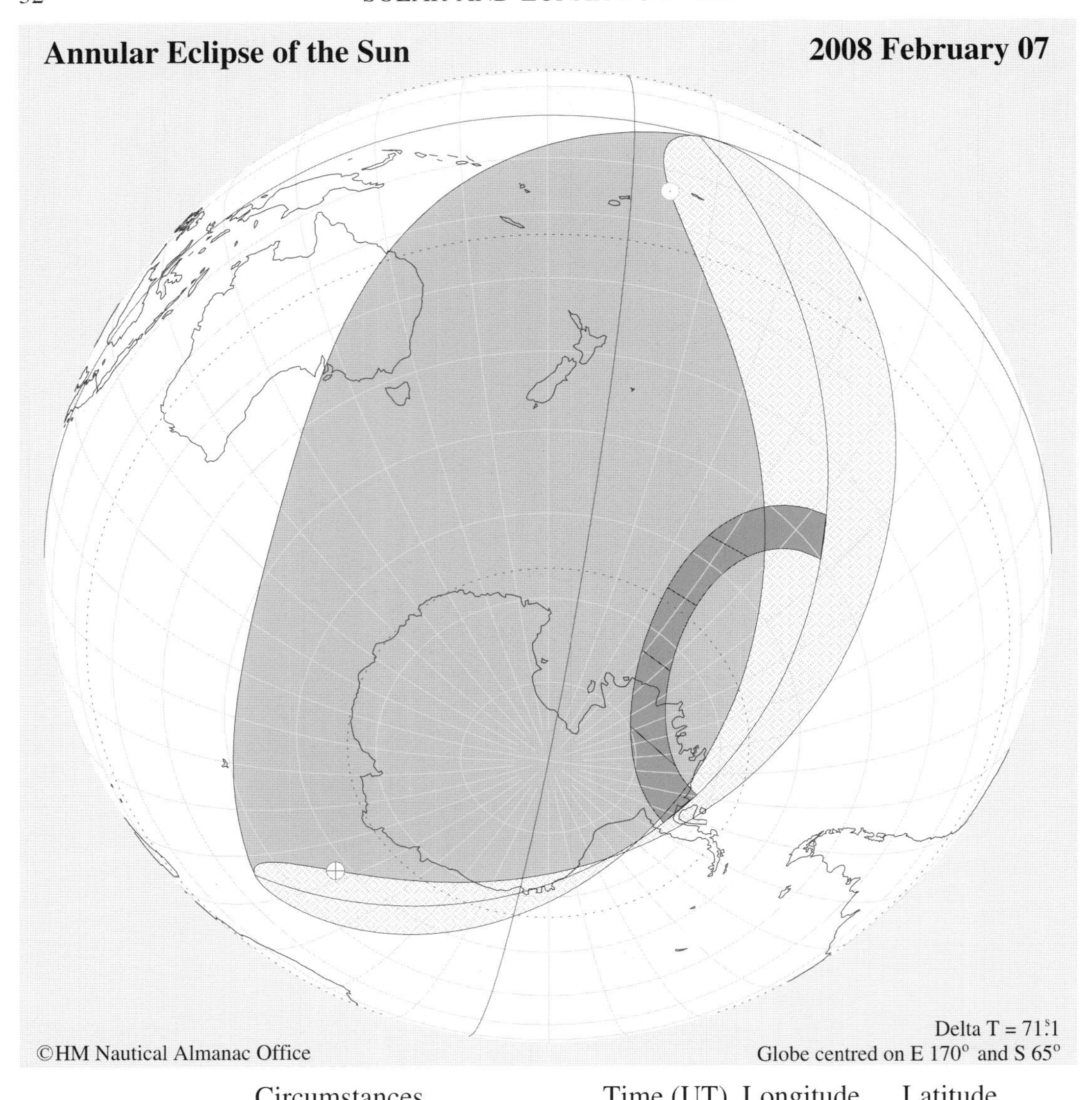

	Circumstances	Time (UT)		Longitude			Latitude	
		h	m		°	′	°	′
⊕	Eclipse begins; first contact with Earth	1	38.4	E	42	36.9	S 57	53.8
	Beginning of northern limit of penumbra	2	06.1	E	45	44.7	S 45	11.4
	Beginning of northern limit of umbra	3	19.9	W	64	39.7	S 73	42.2
	Beginning of centre line; central eclipse begins	3	23.9	W	72	34.2	S 72	56.9
	Beginning of southern limit of umbra	3	28.5	W	80	50.5	S 71	49.5
	End of southern limit of umbra	4	22.0	W	132	41.0	S 49	58.5
	End of centre line; central eclipse ends	4	26.5	W	135	18.6	S 47	46.6
	End of northern limit of umbra	4	30.5	W	137	29.6	S 45	50.6
	End of northern limit of penumbra	5	44.0	W	172	23.5	S 0	18.5
○	Eclipse ends; last contact with Earth	6	11.8	W	175	27.8	S 13	59.4

Total Eclipse of the Moon 2008 February 21

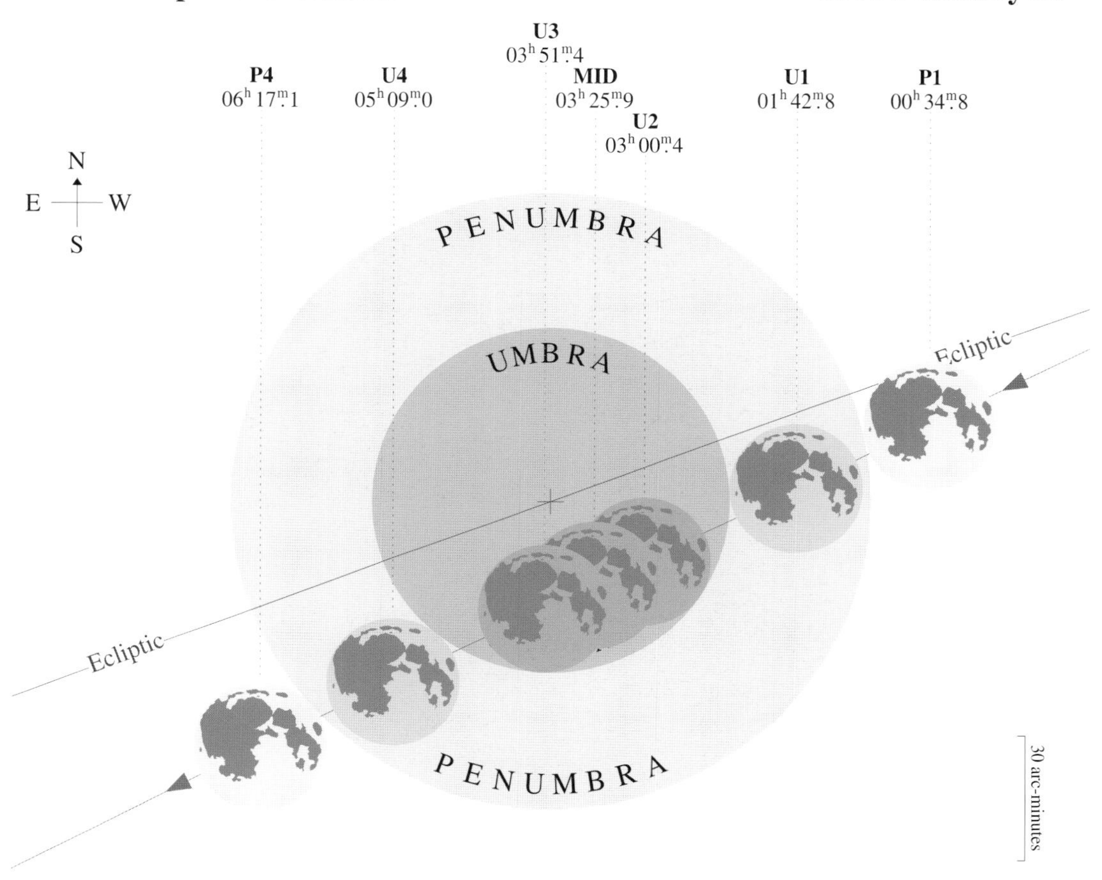

Delta T = $71^{s}.1$

Areas of visibility of the Moon at different stages

U1	U2	U3	U4
Enters umbra	Enters totality	Leaves totality	Leaves umbra
$01^{h}\,42^{m}.8$ UT	$03^{h}\,00^{m}.4$ UT	$03^{h}\,51^{m}.4$ UT	$05^{h}\,09^{m}.0$ UT

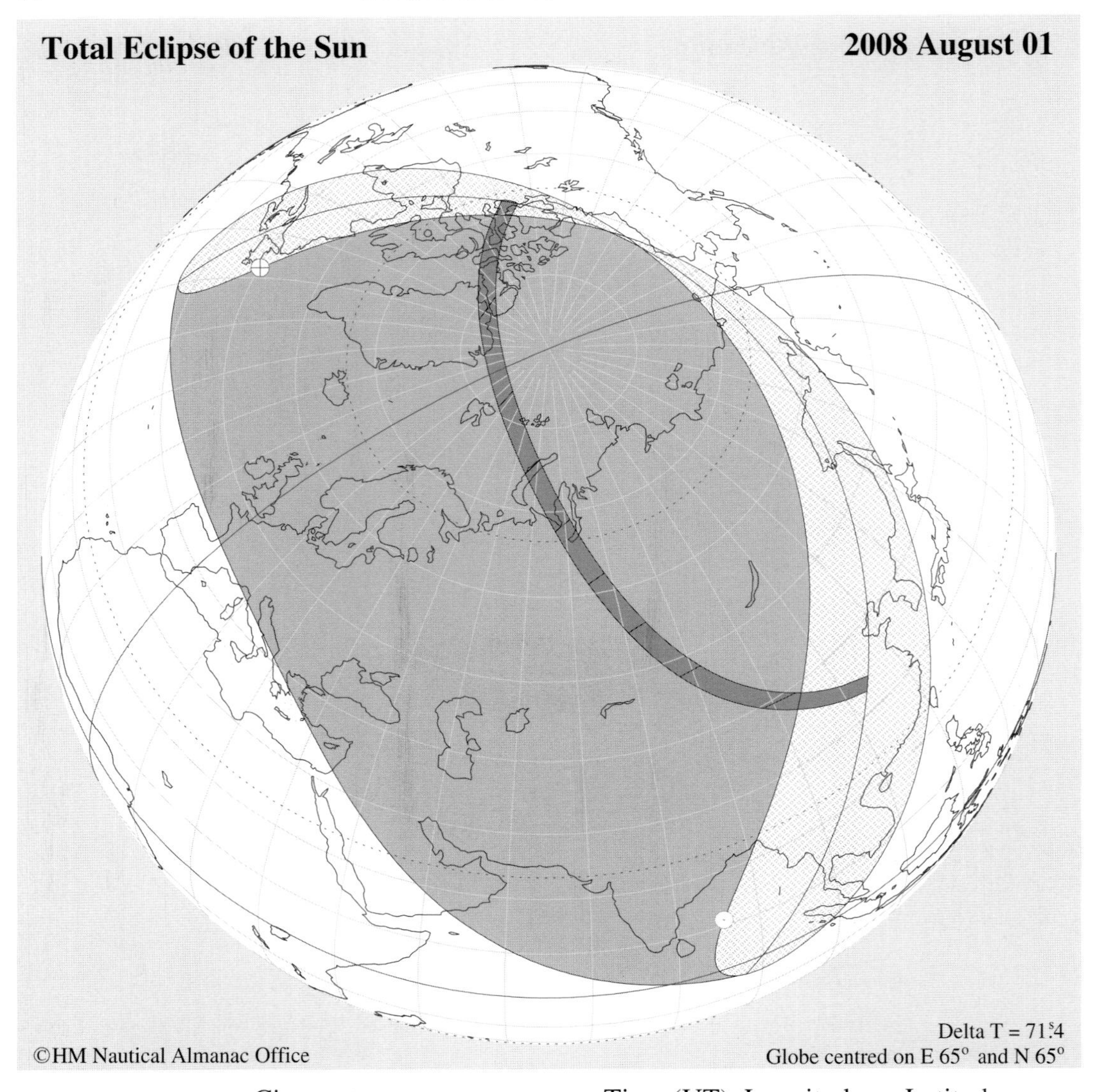

Circumstances	Time (UT)	Longitude	Latitude
	h m	° ′	° ′
⊕ Eclipse begins; first contact with Earth	8 04.0	W 52 12.7	N 50 12.4
Beginning of southern limit of penumbra	8 32.4	W 50 13.0	N 36 15.7
Beginning of southern limit of umbra	9 21.2	W 100 55.6	N 67 48.1
Beginning of centre line; central eclipse begins	9 22.5	W 103 05.6	N 68 16.7
Beginning of northern limit of umbra	9 23.9	W 105 23.4	N 68 44.6
Central eclipse at local apparent noon	9 47.2	E 34 45.9	N 81 06.9
End of northern limit of umbra	11 18.5	E 114 37.3	N 34 15.6
End of centre line; central eclipse ends	11 19.8	E 113 54.7	N 33 29.1
End of southern limit of umbra	11 21.2	E 113 13.7	N 32 43.5
End of southern limit of penumbra	12 09.8	E 87 58.5	S 3 34.7
⊙ Eclipse ends; last contact with Earth	12 38.3	E 85 37.9	N 11 09.9

Partial Eclipse of the Moon 2008 August 16

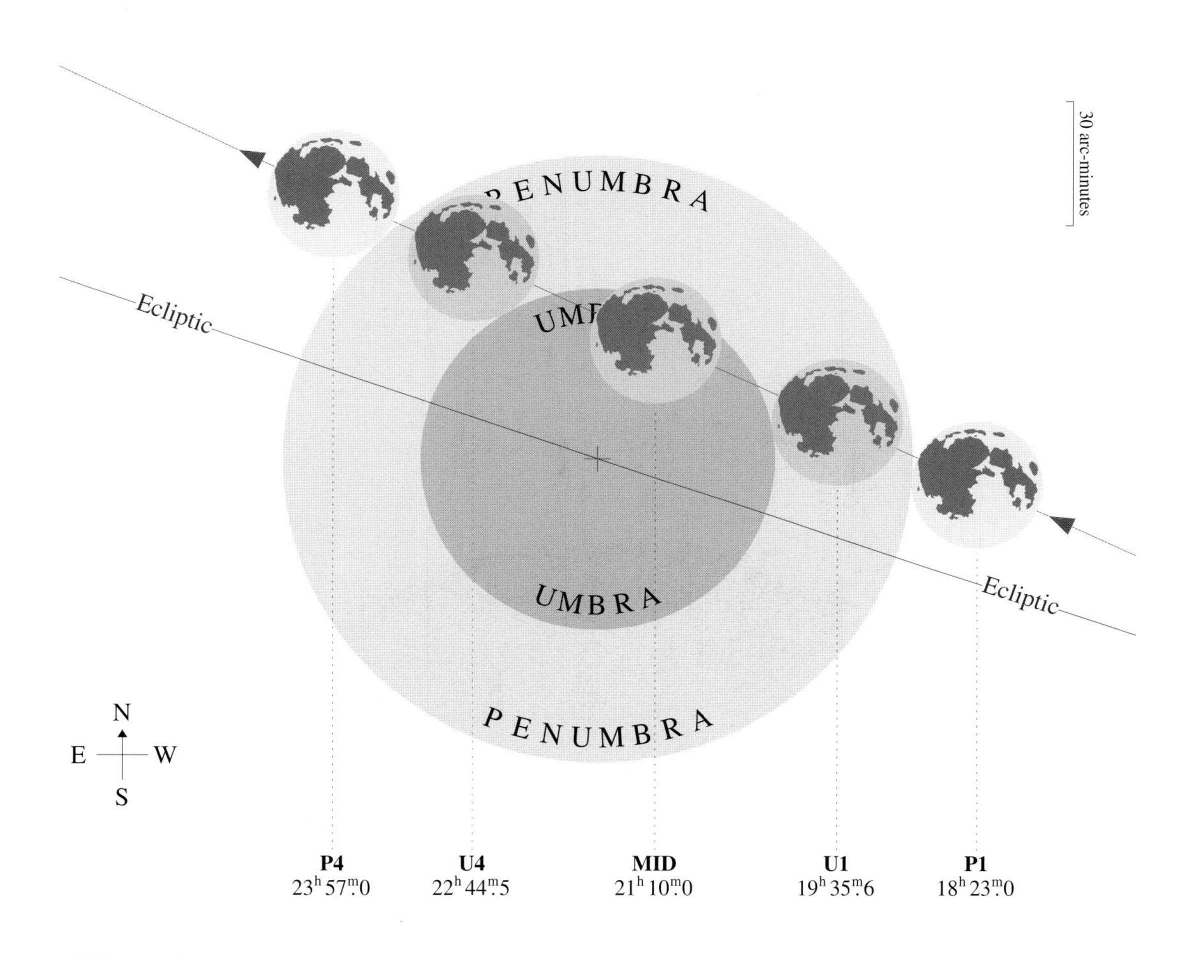

Delta T = 71s.5

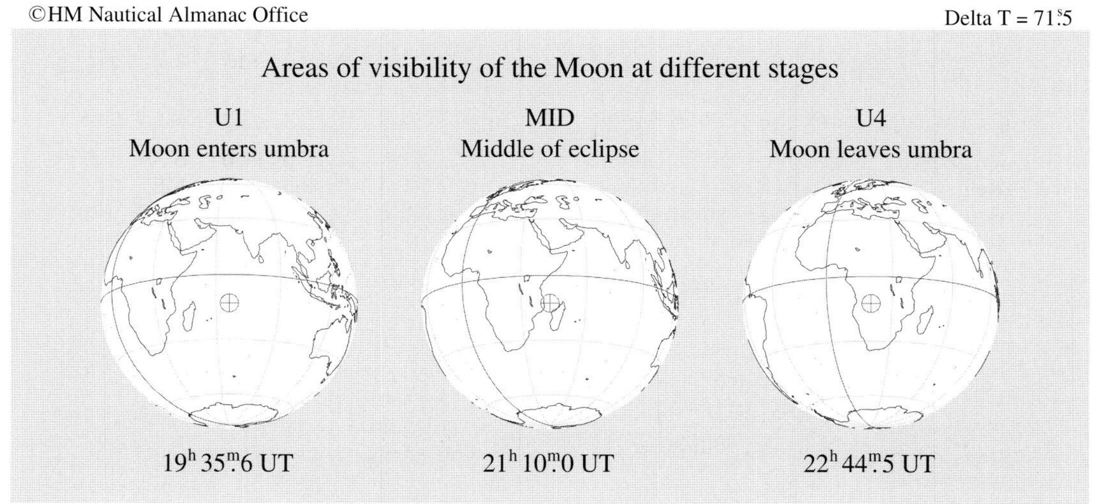

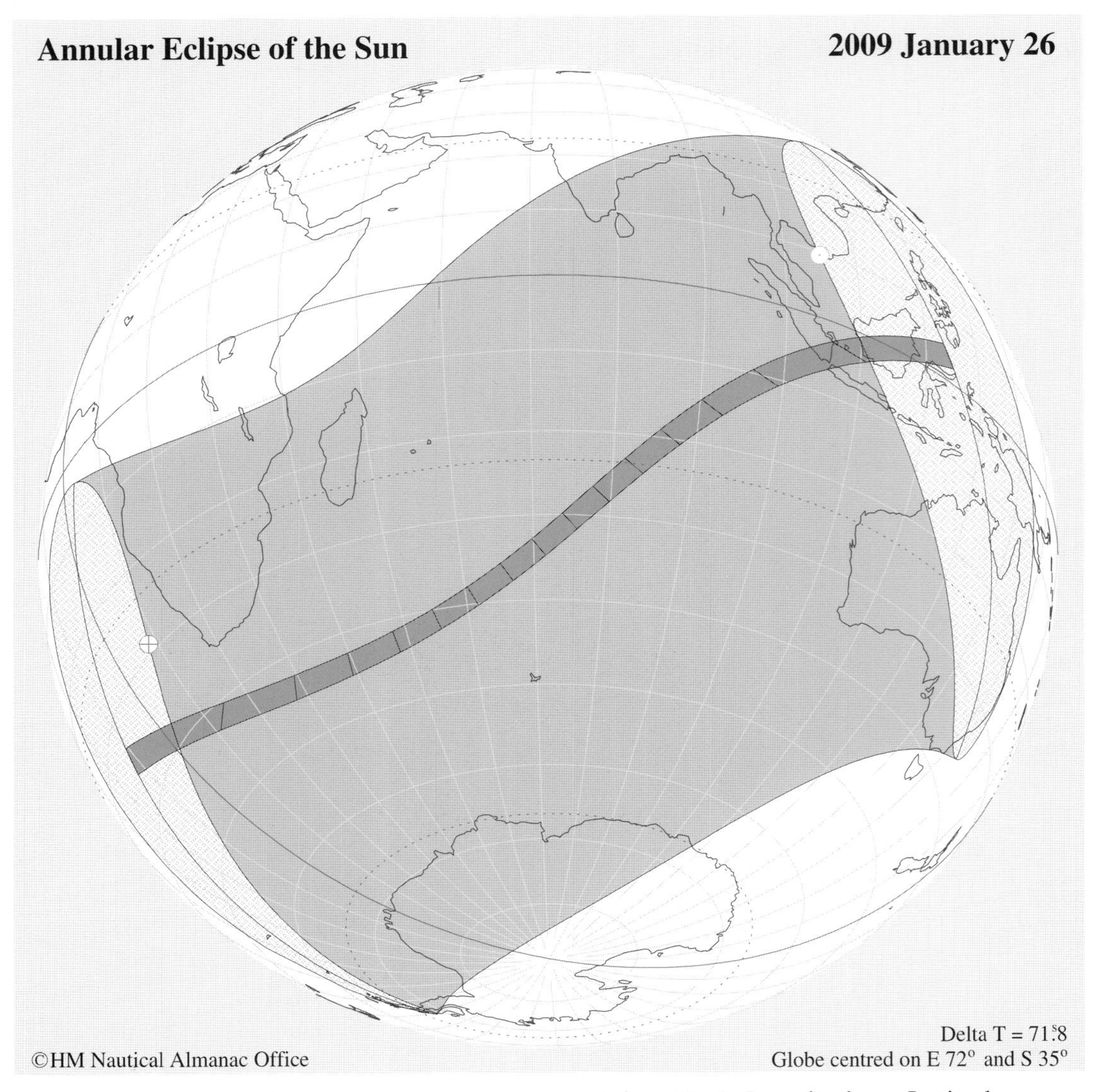

	Circumstances	Time (UT)	Longitude	Latitude
		h m	° ′	° ′
⊕	Eclipse begins; first contact with Earth	4 56.5	E 8 15.0	S 28 54.7
	Beginning of northern limit of penumbra	5 54.4	E 3 18.7	S 3 38.9
	Beginning of northern limit of umbra	6 04.4	W 10 43.0	S 33 10.0
	Beginning of centre line; central eclipse begins	6 05.7	W 11 44.5	S 34 33.4
	Beginning of southern limit of umbra	6 07.1	W 12 48.7	S 35 57.3
	Beginning of southern limit of penumbra	7 02.6	W 71 20.5	S 68 27.8
	Central eclipse at local apparent noon	7 46.3	E 66 34.0	S 36 21.0
	End of southern limit of penumbra	8 54.9	E 153 40.8	S 36 08.6
	End of southern limit of umbra	9 50.1	E 124 50.2	N 2 16.3
	End of centre line; central eclipse ends	9 51.5	E 124 01.2	N 3 42.0
	End of northern limit of umbra	9 52.8	E 123 13.2	N 5 06.9
	End of northern limit of penumbra	10 02.5	E 109 04.5	N 34 36.5
⊙	Eclipse ends; last contact with Earth	11 00.6	E 104 46.6	N 9 28.4

Penumbral Eclipse of the Moon — 2009 February 09

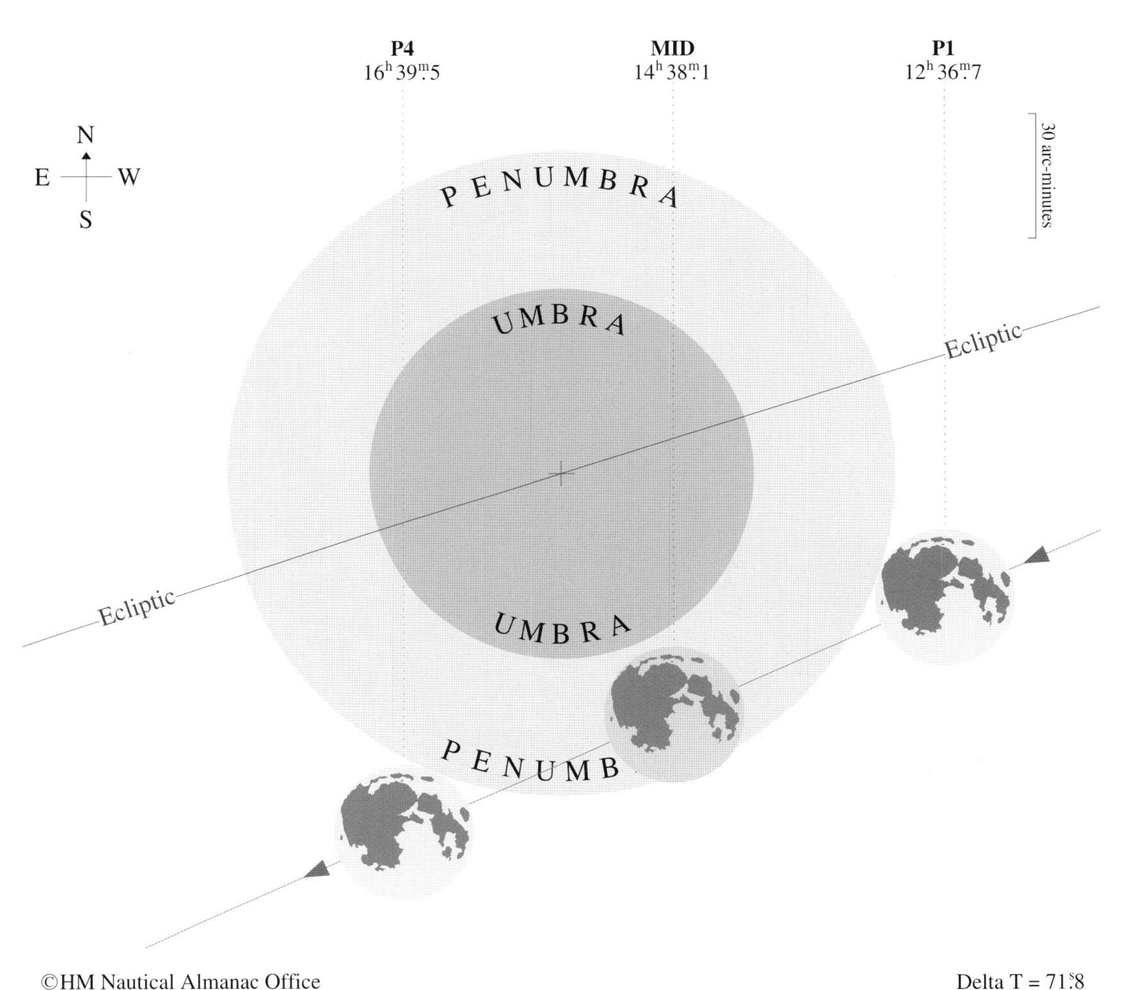

Areas of visibility of the Moon at different stages

P1 Moon enters penumbra	MID Middle of eclipse	P4 Moon leaves penumbra
$12^{h} 36^{m}.7$ UT	$14^{h} 38^{m}.1$ UT	$16^{h} 39^{m}.5$ UT

Penumbral Eclipse of the Moon 2009 July 07

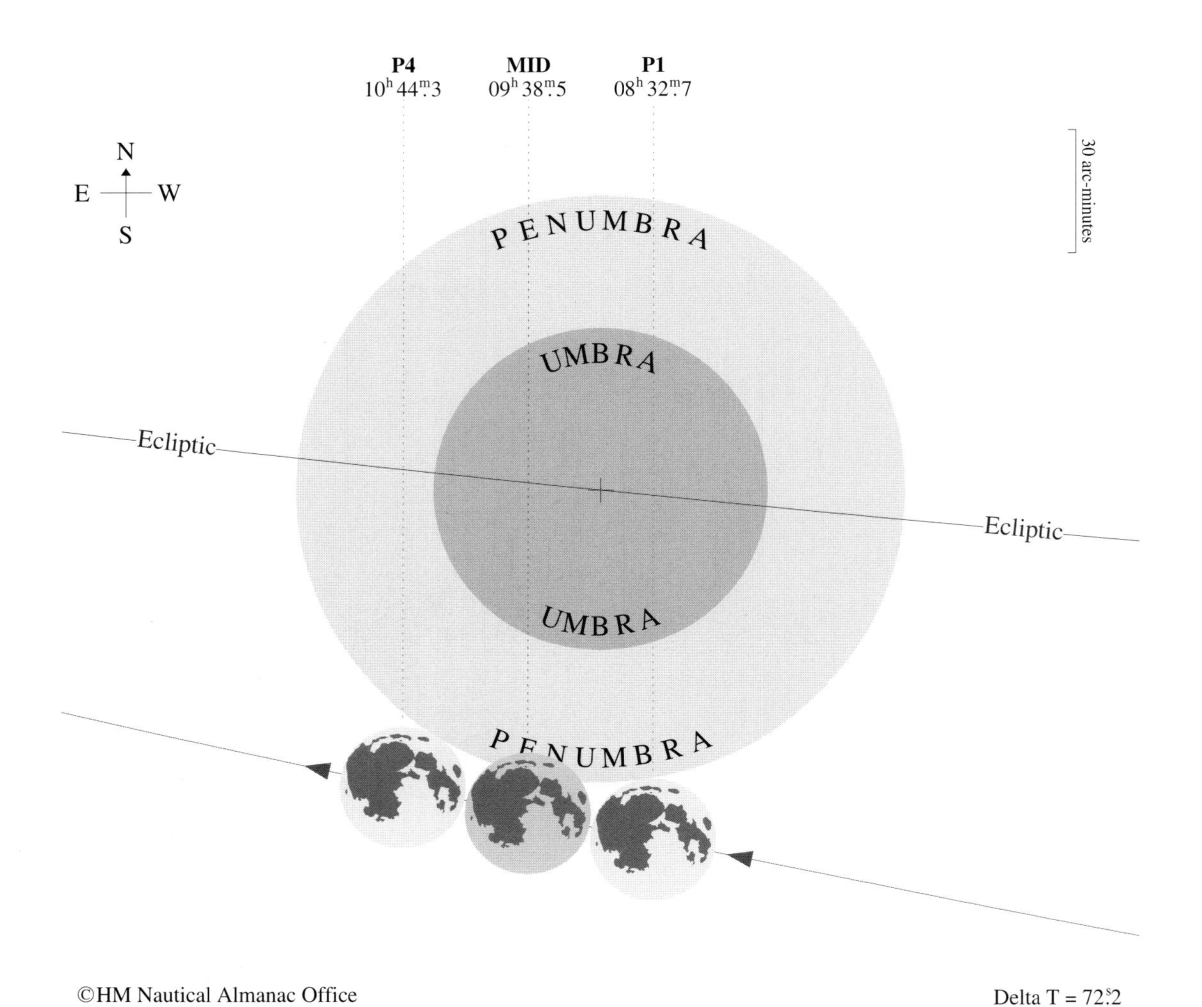

Delta T = $72^s.2$

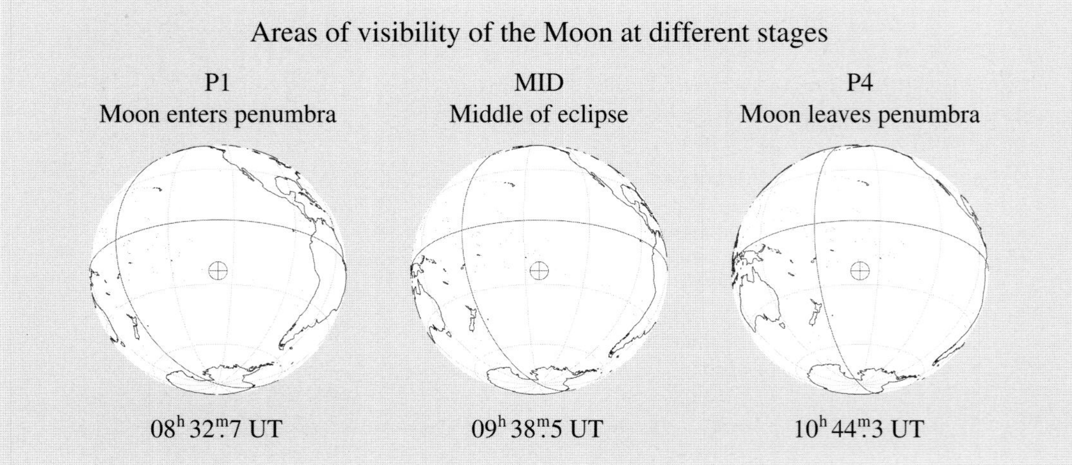

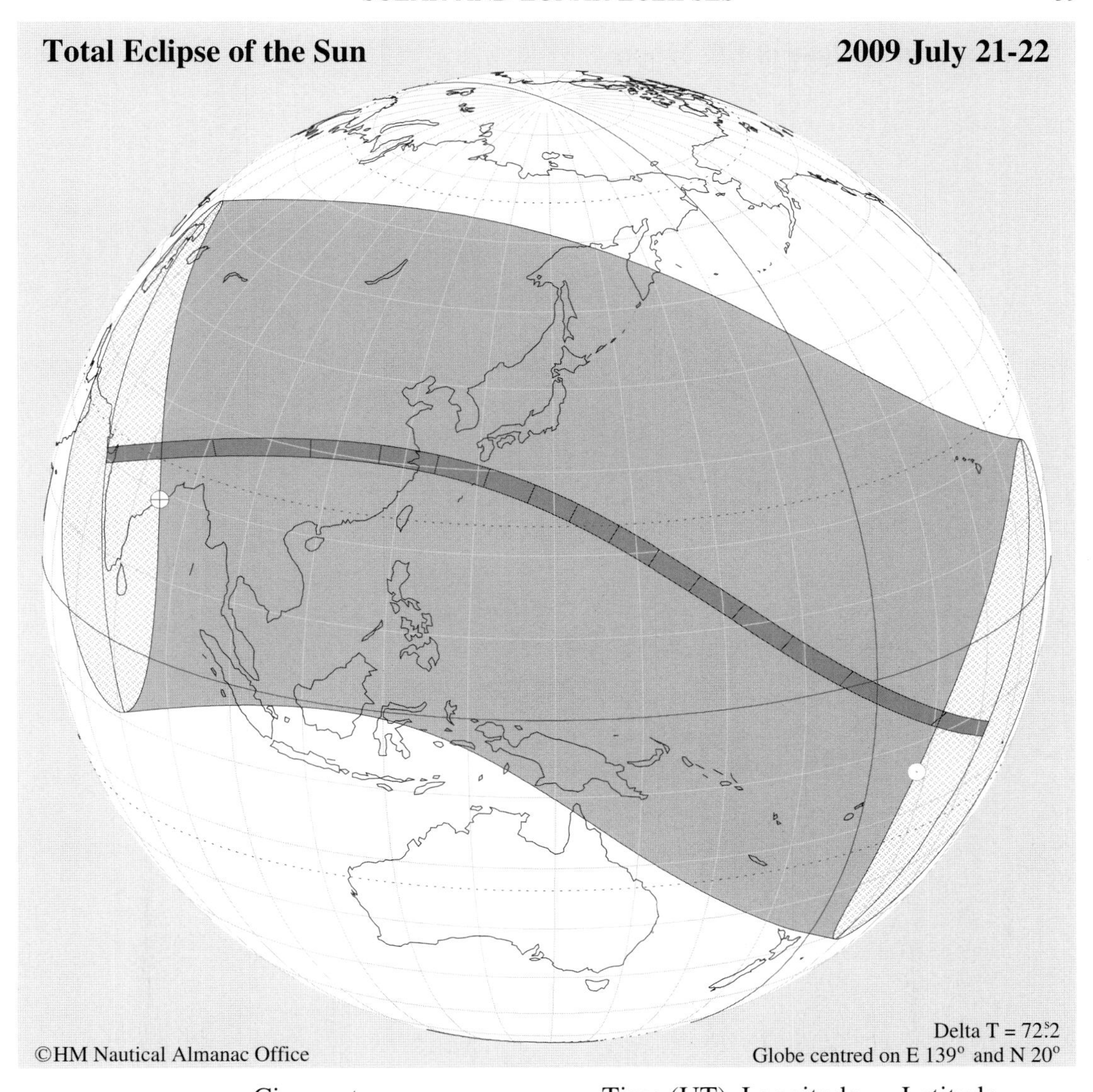

Circumstances	Time (UT)	Longitude	Latitude
	h m	° ′	° ′
⊕ Eclipse begins; first contact with Earth	23 58.1	E 84 44.7	N 19 02.9
Beginning of southern limit of umbra	0 52.4	E 70 59.6	N 19 30.8
Beginning of centre line; central eclipse begins	0 52.7	E 70 33.0	N 20 21.5
Beginning of northern limit of umbra	0 53.1	E 70 05.7	N 21 12.1
Beginning of southern limit of penumbra	0 55.5	E 81 01.9	S 8 51.4
Beginning of northern limit of penumbra	1 19.1	E 45 53.1	N 49 49.4
Central eclipse at local apparent noon	2 32.9	E 143 23.2	N 24 36.7
End of northern limit of penumbra	3 51.5	W 139 26.2	N 17 50.6
End of southern limit of penumbra	4 14.6	W 171 07.2	S 41 31.8
End of northern limit of umbra	4 17.3	W 157 14.2	S 12 04.1
End of centre line; central eclipse ends	4 17.7	W 157 39.6	S 12 54.9
End of southern limit of umbra	4 18.0	W 158 04.6	S 13 45.7
○ Eclipse ends; last contact with Earth	5 12.3	W 171 49.3	S 14 13.8

Penumbral Eclipse of the Moon — 2009 August 05-06

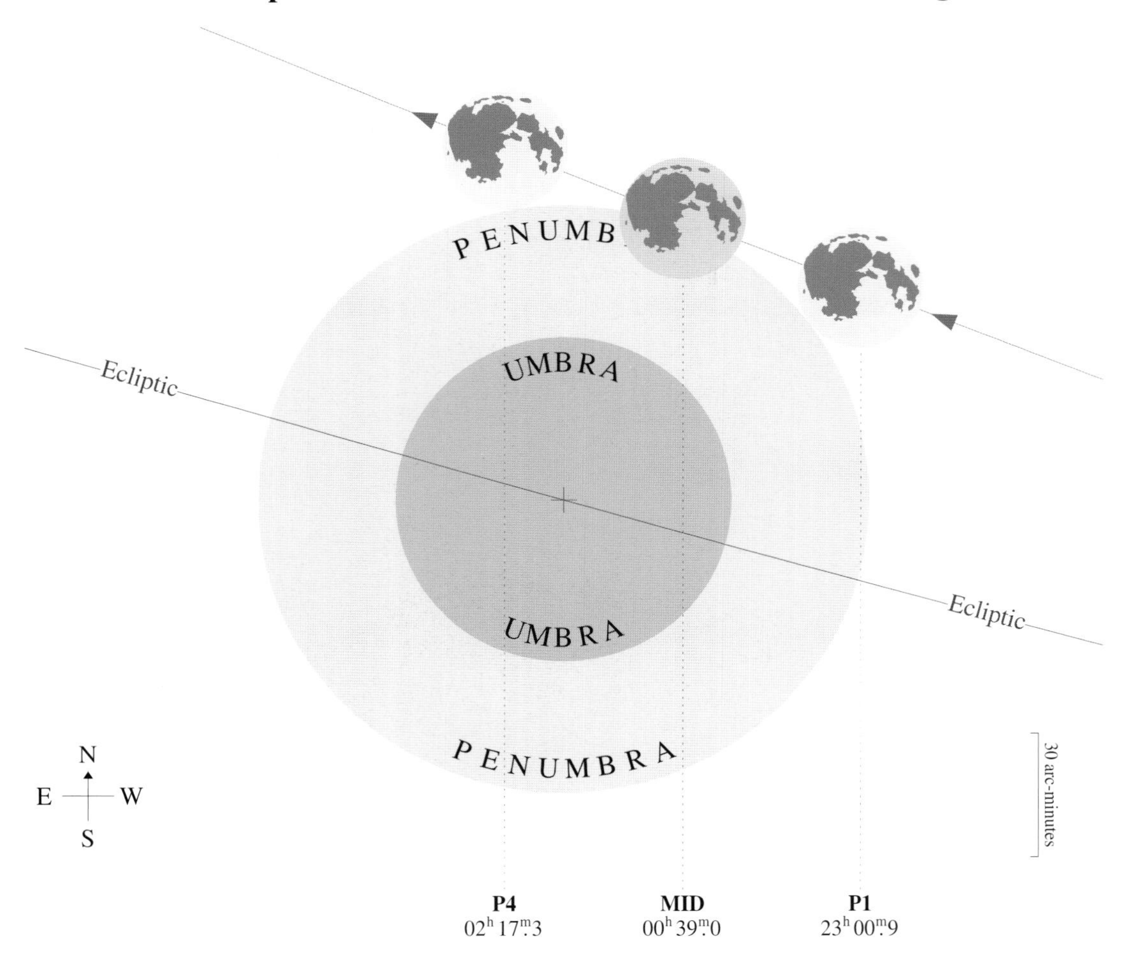

Delta T = $72^s.2$

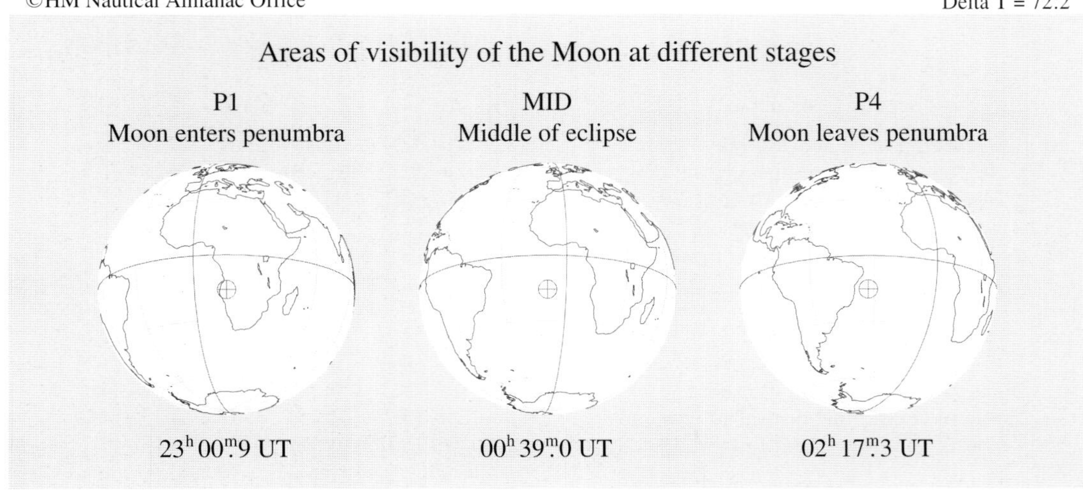

Partial Eclipse of the Moon **2009 December 31**

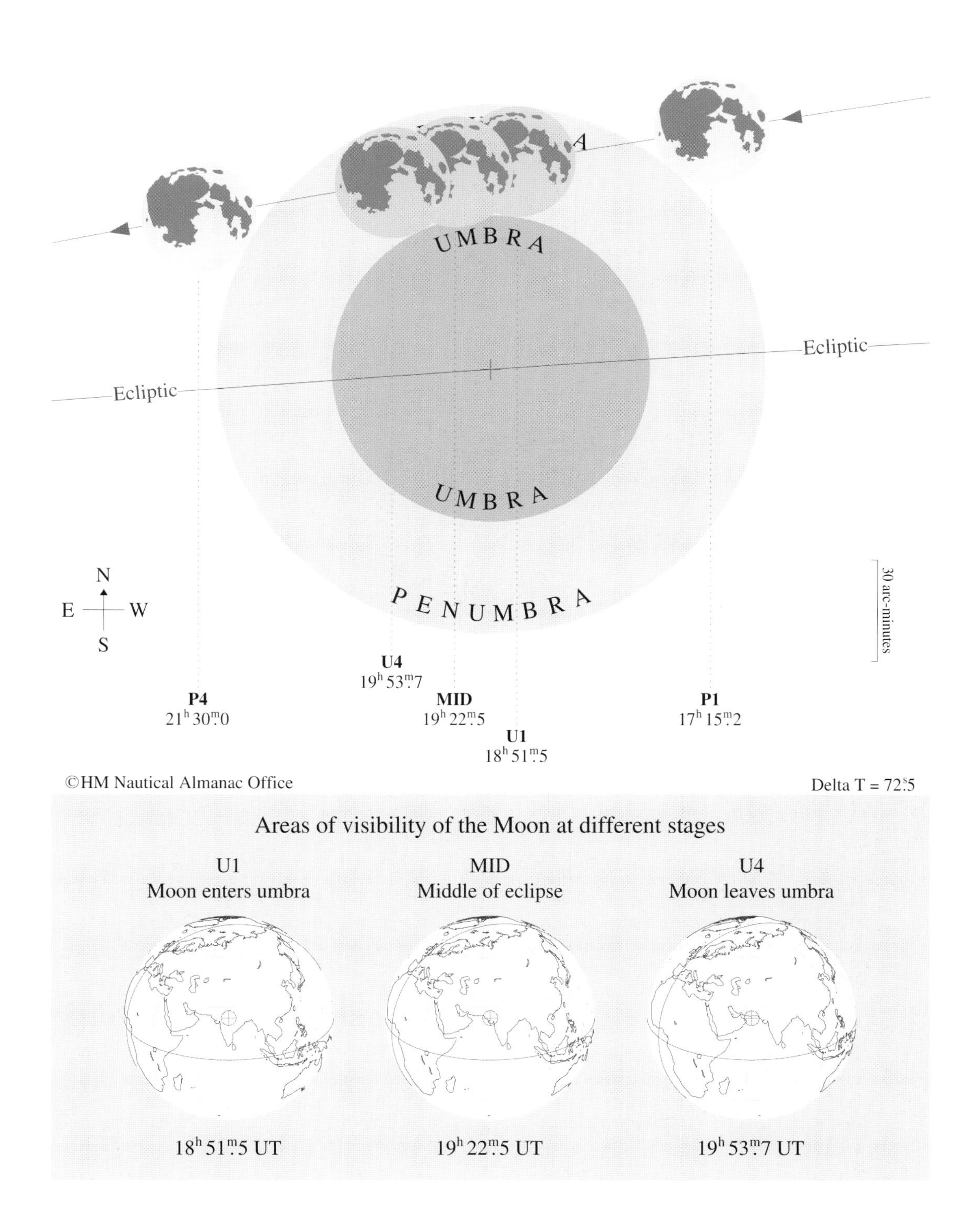

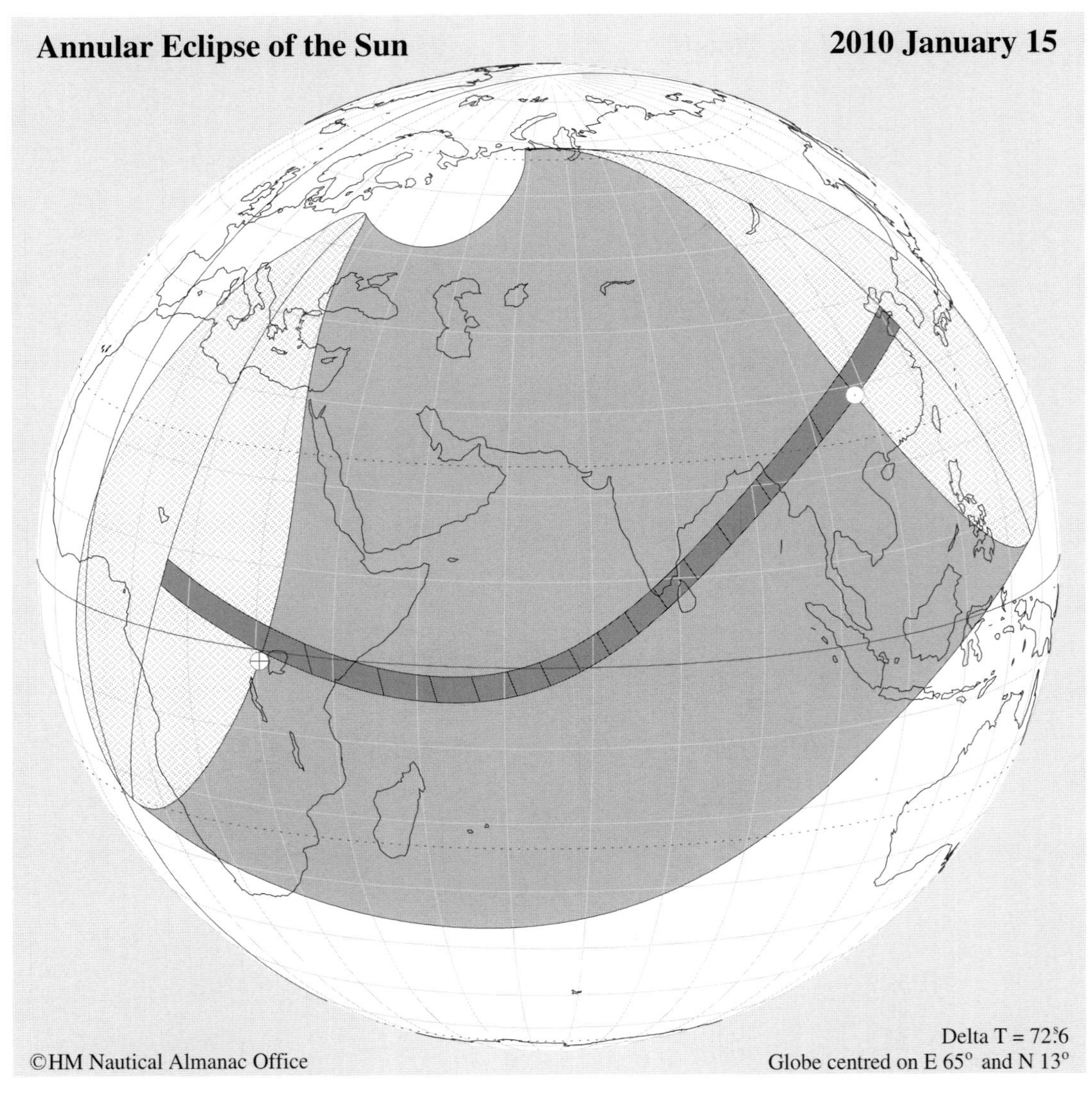

Circumstances	Time (UT)	Longitude	Latitude
	h m	° ′	° ′
⊕ Eclipse begins; first contact with Earth	4 05.3	E 30 28.7	S 1 19.5
Beginning of southern limit of umbra	5 16.7	E 15 14.3	N 5 21.9
Beginning of centre line; central eclipse begins	5 17.4	E 15 40.5	N 6 58.3
Beginning of northern limit of umbra	5 18.3	E 16 05.8	N 8 36.0
Beginning of southern limit of penumbra	5 21.2	E 2 08.2	S 23 58.3
Beginning of northern limit of penumbra	6 37.2	E 26 33.0	N 55 01.0
Central eclipse at local apparent noon	7 20.2	E 72 16.5	N 3 30.9
End of northern limit of penumbra	7 35.1	E 58 02.7	N 68 32.8
End of southern limit of penumbra	8 51.8	E 136 58.2	N 6 14.2
End of northern limit of umbra	8 54.4	E 120 56.1	N 38 22.9
End of centre line; central eclipse ends	8 55.3	E 121 42.9	N 36 49.3
End of southern limit of umbra	8 56.0	E 122 28.3	N 35 16.4
⊙ Eclipse ends; last contact with Earth	10 07.5	E 108 13.3	N 28 47.9

Partial Eclipse of the Moon 2010 June 26

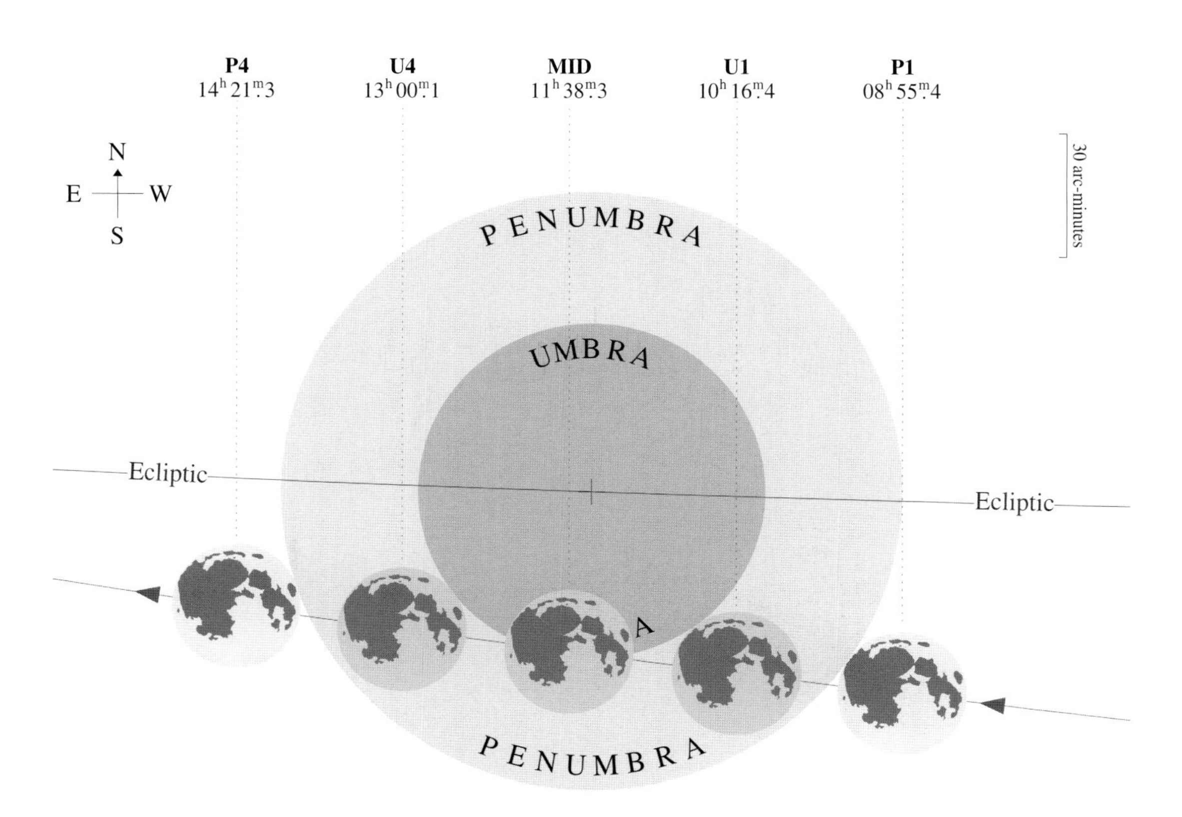

Delta T = $72^s.9$

Areas of visibility of the Moon at different stages

U1	MID	U4
Moon enters umbra	Middle of eclipse	Moon leaves umbra
$10^h 16^m.4$ UT	$11^h 38^m.3$ UT	$13^h 00^m.1$ UT

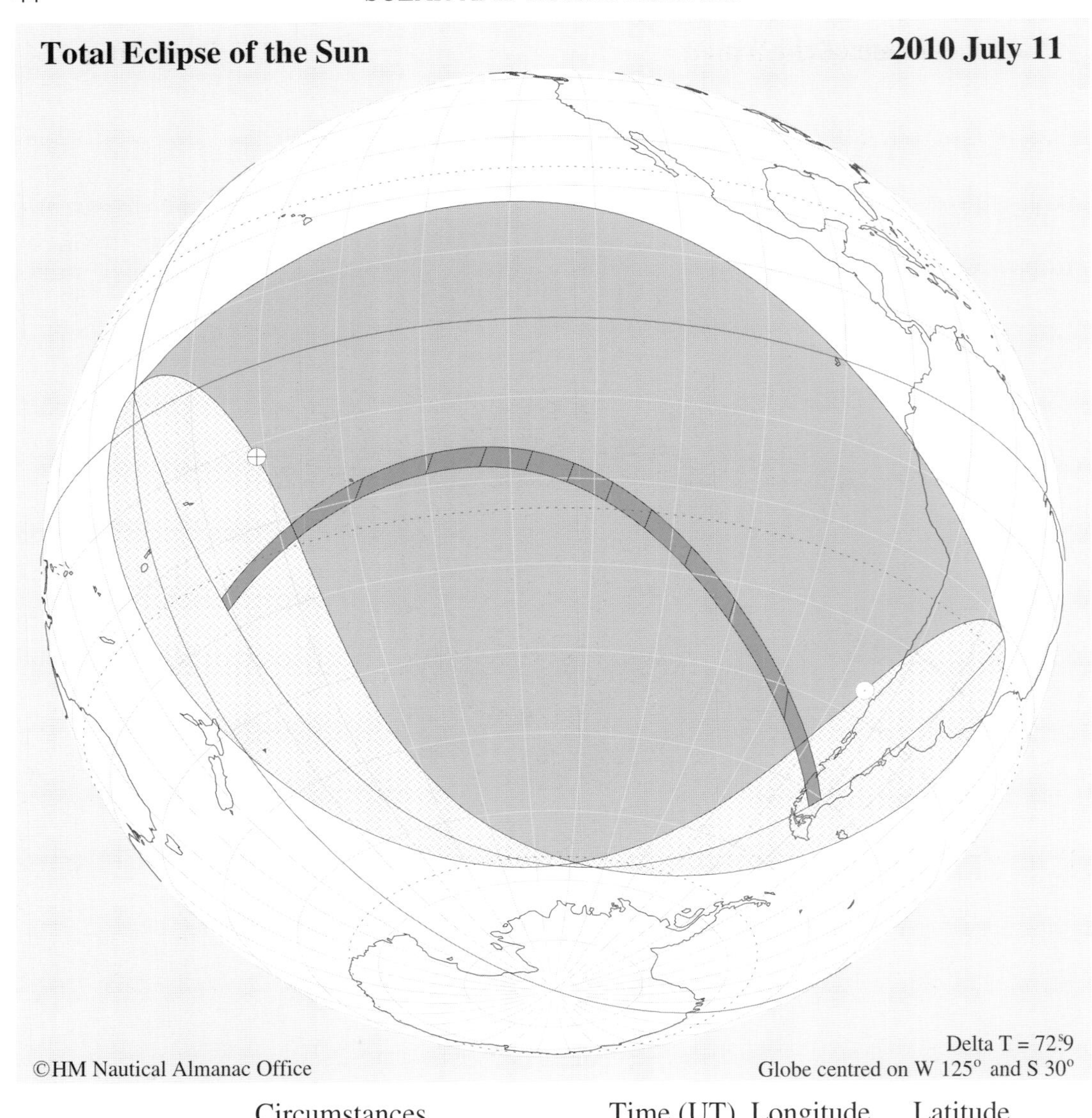

Circumstances	Time (UT)	Longitude	Latitude
	h m	° ′	° ′
⊕ Eclipse begins; first contact with Earth	17 09.5	W161 11.9	S 11 39.0
Beginning of northern limit of penumbra	17 59.5	E 179 36.5	N 4 40.1
Beginning of northern limit of umbra	18 15.8	W171 09.0	S 26 02.2
Beginning of centre line; central eclipse begins	18 16.7	W170 57.5	S 26 51.6
Beginning of southern limit of umbra	18 17.7	W170 46.0	S 27 41.8
Central eclipse at local apparent noon	19 50.8	W116 18.7	S 22 28.0
End of southern limit of umbra	20 48.9	W 71 32.7	S 51 36.6
End of centre line; central eclipse ends	20 49.9	W 70 54.0	S 50 51.6
End of northern limit of umbra	20 50.8	W 70 17.2	S 50 07.0
End of northern limit of penumbra	21 07.2	W 54 18.9	S 20 55.8
○ Eclipse ends; last contact with Earth	21 57.1	W 75 30.0	S 36 47.4

Total Eclipse of the Moon 2010 December 21

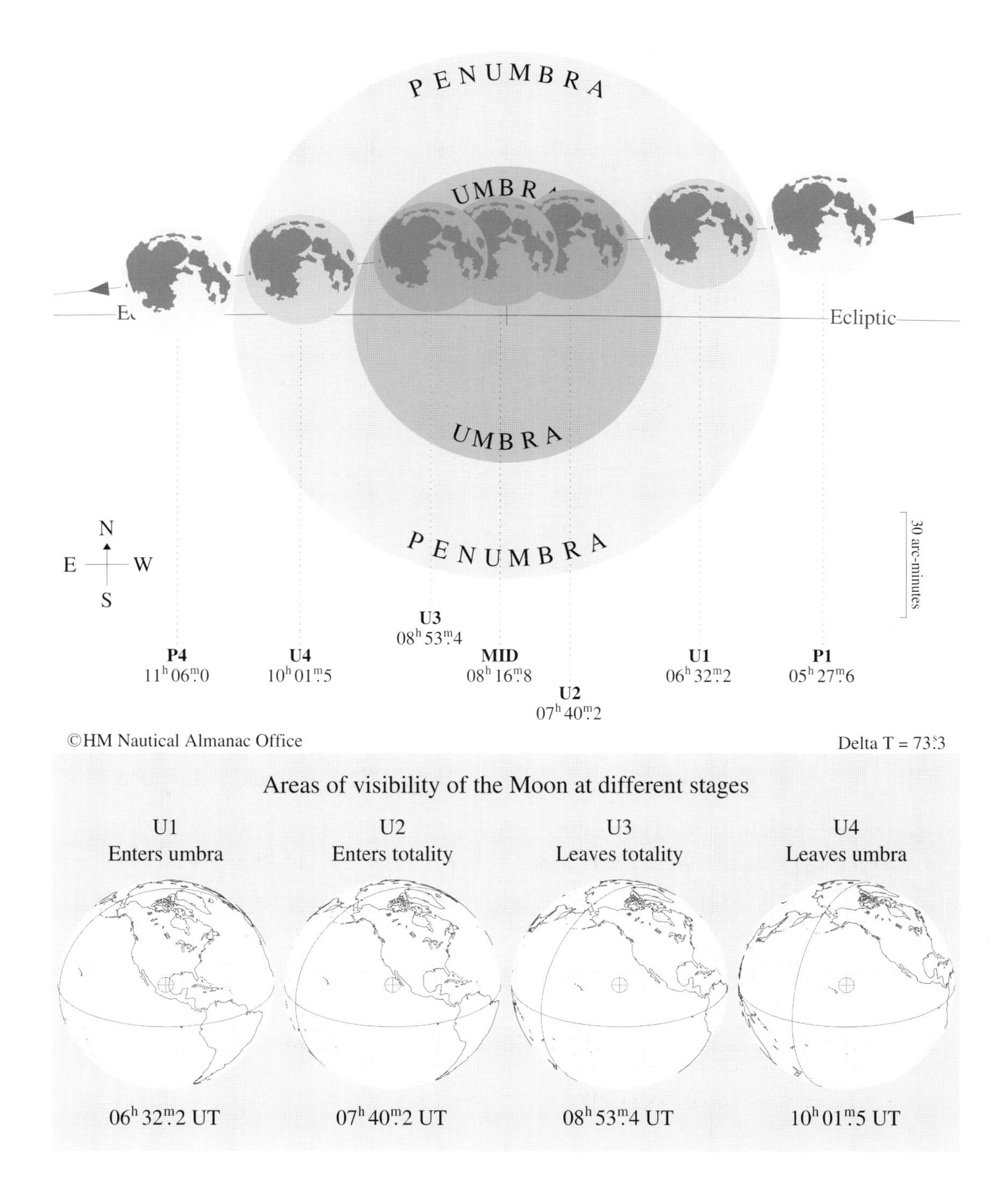

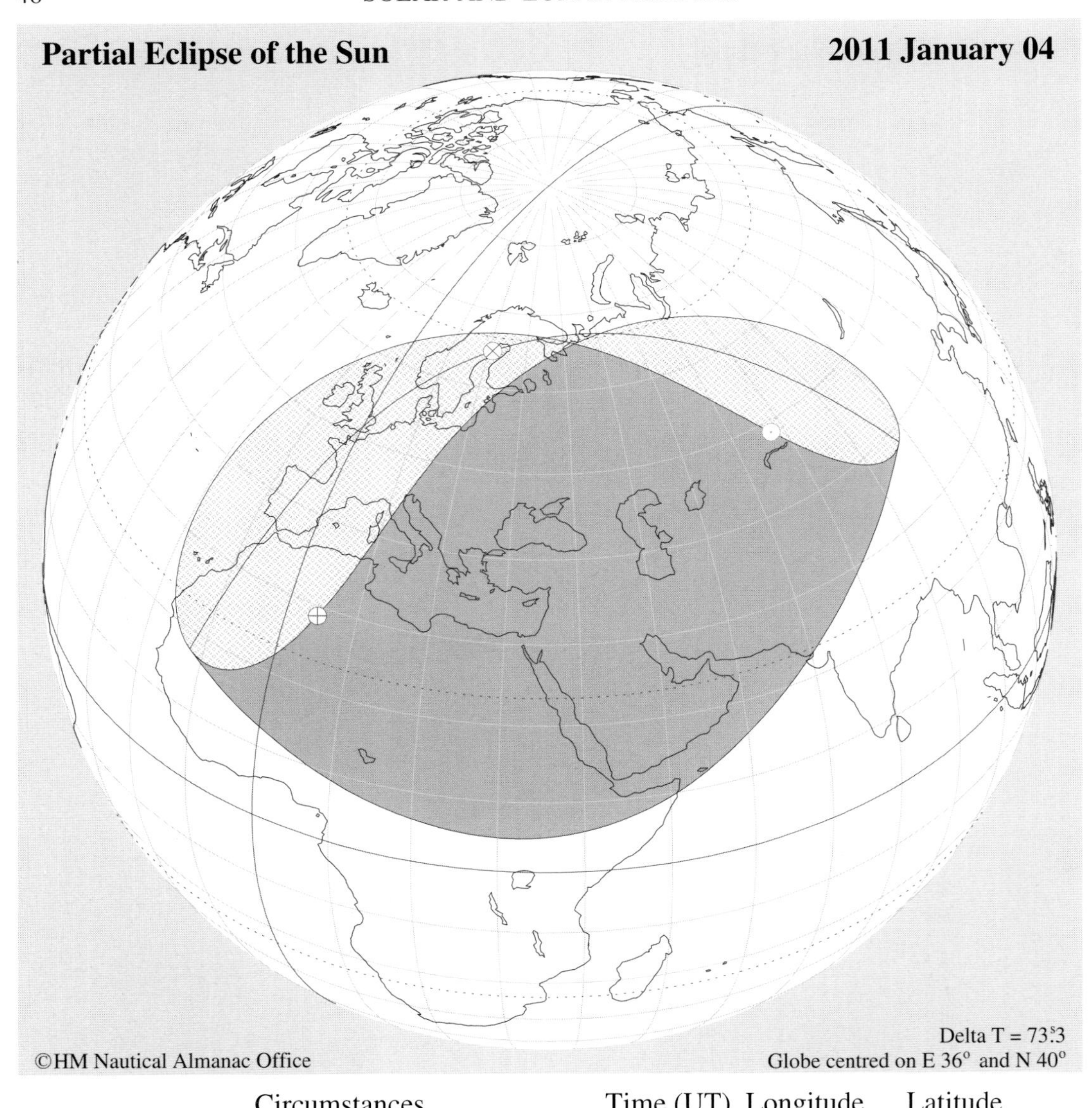

Circumstances	Time (UT)	Longitude	Latitude
	h m	° ′	° ′
⊕ Eclipse begins; first contact with Earth	6 40.1	E 4 30.2	N 28 48.9
Beginning of southern limit of penumbra	7 23.9	W 12 08.6	N 17 37.5
⊗ Greatest eclipse (magnitude = 0.8581)	8 50.5	E 20 56.1	N 64 40.7
End of southern limit of penumbra	10 16.9	E 97 37.2	N 38 19.7
⊙ Eclipse ends; last contact with Earth	11 00.8	E 77 30.1	N 48 42.7

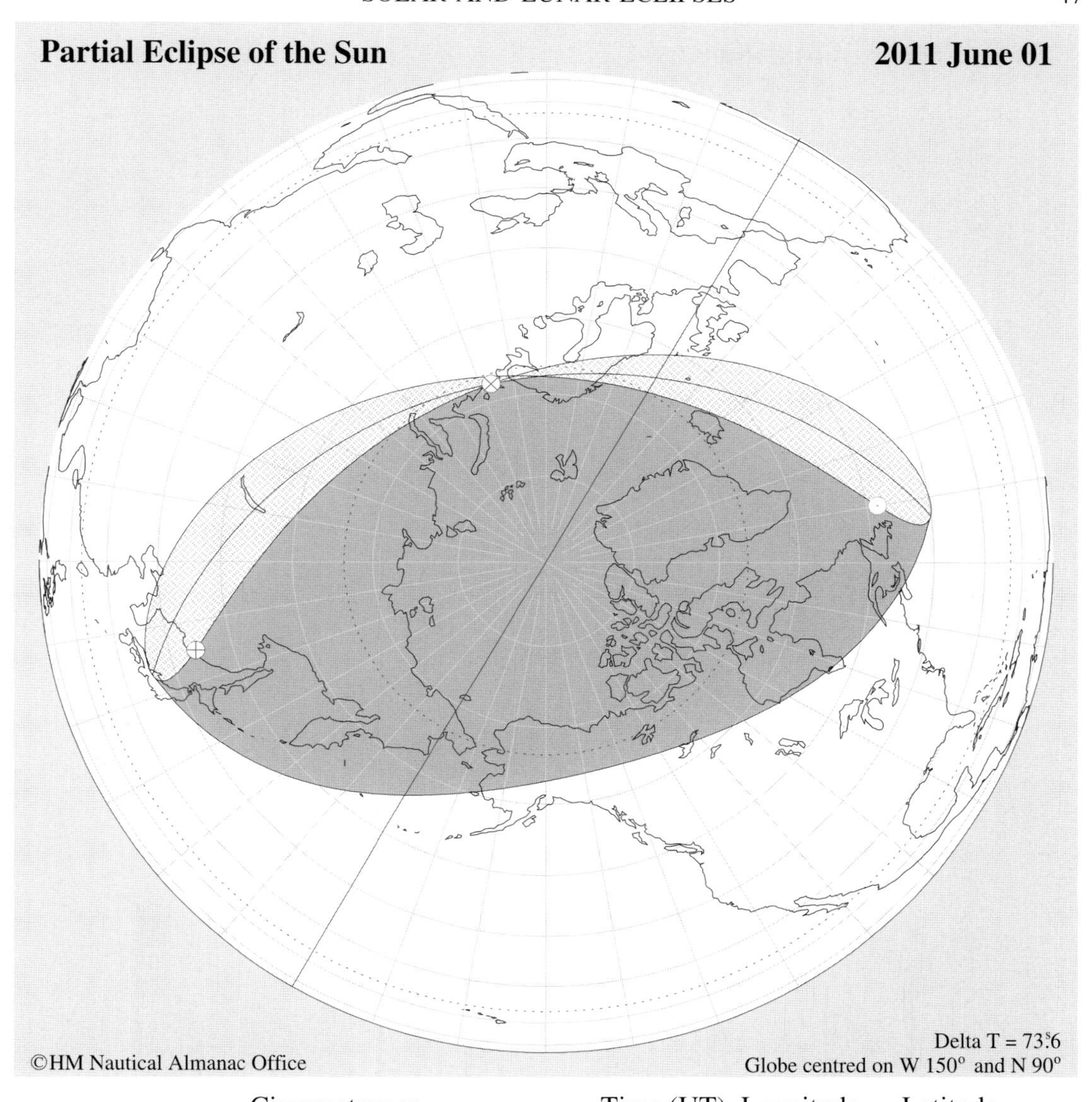

Circumstances	Time (UT)	Longitude	Latitude
	h m	° ′	° ′
⊕ Eclipse begins; first contact with Earth	19 25.2	E 134 47.1	N 44 22.3
Beginning of southern limit of penumbra	19 42.7	E 136 45.6	N 35 48.9
⊗ Greatest eclipse (magnitude = 0.6014)	21 16.1	E 46 46.7	N 67 47.0
End of southern limit of penumbra	22 49.3	W 52 53.3	N 40 04.0
⊙ Eclipse ends; last contact with Earth	23 06.8	W 50 00.6	N 48 24.5

Total Eclipse of the Moon

2011 June 15

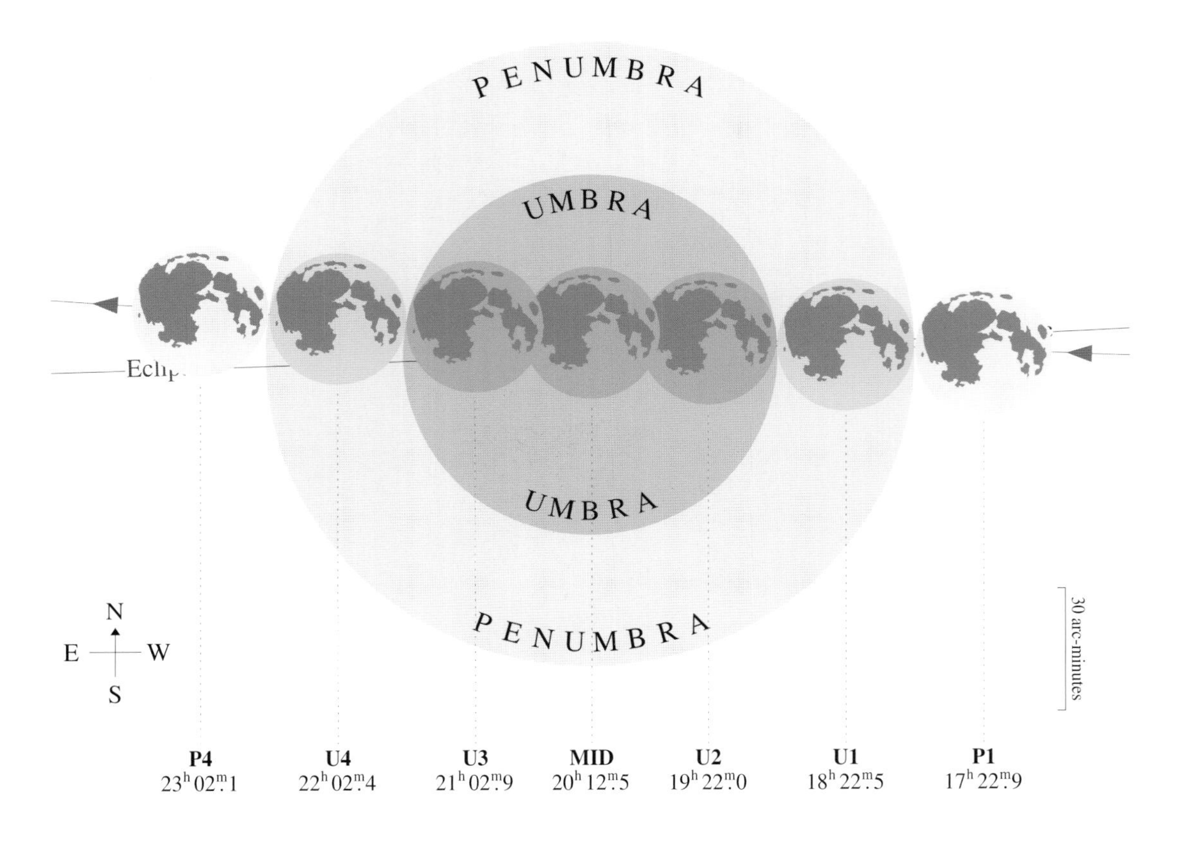

Delta T = $73^{s}.6$

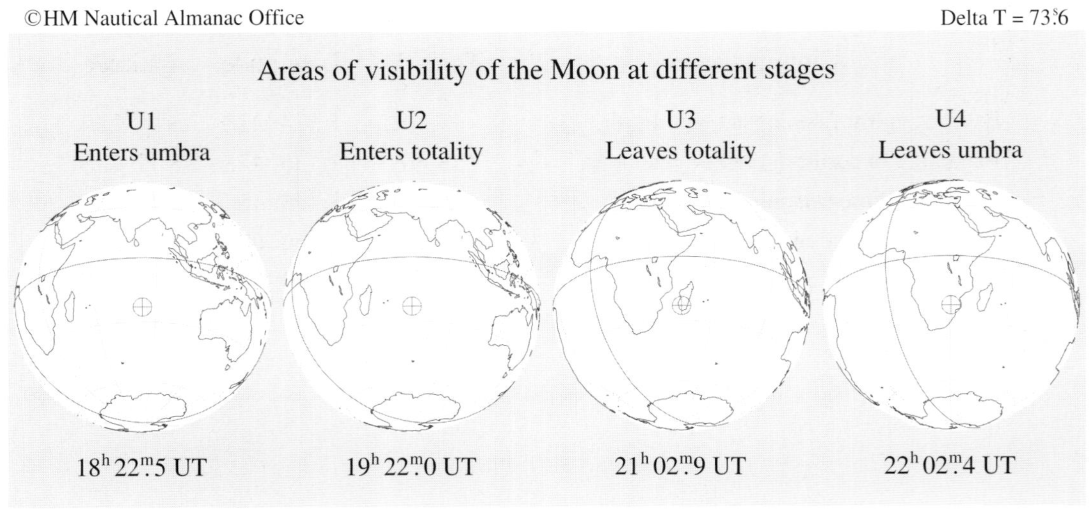

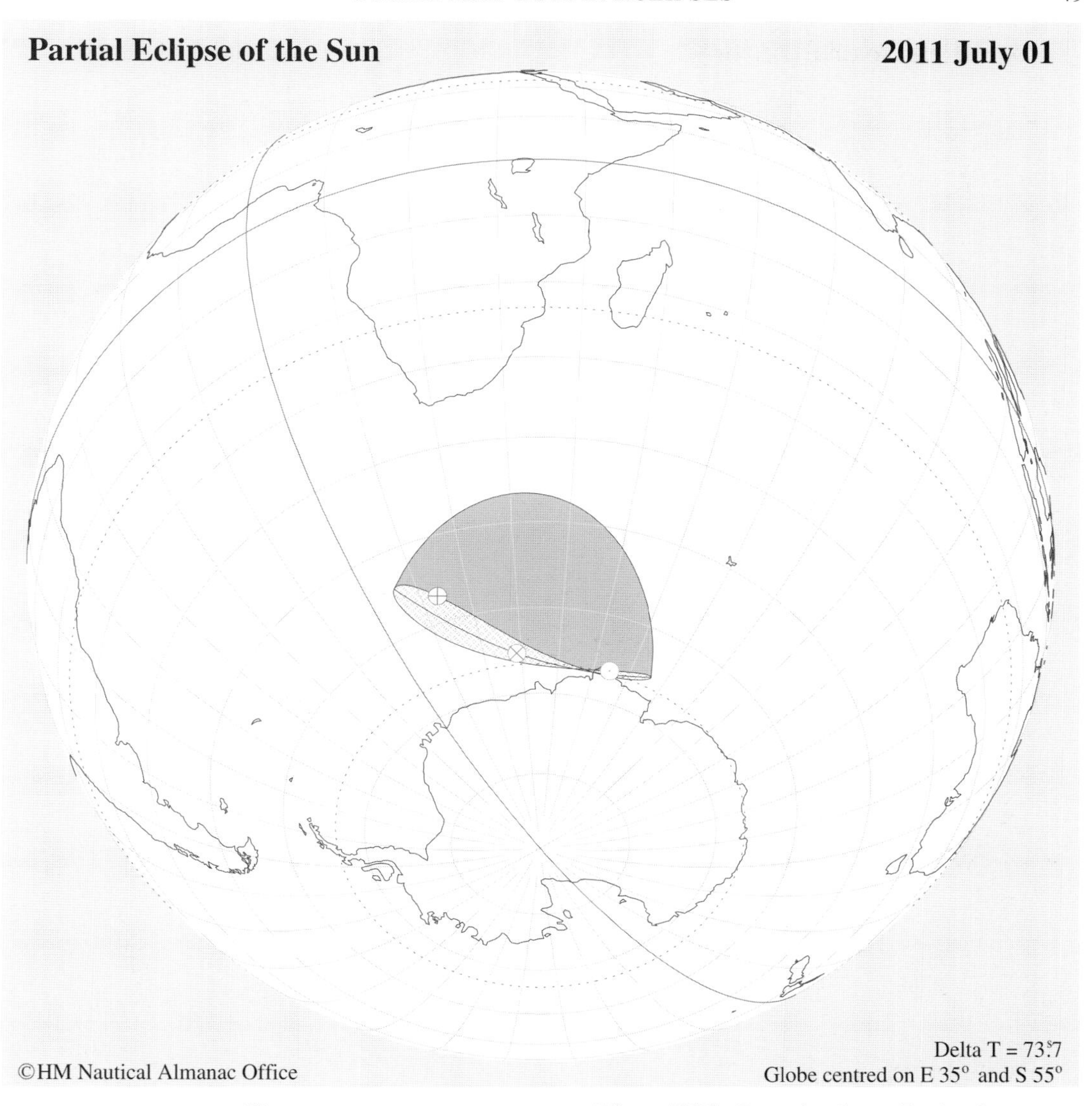

Circumstances	Time (UT)	Longitude	Latitude
	h m	° ′	° ′
⊕ Eclipse begins; first contact with Earth	7 53.6	E 13 27.9	S 56 53.7
Beginning of northern limit of penumbra	8 07.3	E 5 42.7	S 54 23.3
⊗ Greatest eclipse (magnitude = 0.0970)	8 38.3	E 28 45.7	S 65 10.6
End of northern limit of penumbra	9 08.9	E 66 00.7	S 65 13.8
○ Eclipse ends; last contact with Earth	9 22.7	E 54 31.3	S 66 13.7

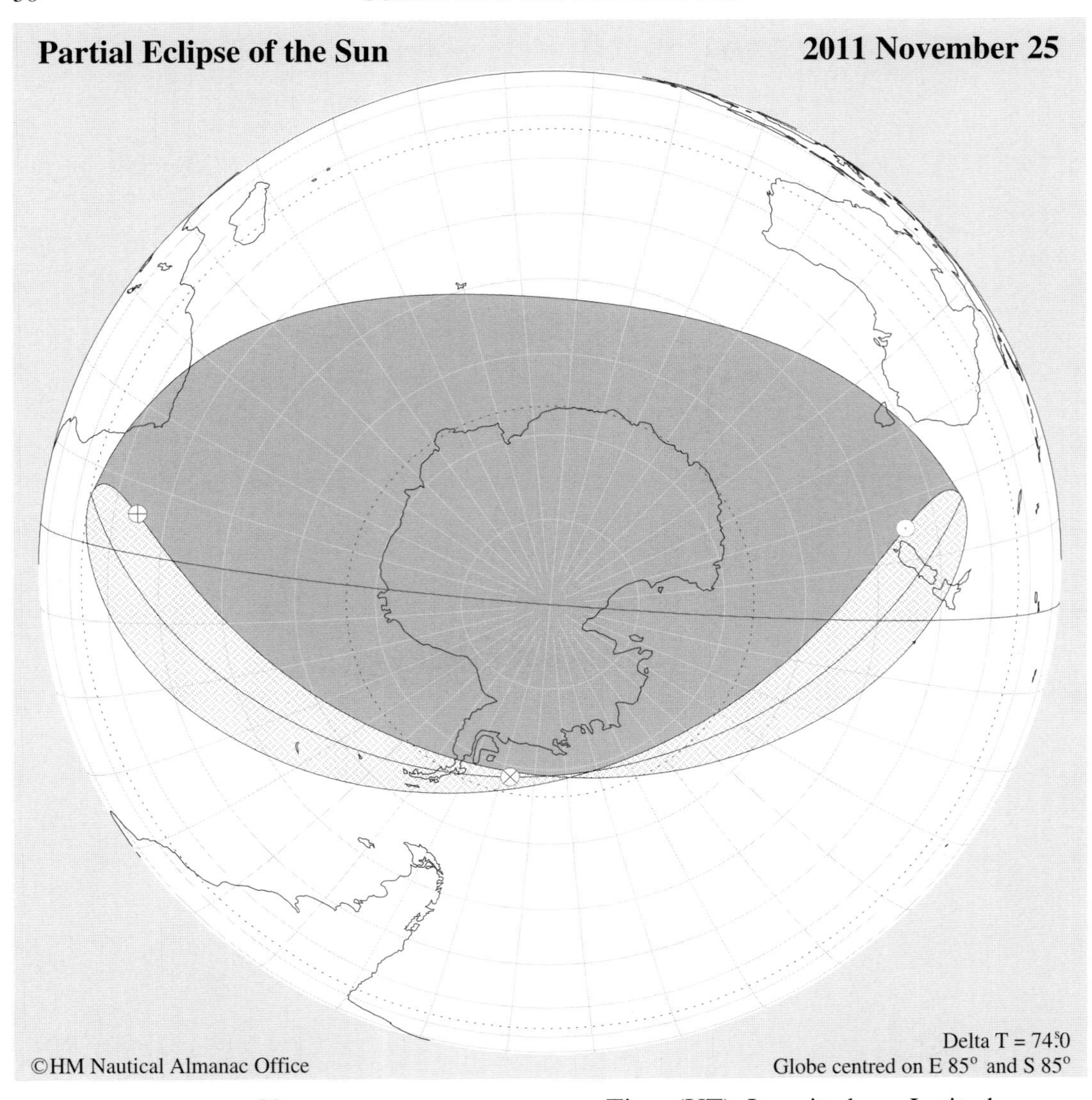

Circumstances	Time (UT)	Longitude	Latitude
	h m	° ′	° ′
⊕ Eclipse begins; first contact with Earth	4 23.1	E 5 43.8	S 34 46.7
Beginning of northern limit of penumbra	4 43.8	E 6 10.6	S 23 47.0
⊗ Greatest eclipse (magnitude = 0.9049)	6 20.2	W 82 30.0	S 68 34.6
End of northern limit of penumbra	7 56.5	E 162 32.2	S 34 19.8
⊙ Eclipse ends; last contact with Earth	8 17.1	E 164 36.2	S 44 59.1

Total Eclipse of the Moon 2011 December 10

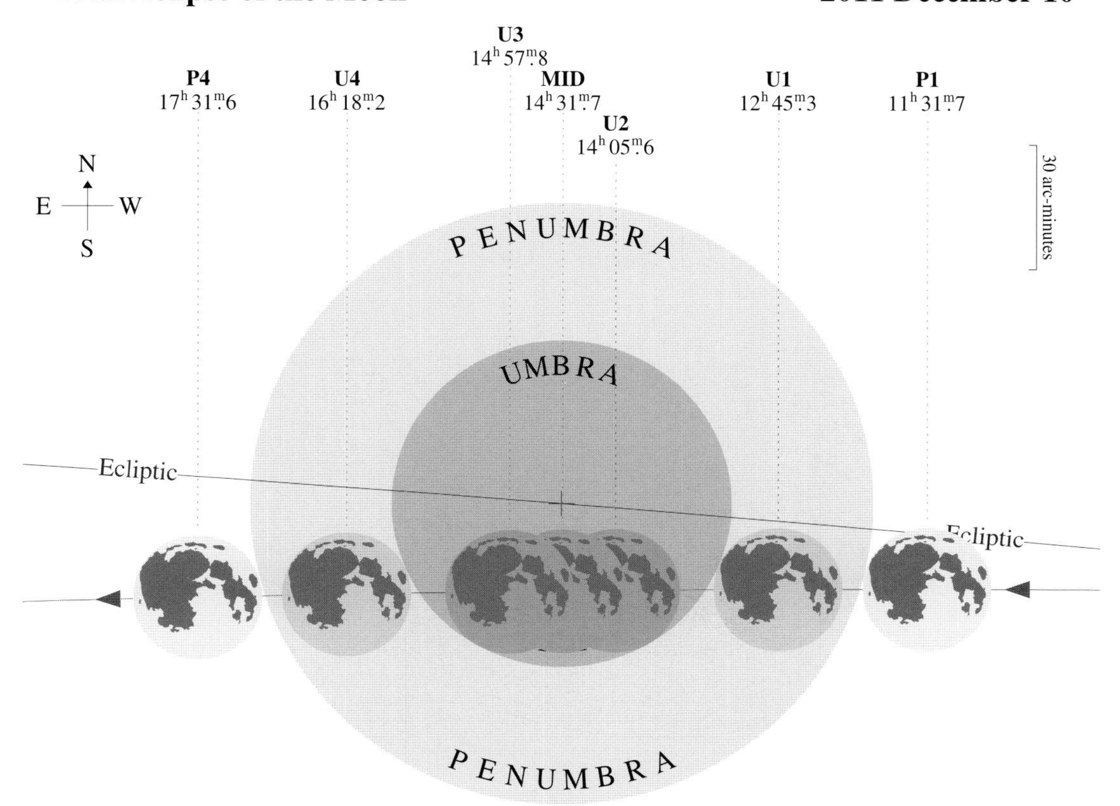

Delta T = 74.s0

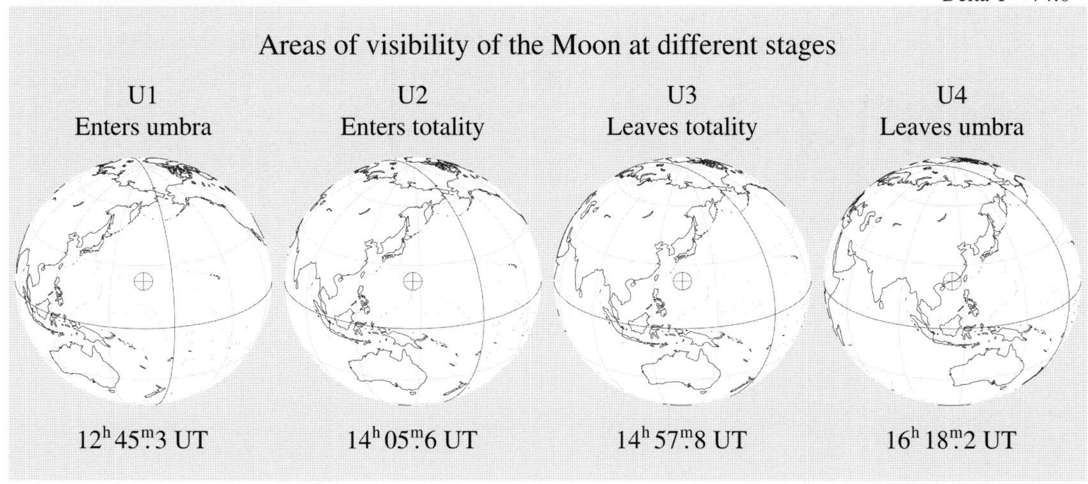

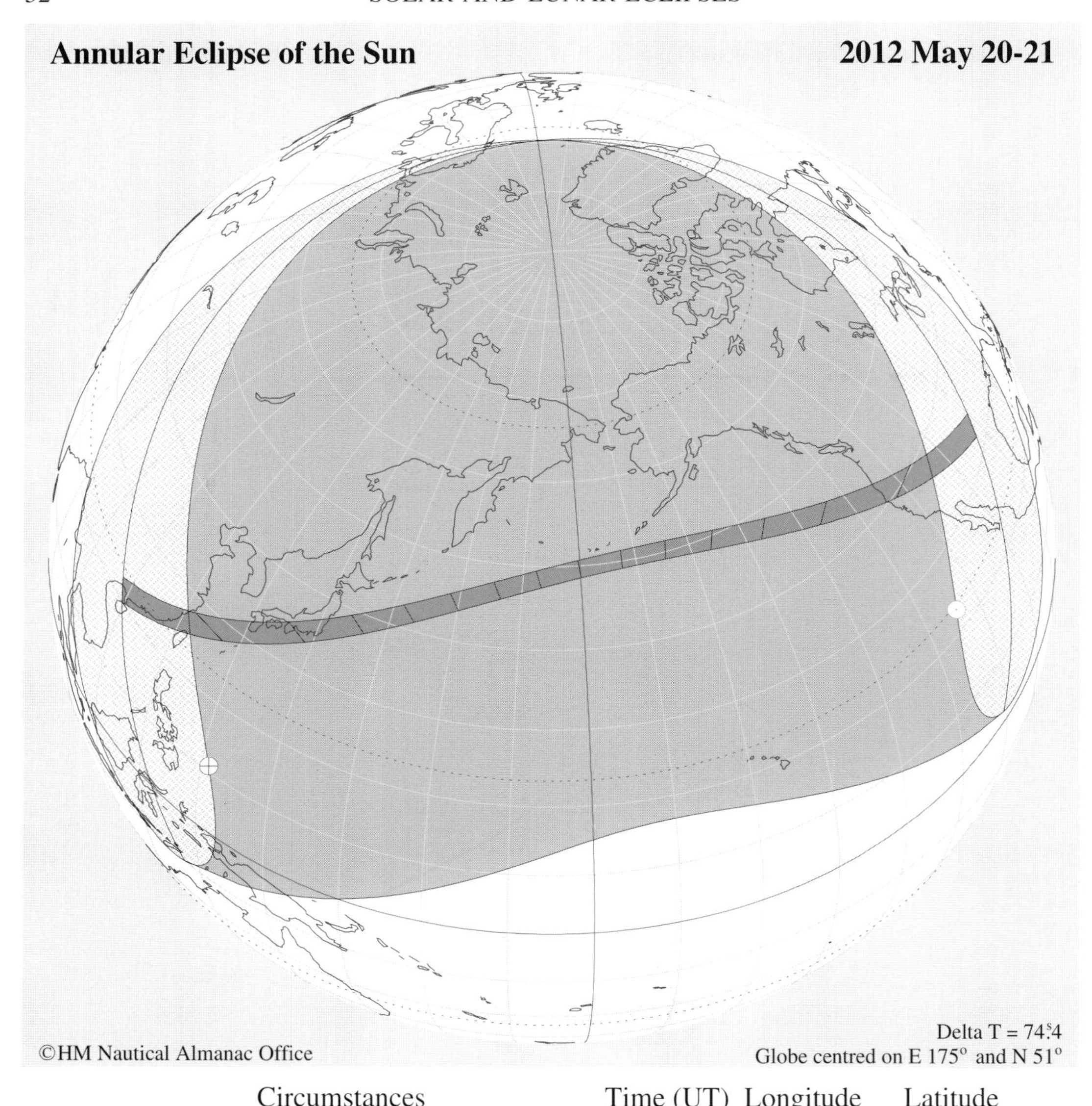

Circumstances	Time (UT)		Longitude		Latitude	
	h	m	°	′	°	′
⊕ Eclipse begins; first contact with Earth	20	56.0	E 131	05.9	N 10	53.2
Beginning of southern limit of penumbra	21	42.3	E 127	18.7	S 10	03.8
Beginning of southern limit of umbra	22	07.3	E 109	39.5	N 19	56.6
Beginning of centre line; central eclipse begins	22	08.9	E 108	44.5	N 21	08.9
Beginning of northern limit of umbra	22	10.5	E 107	47.9	N 22	22.2
Central eclipse at local apparent noon	23	59.0	E 179	23.5	N 49	31.9
End of northern limit of umbra	1	34.6	W 100	04.0	N 34	06.1
End of centre line; central eclipse ends	1	36.3	W 101	08.0	N 32	54.7
End of southern limit of umbra	1	37.9	W 102	09.6	N 31	44.1
End of southern limit of penumbra	2	03.0	W 120	52.9	N 1	56.7
⊙ Eclipse ends; last contact with Earth	2	49.2	W 124	14.6	N 22	48.0

Partial Eclipse of the Moon 2012 June 04

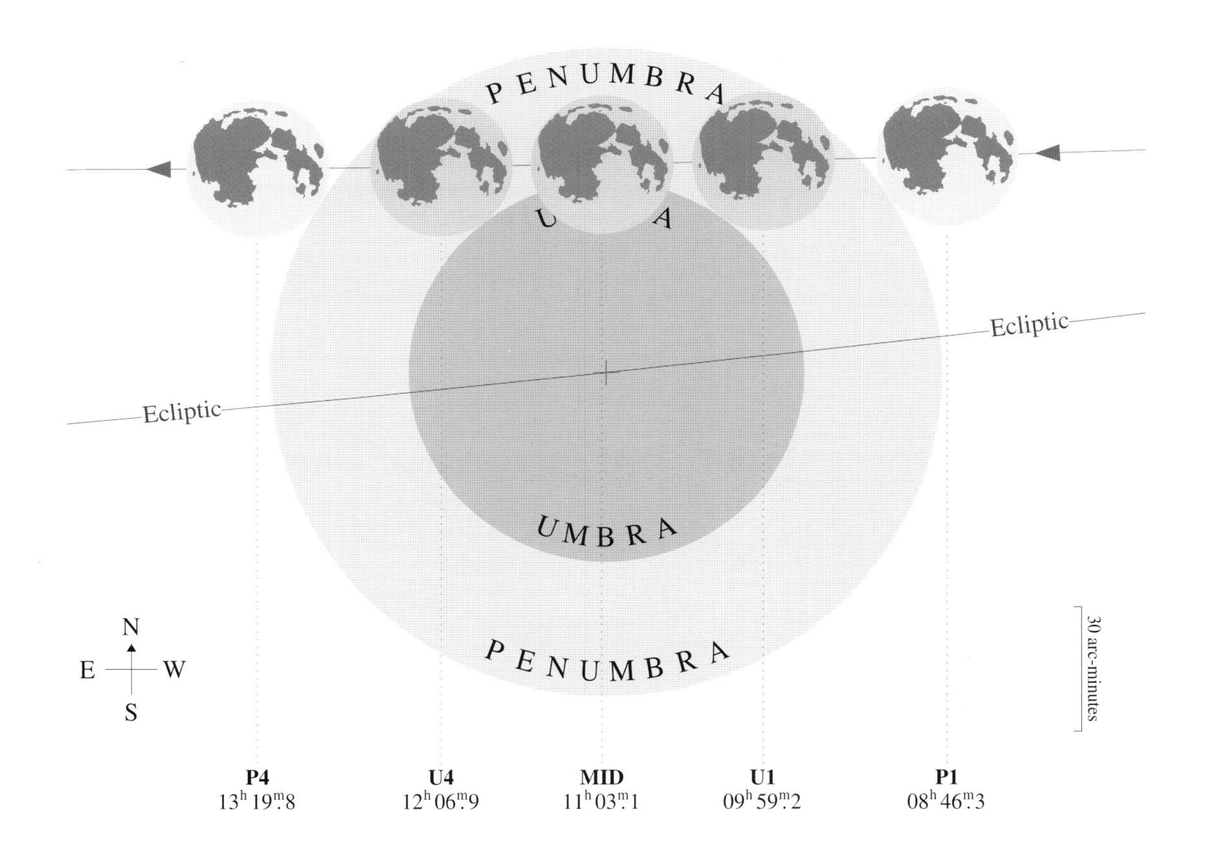

Delta T = $74^{s}.4$

Areas of visibility of the Moon at different stages

U1	MID	U4
Moon enters umbra	Middle of eclipse	Moon leaves umbra
$09^{h}59^{m}.2$ UT	$11^{h}03^{m}.1$ UT	$12^{h}06^{m}.9$ UT

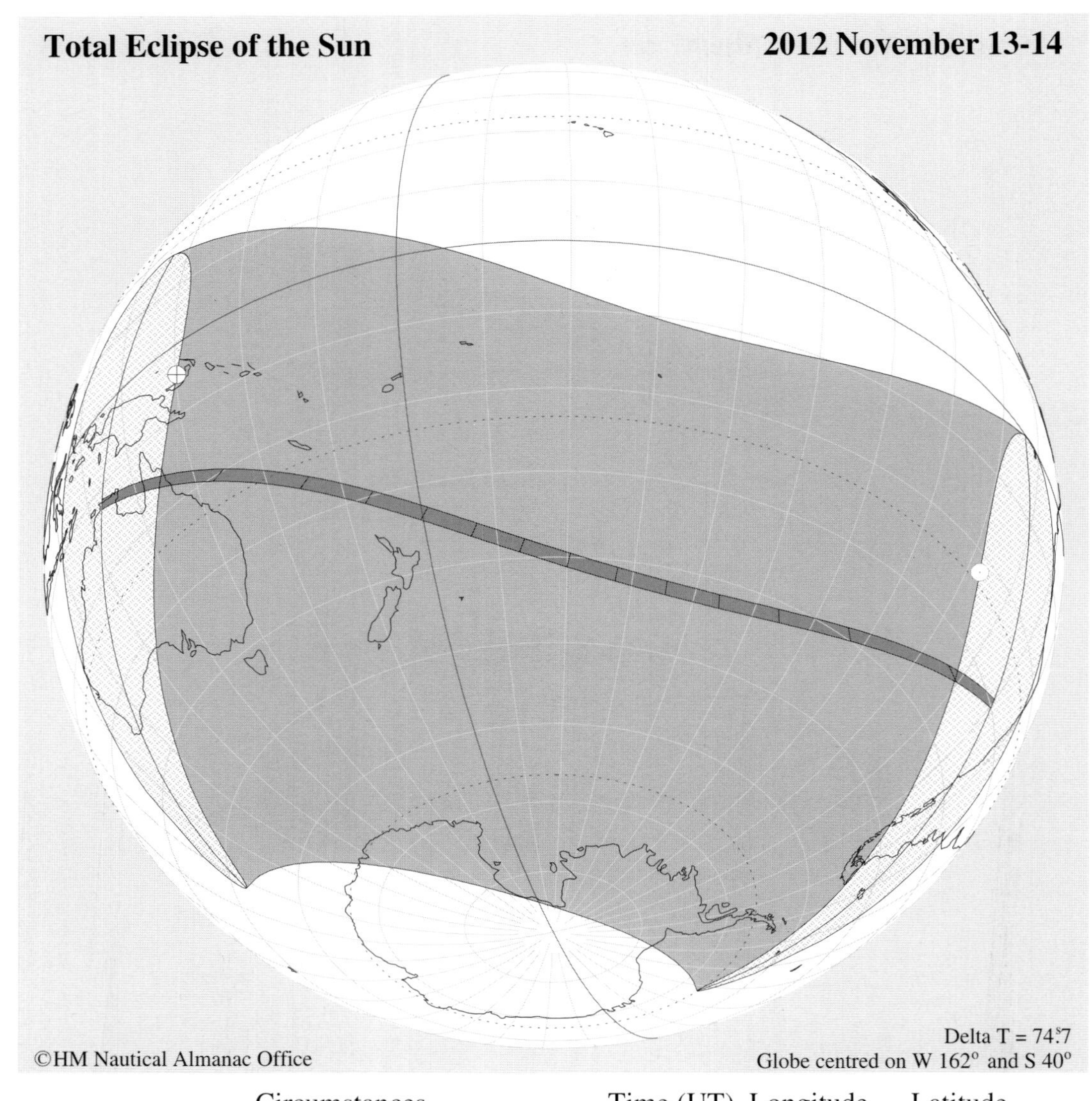

Circumstances	Time (UT)	Longitude	Latitude
	h m	° ′	° ′
⊕ Eclipse begins; first contact with Earth	19 37.8	E 150 10.6	S 4 27.6
Beginning of northern limit of penumbra	20 22.0	E 146 37.9	N 17 40.1
Beginning of northern limit of umbra	20 35.5	E 133 24.7	S 11 26.2
Beginning of centre line; central eclipse begins	20 36.0	E 133 06.9	S 11 57.2
Beginning of southern limit of umbra	20 36.4	E 132 48.9	S 12 28.3
Beginning of southern limit of penumbra	21 31.9	E 97 50.6	S 52 19.8
Central eclipse at local apparent noon	22 17.9	W 158 22.8	S 40 37.0
End of southern limit of penumbra	22 51.2	W 27 07.2	S 66 34.2
End of southern limit of umbra	23 46.8	W 79 34.9	S 30 03.5
End of centre line; central eclipse ends	23 47.3	W 79 56.1	S 29 32.6
End of northern limit of umbra	23 47.8	W 80 16.9	S 29 01.8
End of northern limit of penumbra	0 01.4	W 94 13.3	S 0 04.0
○ Eclipse ends; last contact with Earth	0 45.4	W 97 31.6	S 22 08.7

Penumbral Eclipse of the Moon 2012 November 28

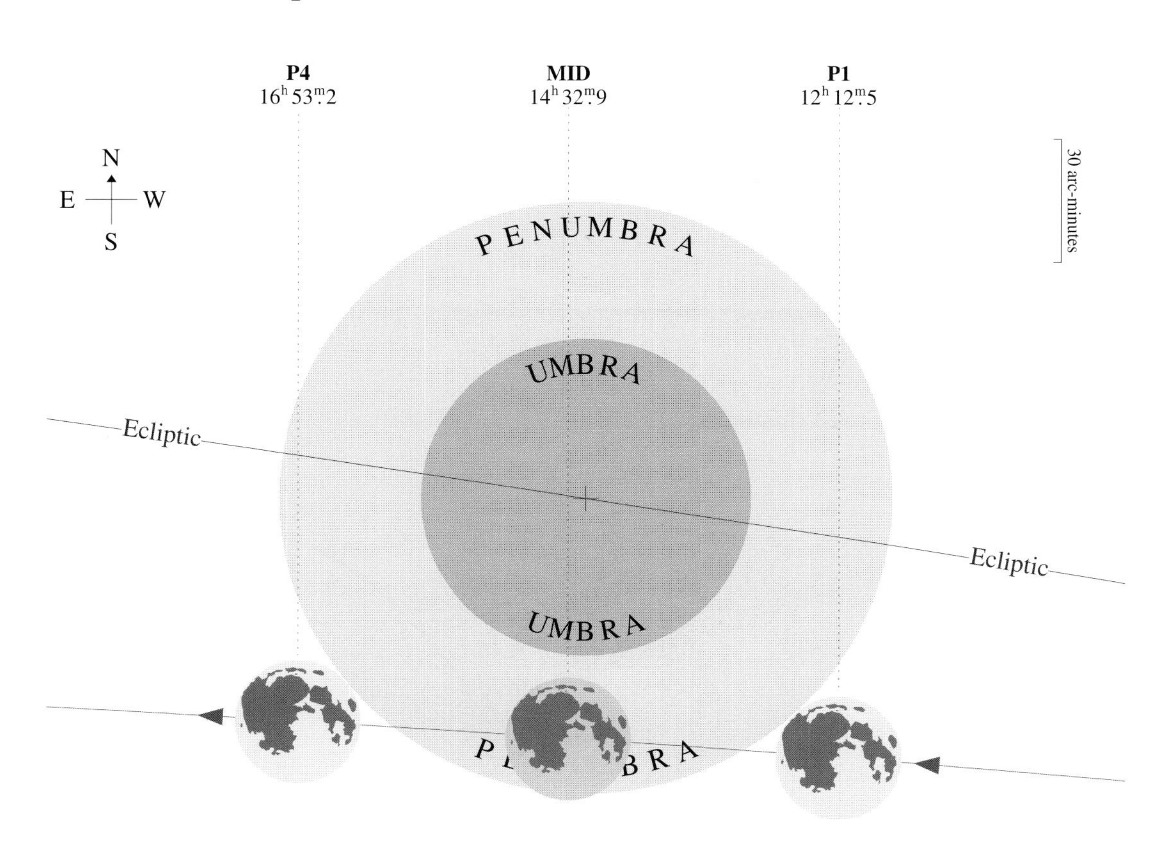

Delta T = 74s.8

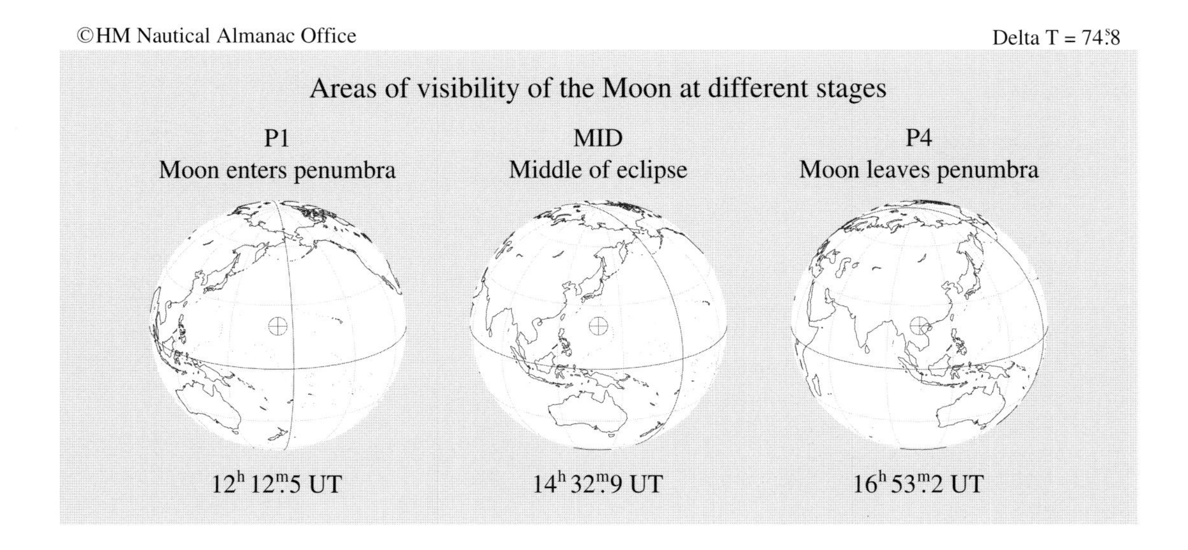

Partial Eclipse of the Moon 2013 April 25

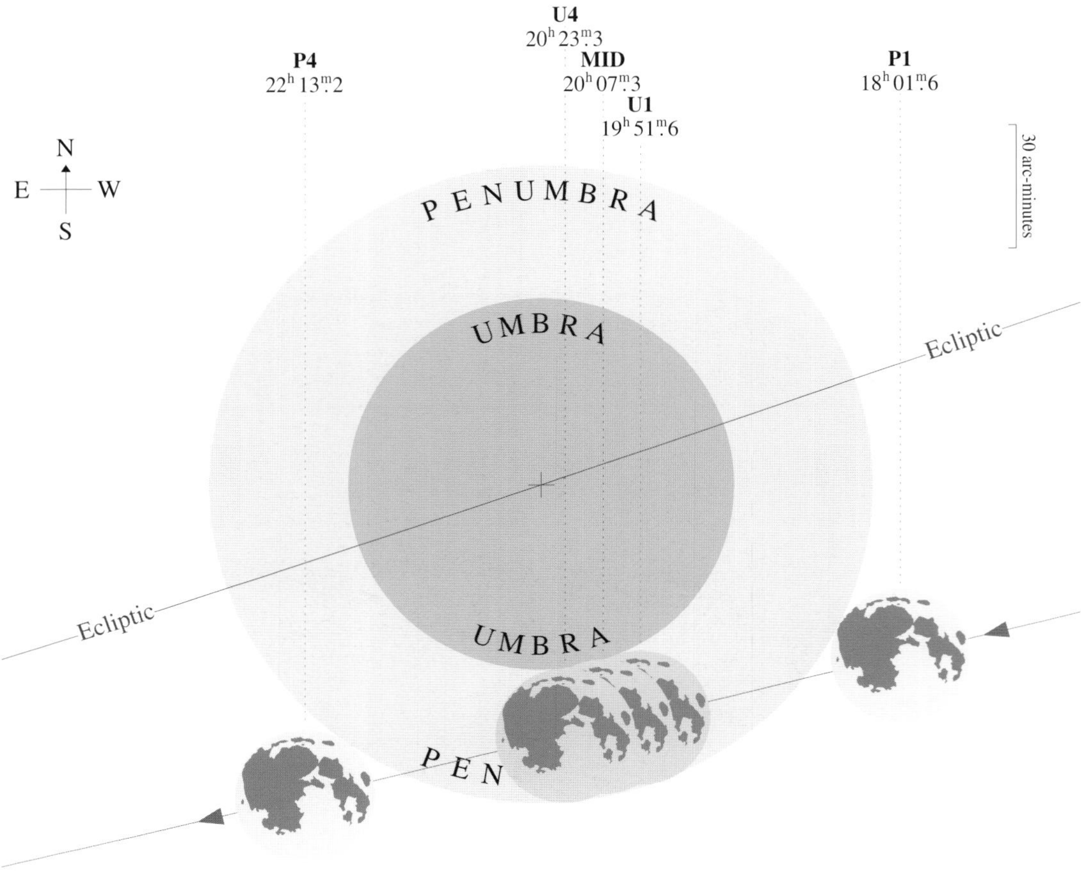

U4 $20^{h} 23^{m}.3$

P4 $22^{h} 13^{m}.2$

MID $20^{h} 07^{m}.3$

U1 $19^{h} 51^{m}.6$

P1 $18^{h} 01^{m}.6$

N E W S

30 arc-minutes

PENUMBRA

UMBRA

Ecliptic

Ecliptic

UMBRA

PEN

Delta T = $75^{s}.1$

Areas of visibility of the Moon at different stages

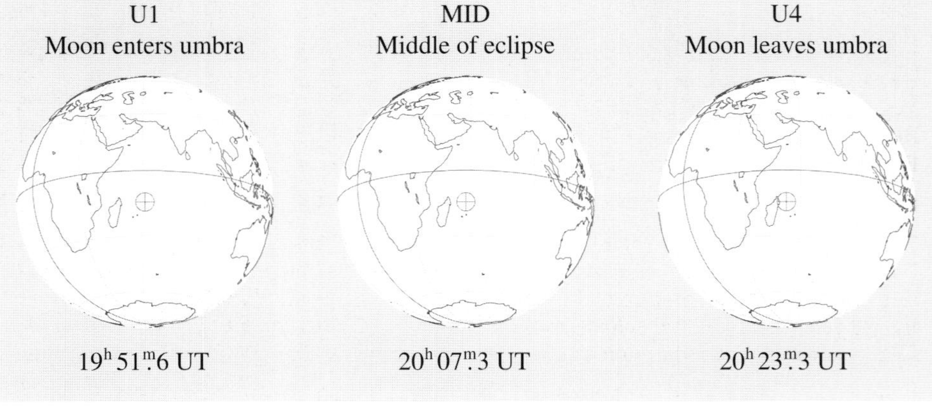

U1
Moon enters umbra

MID
Middle of eclipse

U4
Moon leaves umbra

$19^{h} 51^{m}.6$ UT

$20^{h} 07^{m}.3$ UT

$20^{h} 23^{m}.3$ UT

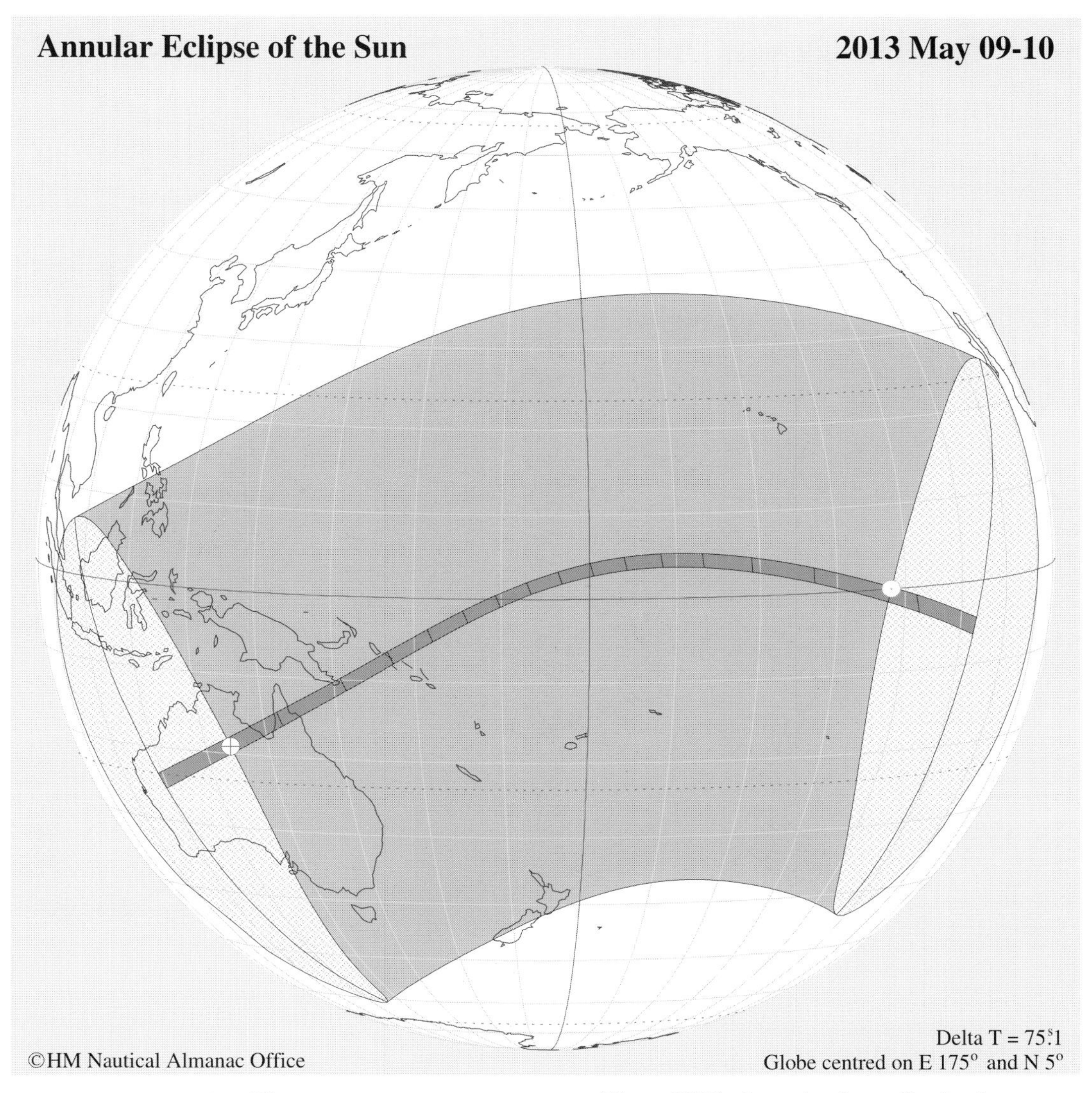

Circumstances	Time (UT)	Longitude	Latitude
	h m	° ′	° ′
⊕ Eclipse begins; first contact with Earth	21 25.0	E 134 08.1	S 19 03.9
Beginning of northern limit of umbra	22 32.3	E 118 57.8	S 23 31.6
Beginning of centre line; central eclipse begins	22 32.5	E 119 16.1	S 24 28.7
Beginning of southern limit of umbra	22 32.8	E 119 34.4	S 25 25.9
Beginning of northern limit of penumbra	22 42.2	E 106 28.0	N 6 31.0
Beginning of southern limit of penumbra	23 13.9	E 135 22.3	S 60 54.8
Central eclipse at local apparent noon	0 19.5	E 174 12.9	N 1 44.7
End of southern limit of penumbra	1 36.5	W 132 42.9	S 43 43.9
End of northern limit of penumbra	2 07.8	W 114 07.5	N 25 32.4
End of southern limit of umbra	2 17.4	W 127 18.7	S 6 25.7
End of centre line; central eclipse ends	2 17.7	W 127 03.8	S 5 26.9
End of northern limit of umbra	2 17.9	W 126 48.6	S 4 28.3
○ Eclipse ends; last contact with Earth	3 25.2	W 142 12.7	N 0 00.9

Penumbral Eclipse of the Moon 2013 May 25

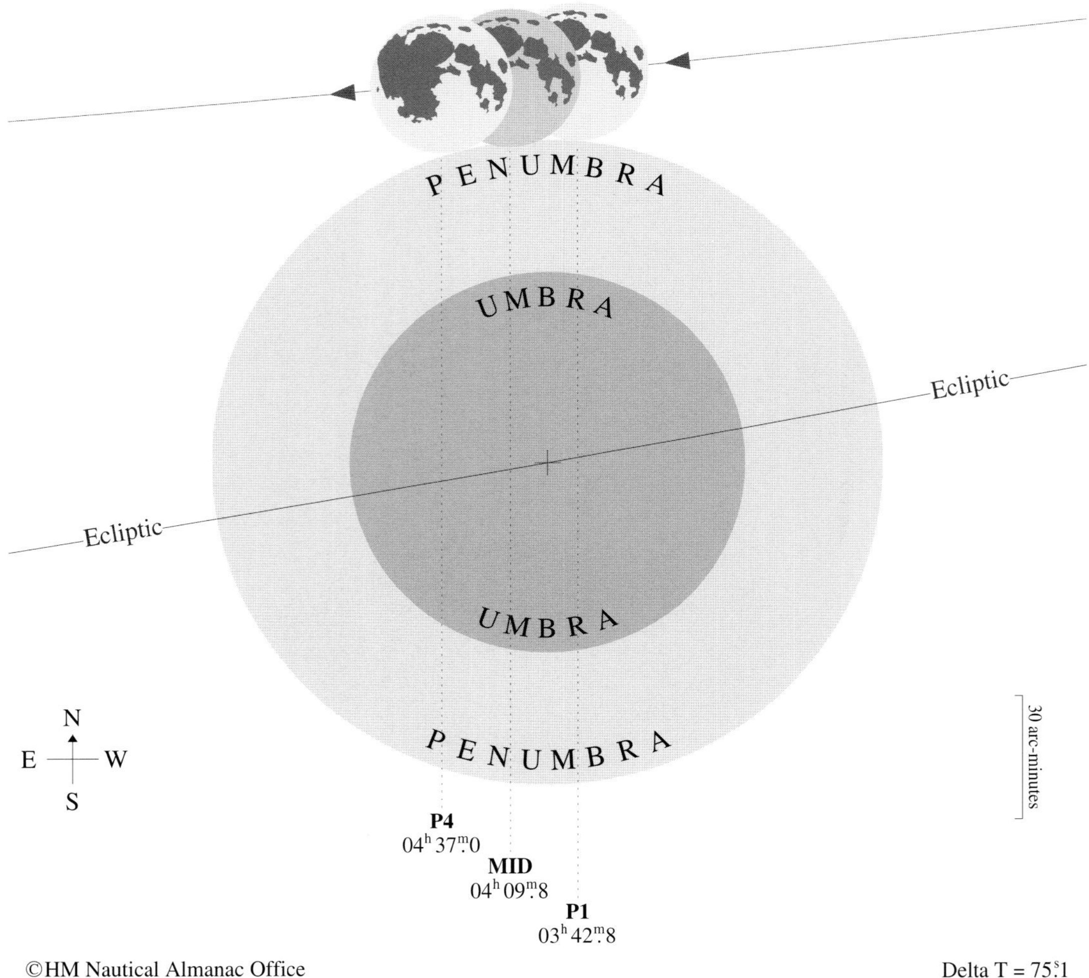

Delta T = 75.ˢ1

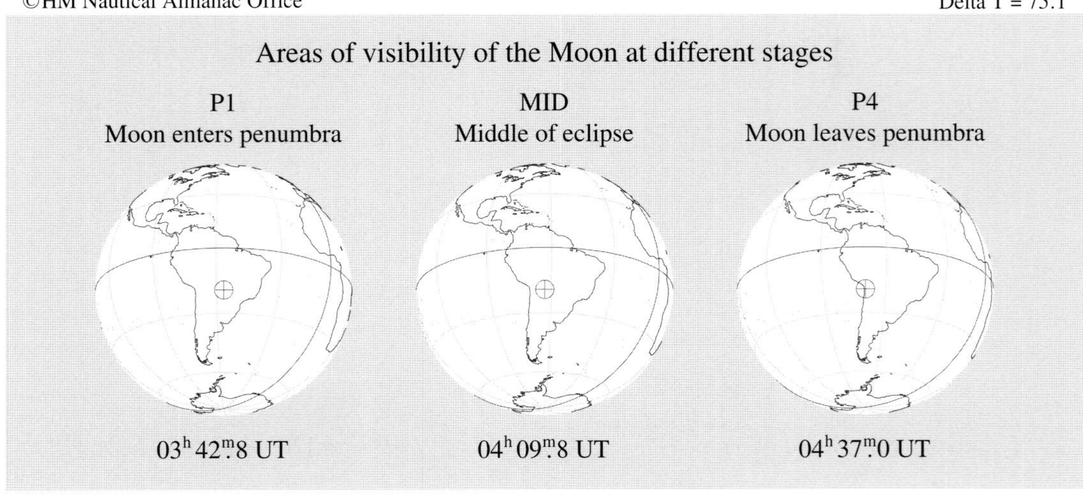

Penumbral Eclipse of the Moon 2013 October 18-19

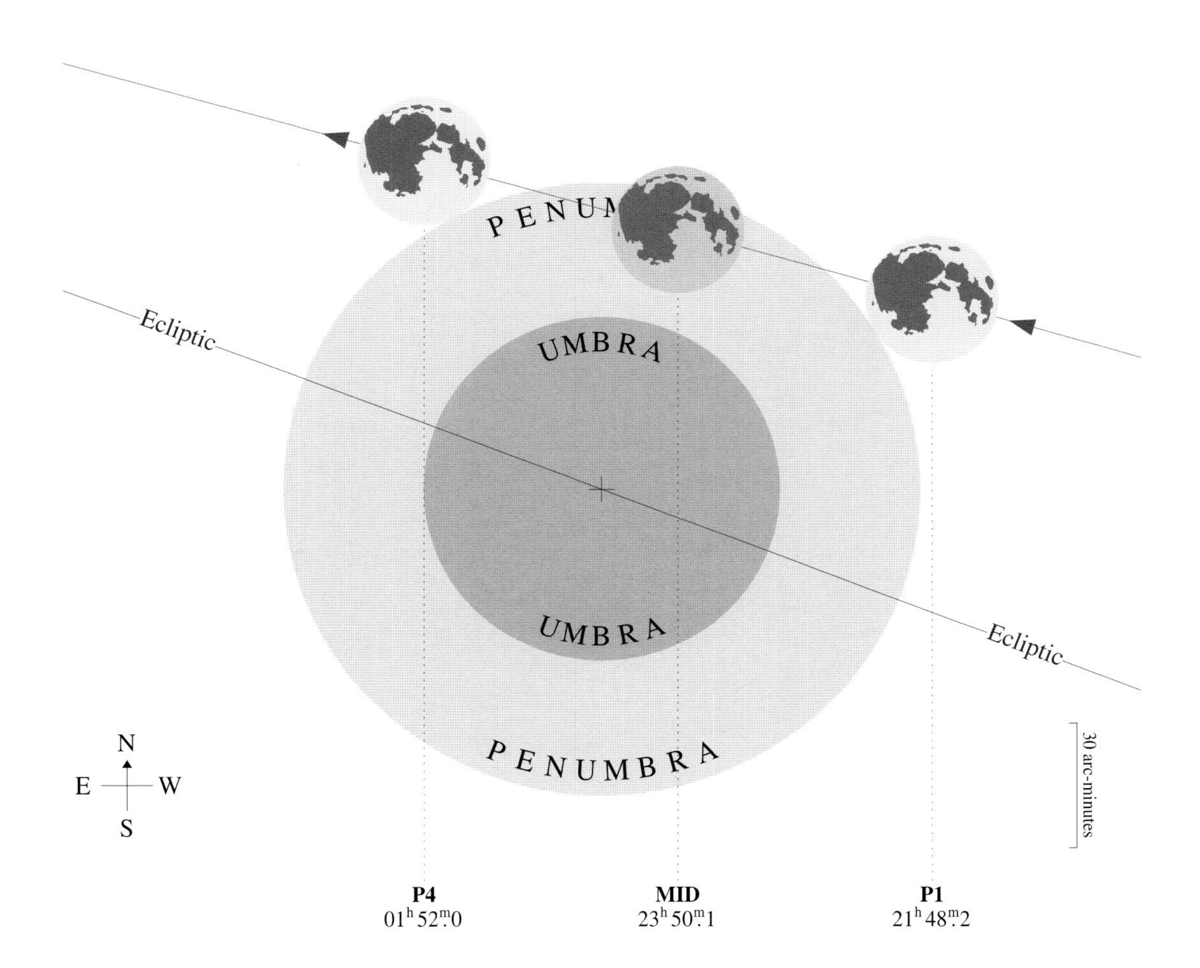

Delta T = $75^{s}.4$

Areas of visibility of the Moon at different stages

P1	MID	P4
Moon enters penumbra	Middle of eclipse	Moon leaves penumbra
$21^{h} 48^{m}.2$ UT	$23^{h} 50^{m}.1$ UT	$01^{h} 52^{m}.0$ UT

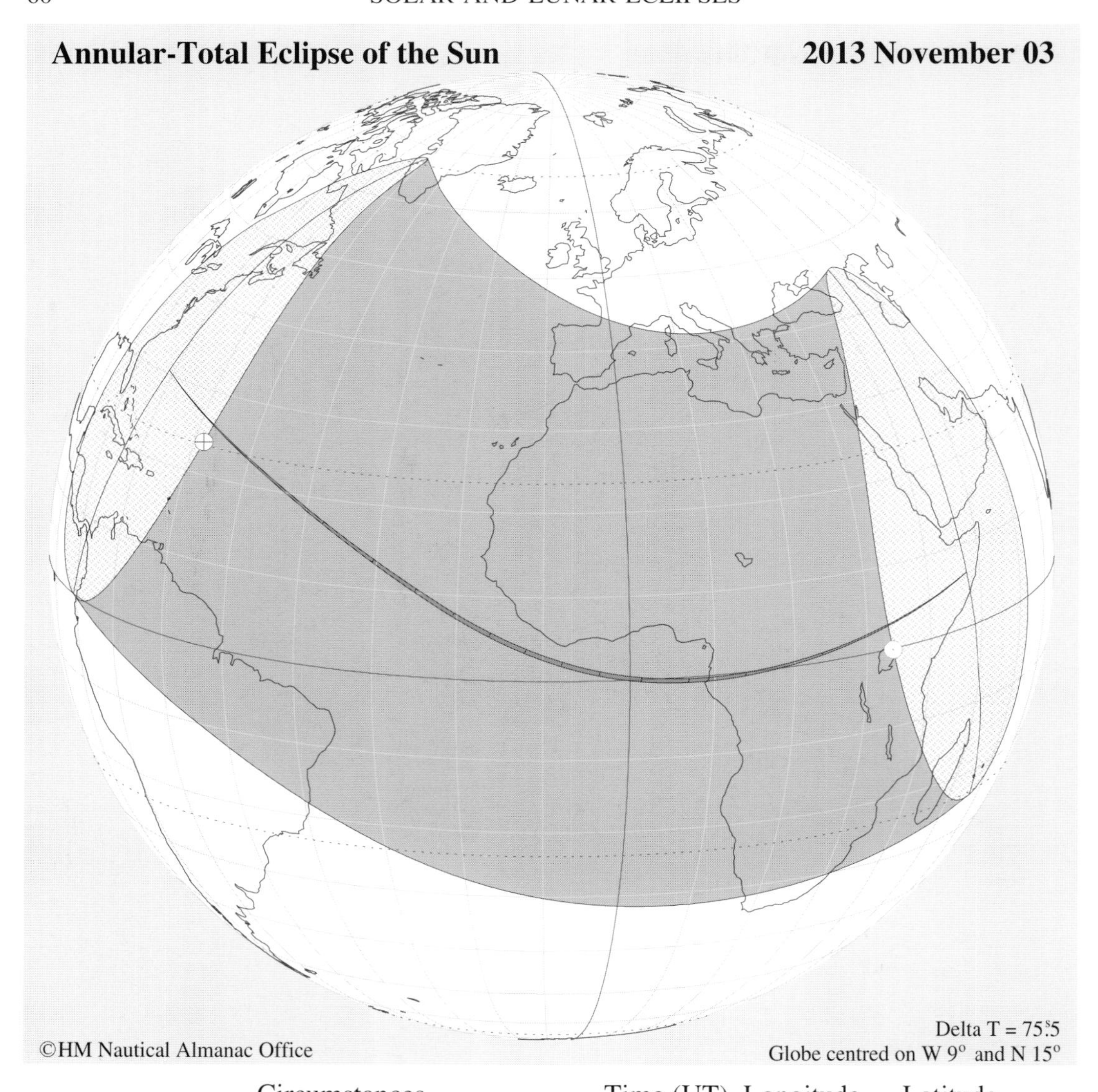

Circumstances	Time (UT)	Longitude	Latitude
	h m	° ′	° ′
⊕ Eclipse begins; first contact with Earth	10 04.4	W 58 19.5	N 23 50.4
Beginning of centre line; central eclipse begins	11 05.2	W 71 13.2	N 30 26.5
Beginning of southern limit of umbra	11 05.2	W 71 13.3	N 30 26.2
Beginning of northern limit of umbra	11 05.2	W 71 13.1	N 30 26.8
Beginning of southern limit of penumbra	11 08.4	W 81 13.0	S 0 03.8
Beginning of northern limit of penumbra	11 50.6	W 50 14.7	N 67 43.1
Central eclipse at local apparent noon	12 38.6	W 13 46.0	N 4 23.3
End of northern limit of penumbra	13 42.4	E 43 40.5	N 46 26.2
End of southern limit of penumbra	14 24.2	E 56 48.8	S 24 01.5
End of southern limit of umbra	14 27.6	E 47 12.6	N 6 32.4
End of northern limit of umbra	14 27.6	E 47 12.9	N 6 30.5
End of centre line; central eclipse ends	14 27.6	E 47 12.8	N 6 31.4
⊙ Eclipse ends; last contact with Earth	15 28.2	E 33 52.4	S 0 08.7

Total Eclipse of the Moon 2014 April 15

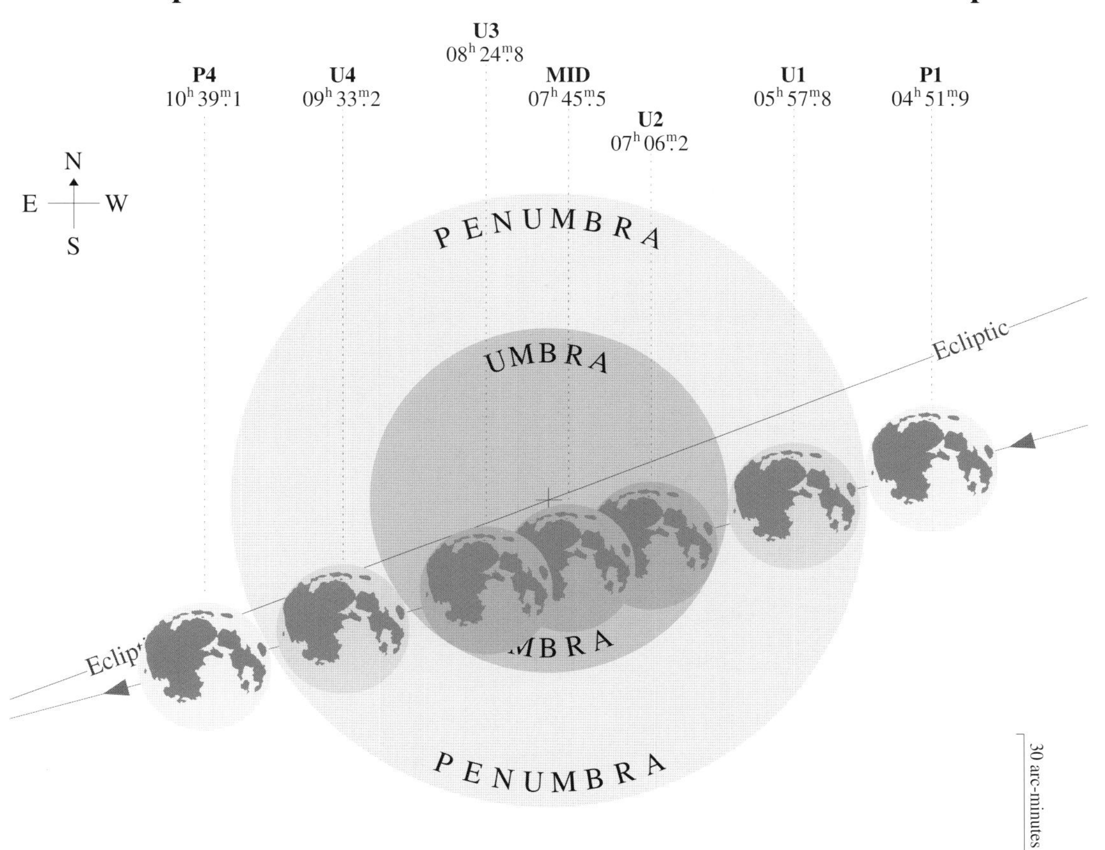

Delta T = 75.s8

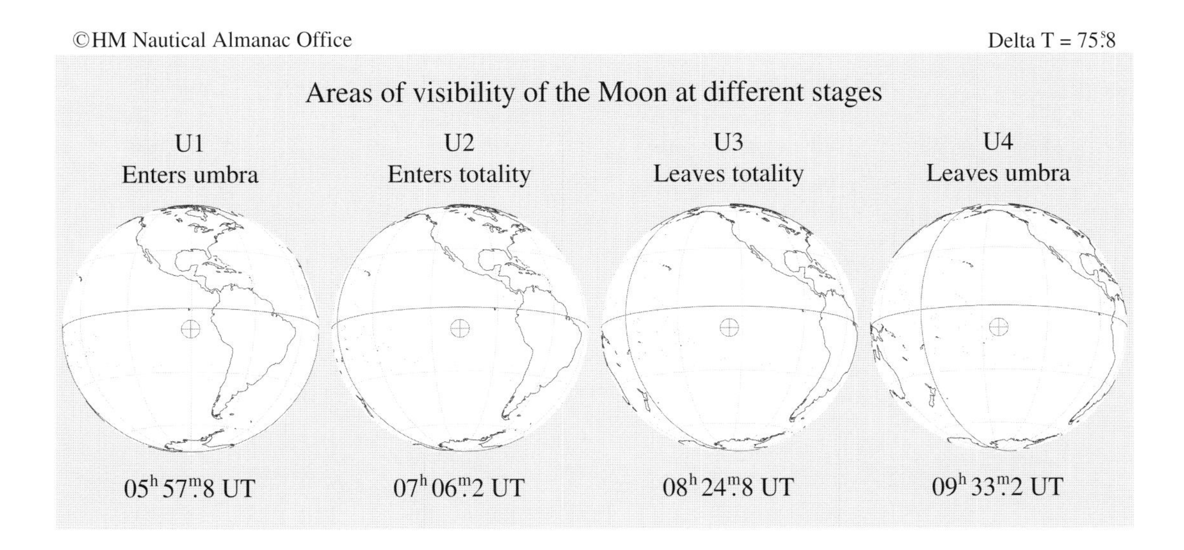

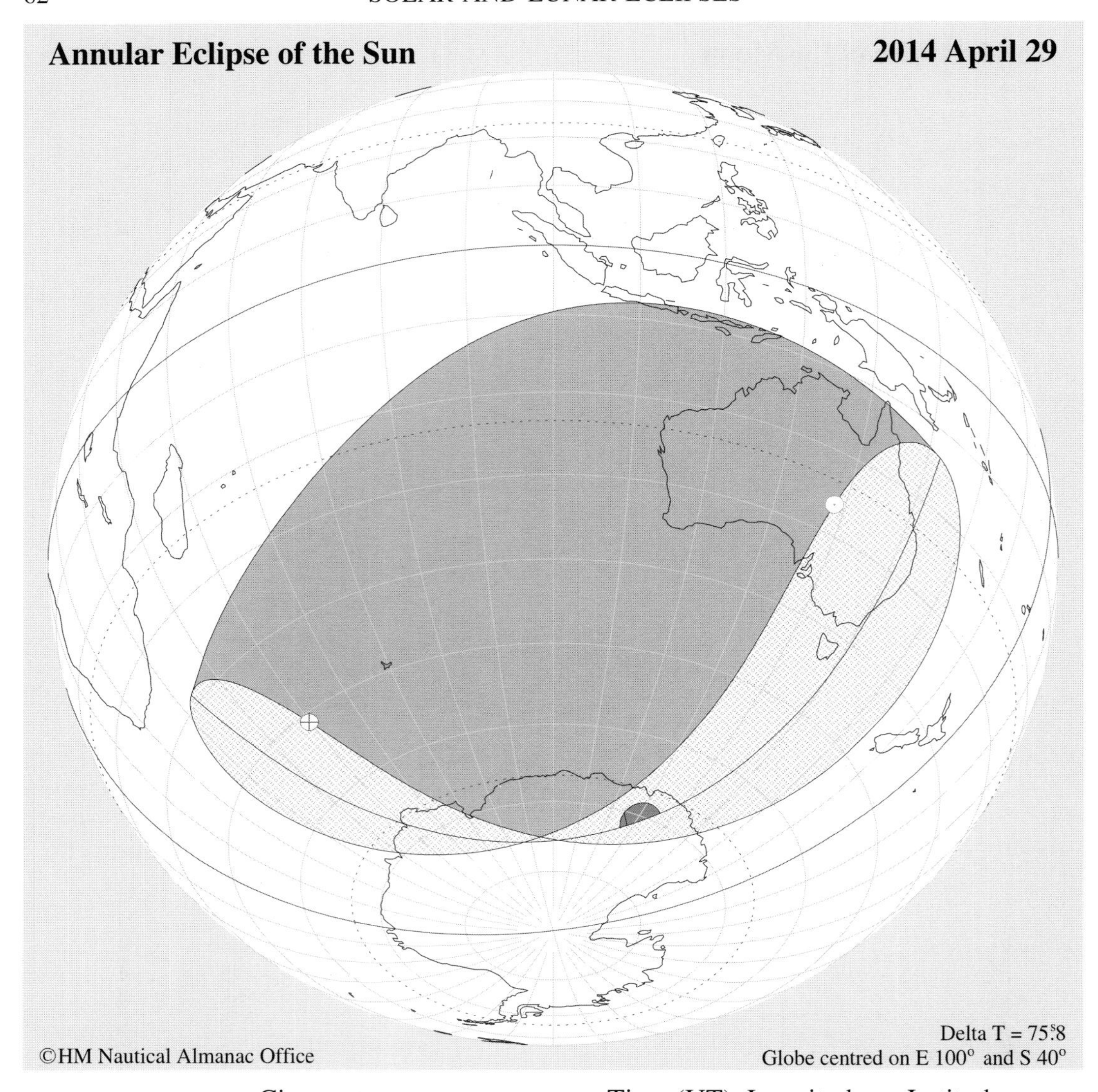

	Circumstances	Time (UT)		Longitude			Latitude	
		h	m		°	′	°	′
⊕	Eclipse begins; first contact with Earth	3	52.5	E	49	47.4	S 51	03.8
	Beginning of northern limit of penumbra	4	31.6	E	33	20.5	S 38	40.1
	Beginning of northern limit of umbra	5	58.1	E	125	48.3	S 72	20.1
	End of northern limit of umbra	6	08.8	E	135	42.6	S 68	43.4
	End of northern limit of penumbra	7	35.1	E	151	59.3	S 13	32.7
⊙	Eclipse ends; last contact with Earth	8	14.3	E	138	23.9	S 26	22.9

Total Eclipse of the Moon 2014 October 08

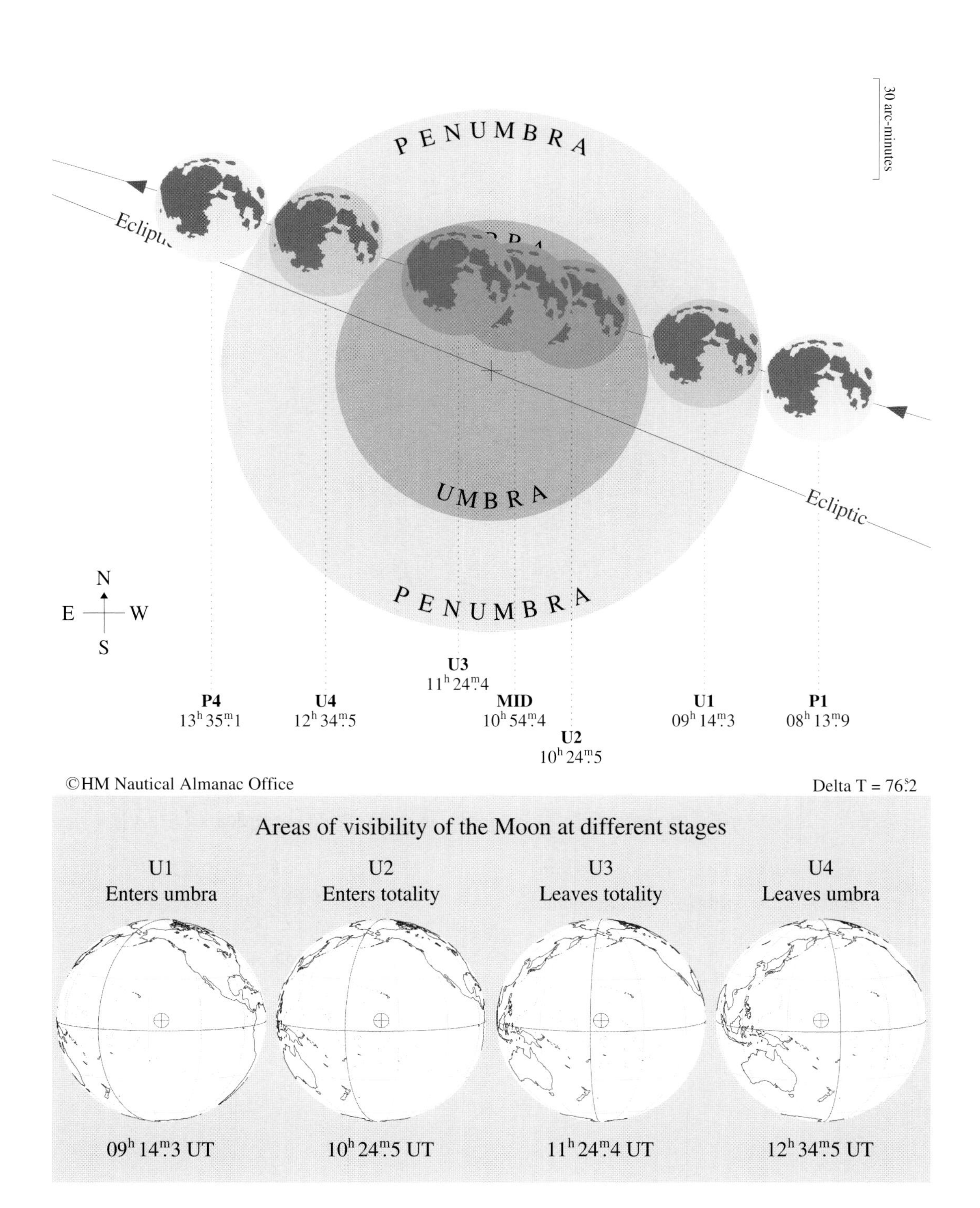

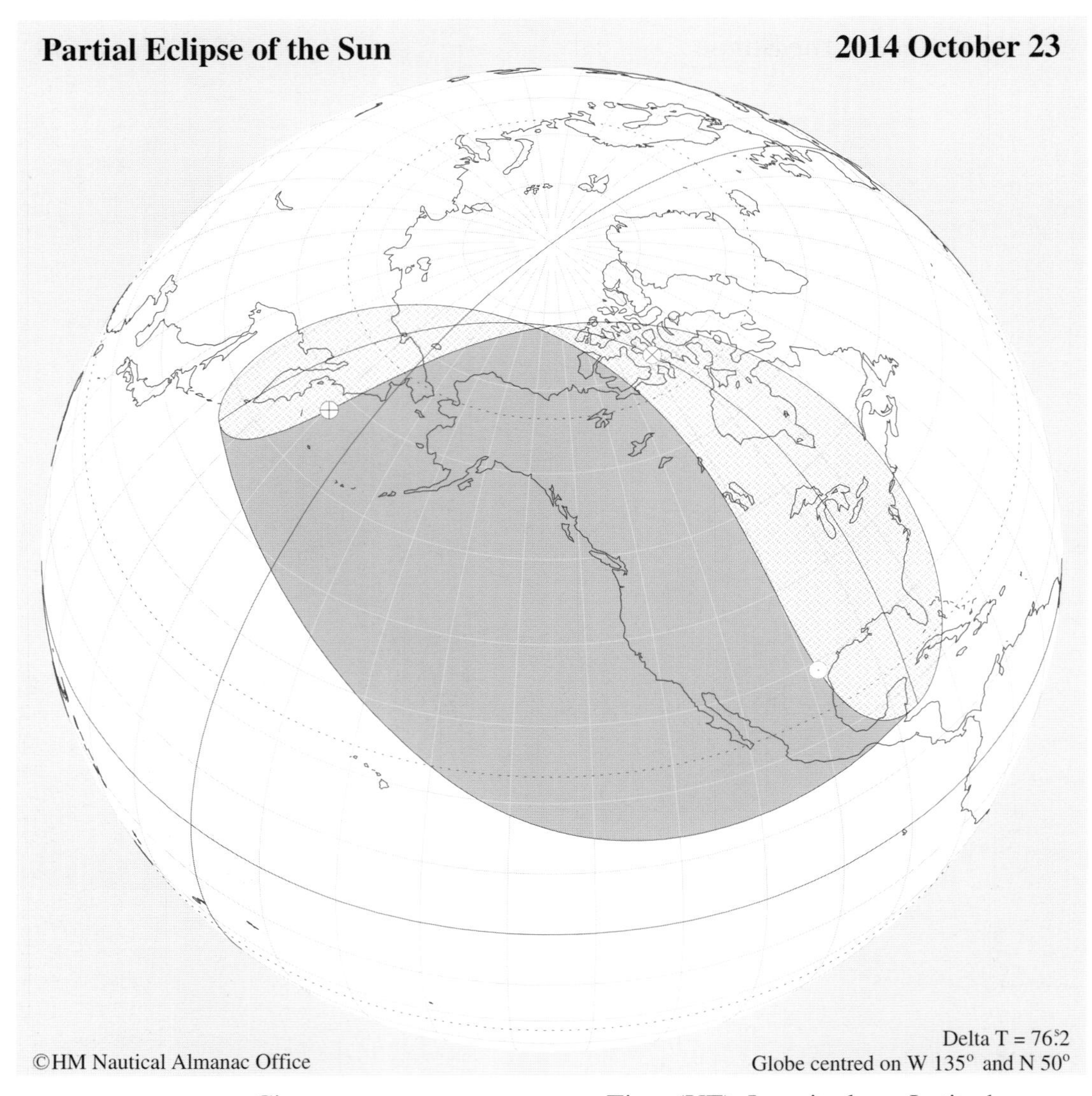

Circumstances	Time (UT)	Longitude	Latitude
	h m	° ′	° ′
⊕ Eclipse begins; first contact with Earth	19 37.4	E 170 33.3	N 57 34.1
Beginning of southern limit of penumbra	20 13.6	E 154 57.4	N 46 02.5
⊗ Greatest eclipse (magnitude = 0.8119)	21 44.4	W 97 15.7	N 71 14.4
End of southern limit of penumbra	23 15.3	W 86 21.2	N 16 57.9
○ Eclipse ends; last contact with Earth	23 51.5	W 98 21.3	N 28 55.8

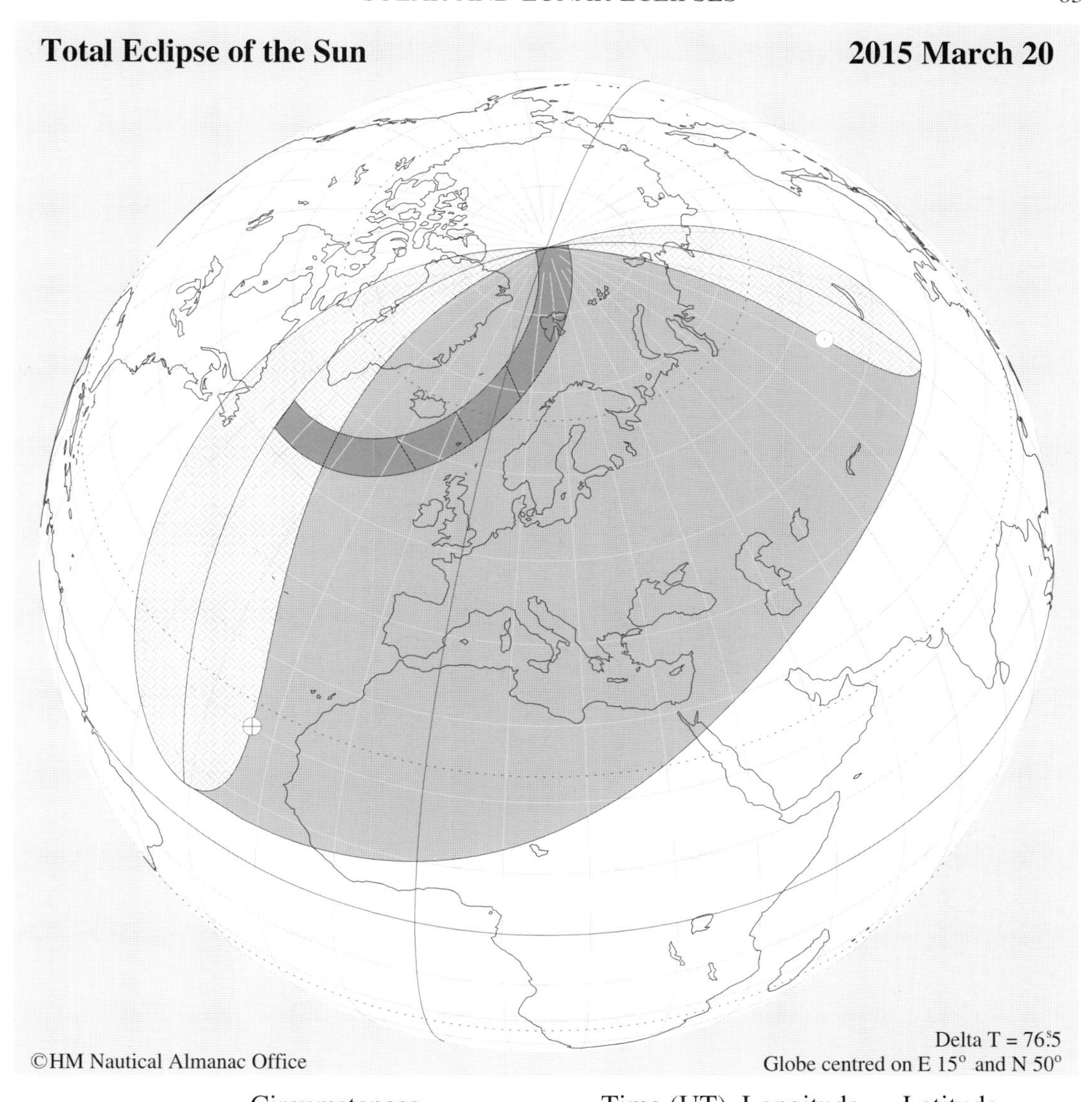

	Circumstances	Time (UT)		Longitude			Latitude	
		h	m		°	′	°	′
⊕	Eclipse begins; first contact with Earth	7	40.7	W	23	10.6	N 20	13.9
	Beginning of southern limit of penumbra	8	11.4	W	30	54.9	N 6	19.3
	Beginning of southern limit of umbra	9	09.5	W	45	11.2	N 51	49.7
	Beginning of centre line; central eclipse begins	9	12.5	W	45	55.9	N 53	37.5
	Beginning of northern limit of umbra	9	15.9	W	46	45.7	N 55	36.5
	End of northern limit of umbra	10	14.7	W	53	13.8	N 88	35.8
	Central eclipse at local apparent noon	10	16.9	E	27	39.5	N 85	06.3
	End of centre line; central eclipse ends	10	18.0	E	97	51.1	N 89	22.8
	End of southern limit of umbra	10	21.1	E	111	38.6	N 87	37.1
	End of southern limit of penumbra	11	19.4	E	101	51.3	N 42	16.1
⊙	Eclipse ends; last contact with Earth	11	50.0	E	94	06.1	N 56	06.1

Total Eclipse of the Moon

2015 April 04

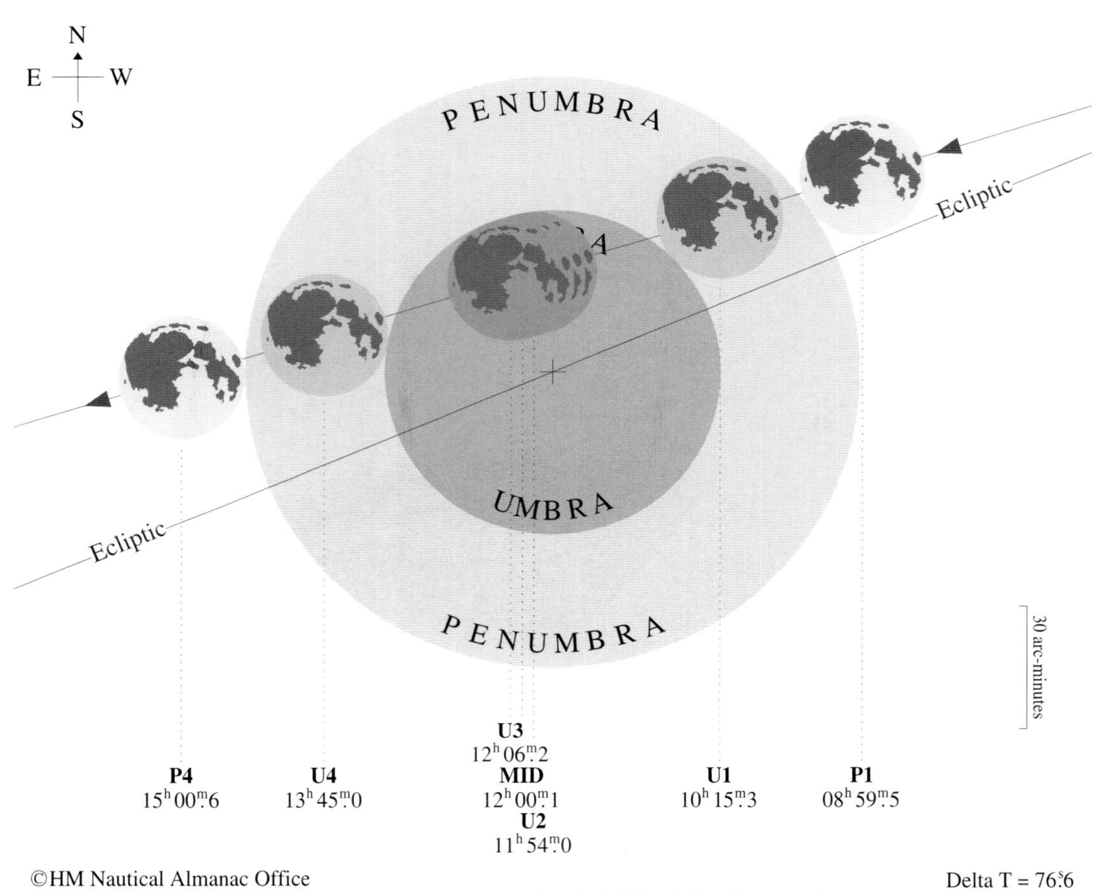

Areas of visibility of the Moon at different stages

U1	U2	U3	U4
Enters umbra	Enters totality	Leaves totality	Leaves umbra
$10^{h}15^{m}.3$ UT	$11^{h}54^{m}.0$ UT	$12^{h}06^{m}.2$ UT	$13^{h}45^{m}.0$ UT

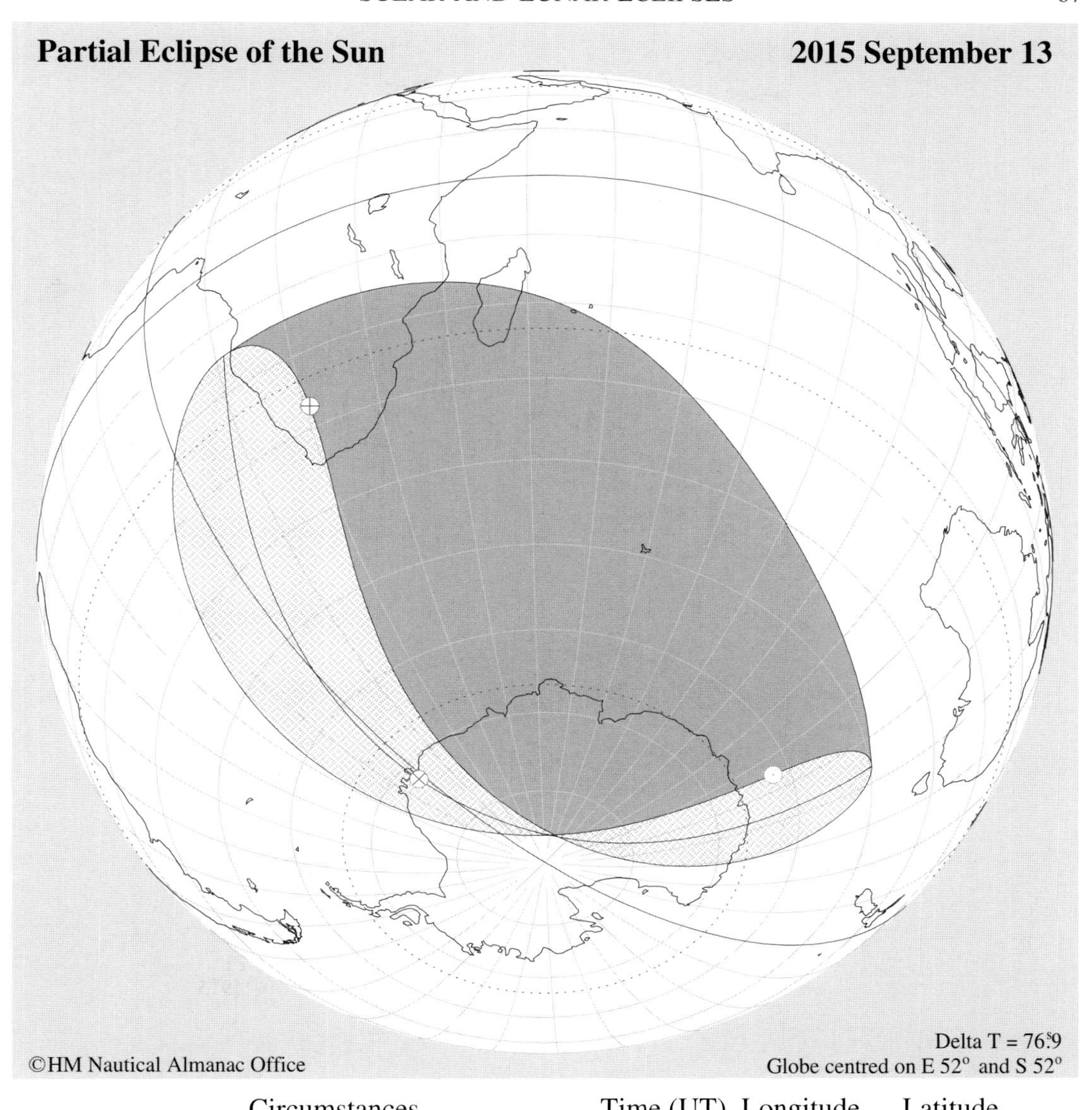

	Circumstances	Time (UT) h m	Longitude ° ′	Latitude ° ′
⊕	Eclipse begins; first contact with Earth	4 41.5	E 20 40.8	S 27 11.7
	Beginning of northern limit of penumbra	5 15.5	E 11 12.0	S 14 45.5
⊗	Greatest eclipse (magnitude = 0.7876)	6 54.0	W 2 14.3	S 72 12.8
	End of northern limit of penumbra	8 32.3	E 136 21.6	S 49 44.9
⊙	Eclipse ends; last contact with Earth	9 06.2	E 125 09.5	S 62 02.6

Total Eclipse of the Moon 2015 September 28

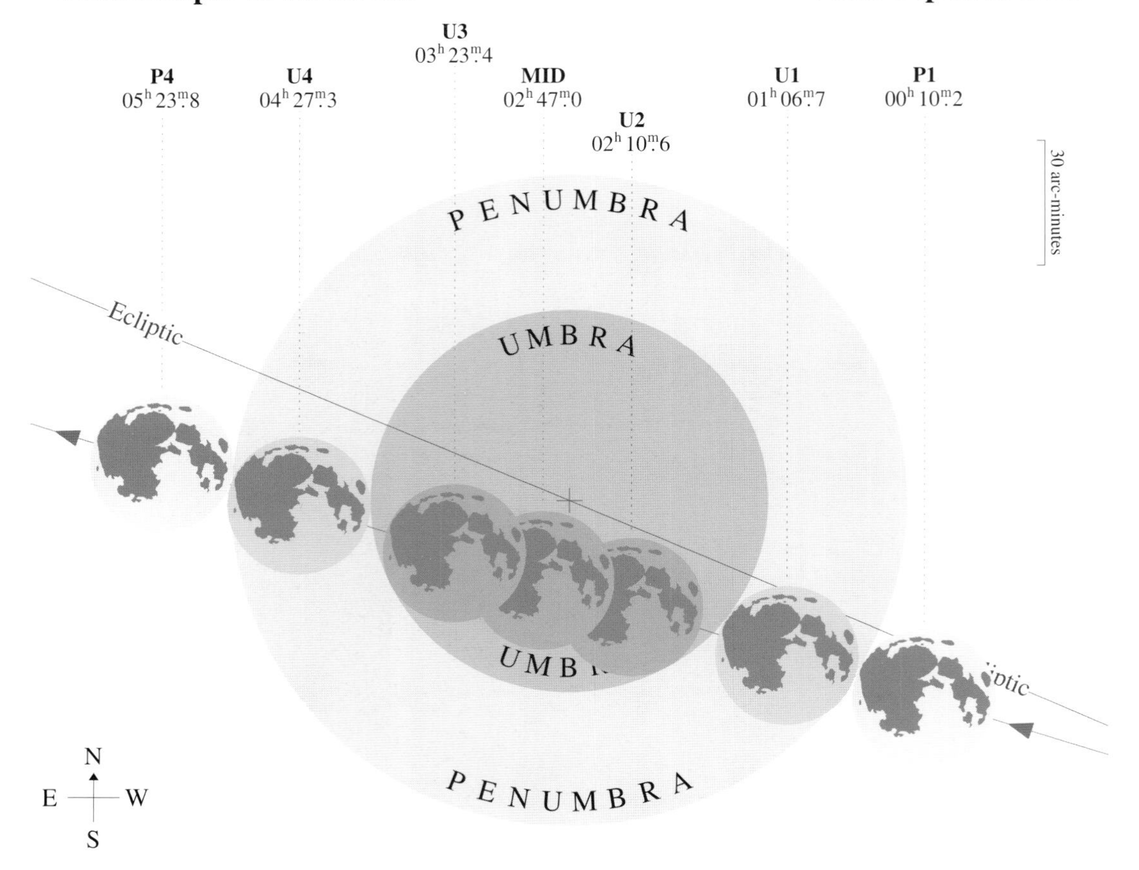

Delta T = 76.s9

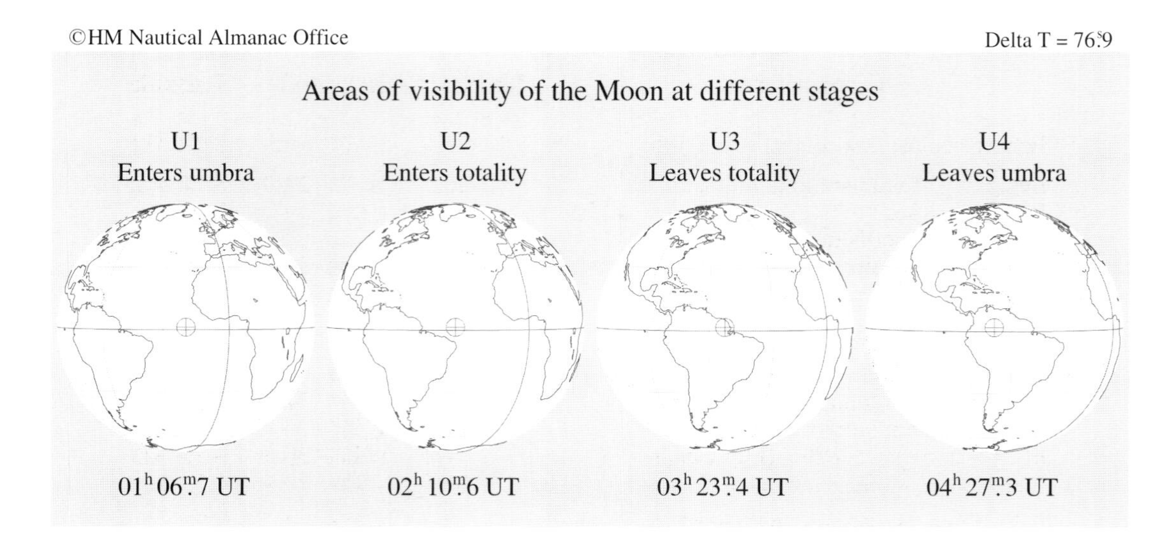

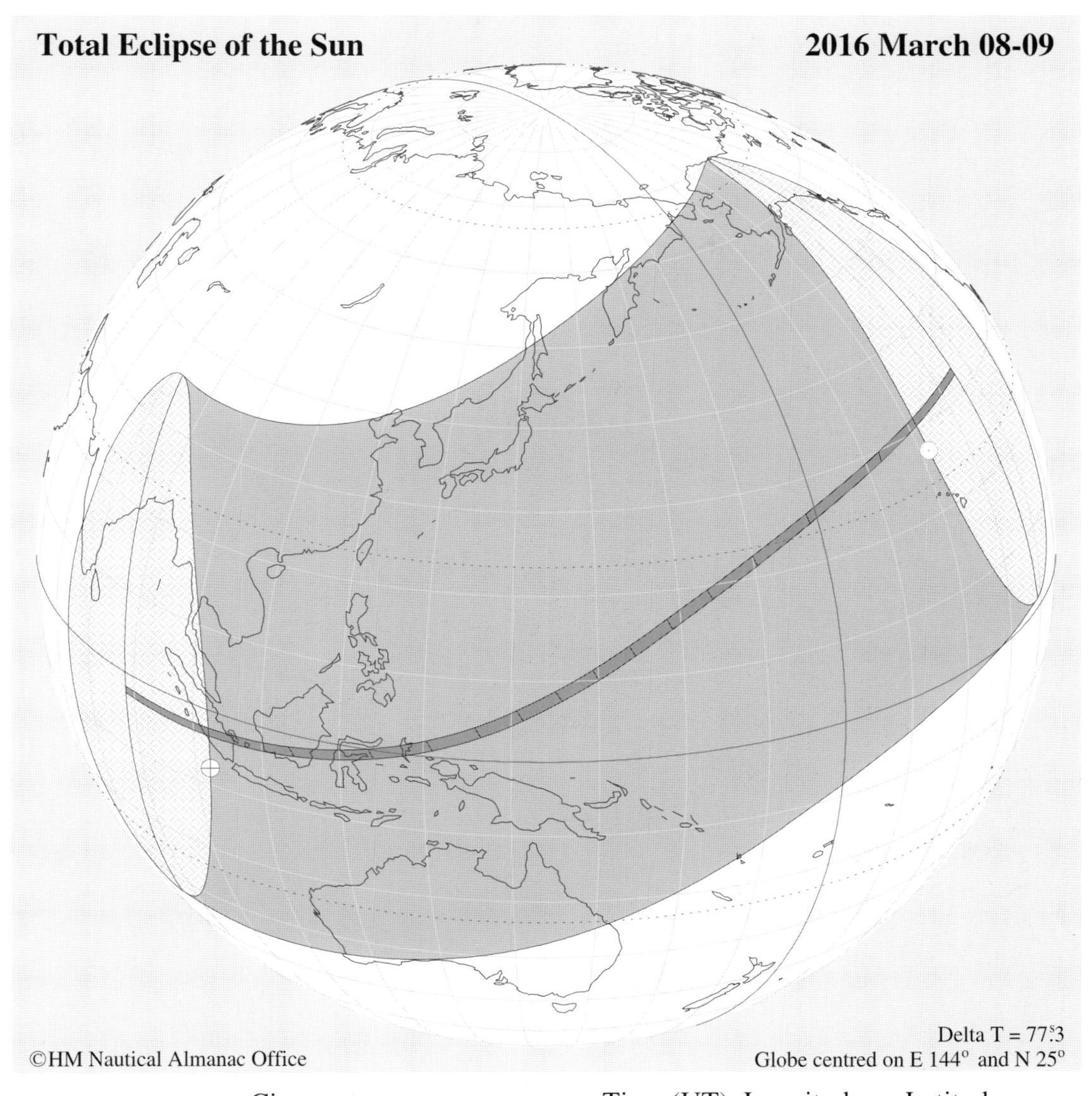

	Circumstances	Time (UT)		Longitude		Latitude	
		h	m	°	′	°	′
⊕	Eclipse begins; first contact with Earth	23	19.2	E 102	15.3	S 7	38.0
	Beginning of southern limit of umbra	0	16.4	E 88	20.1	S 2	41.5
	Beginning of centre line; central eclipse begins	0	16.5	E 88	19.5	S 2	15.2
	Beginning of northern limit of umbra	0	16.7	E 88	18.7	S 1	48.8
	Beginning of southern limit of penumbra	0	18.9	E 84	58.5	S 33	34.7
	Beginning of northern limit of penumbra	0	53.4	E 82	27.9	N 35	52.7
	Central eclipse at local apparent noon	2	05.5	E 151	15.1	N 11	33.8
	End of northern limit of penumbra	3	00.2	W 144	45.5	N 70	20.0
	End of southern limit of penumbra	3	35.3	W 141	17.3	N 1	15.5
	End of northern limit of umbra	3	37.2	W 144	30.6	N 33	01.7
	End of centre line; central eclipse ends	3	37.4	W 144	30.6	N 32	34.6
	End of southern limit of umbra	3	37.6	W 144	30.4	N 32	07.5
⊙	Eclipse ends; last contact with Earth	4	34.7	W 158	18.2	N 27	12.6

Penumbral Eclipse of the Moon 2016 March 23

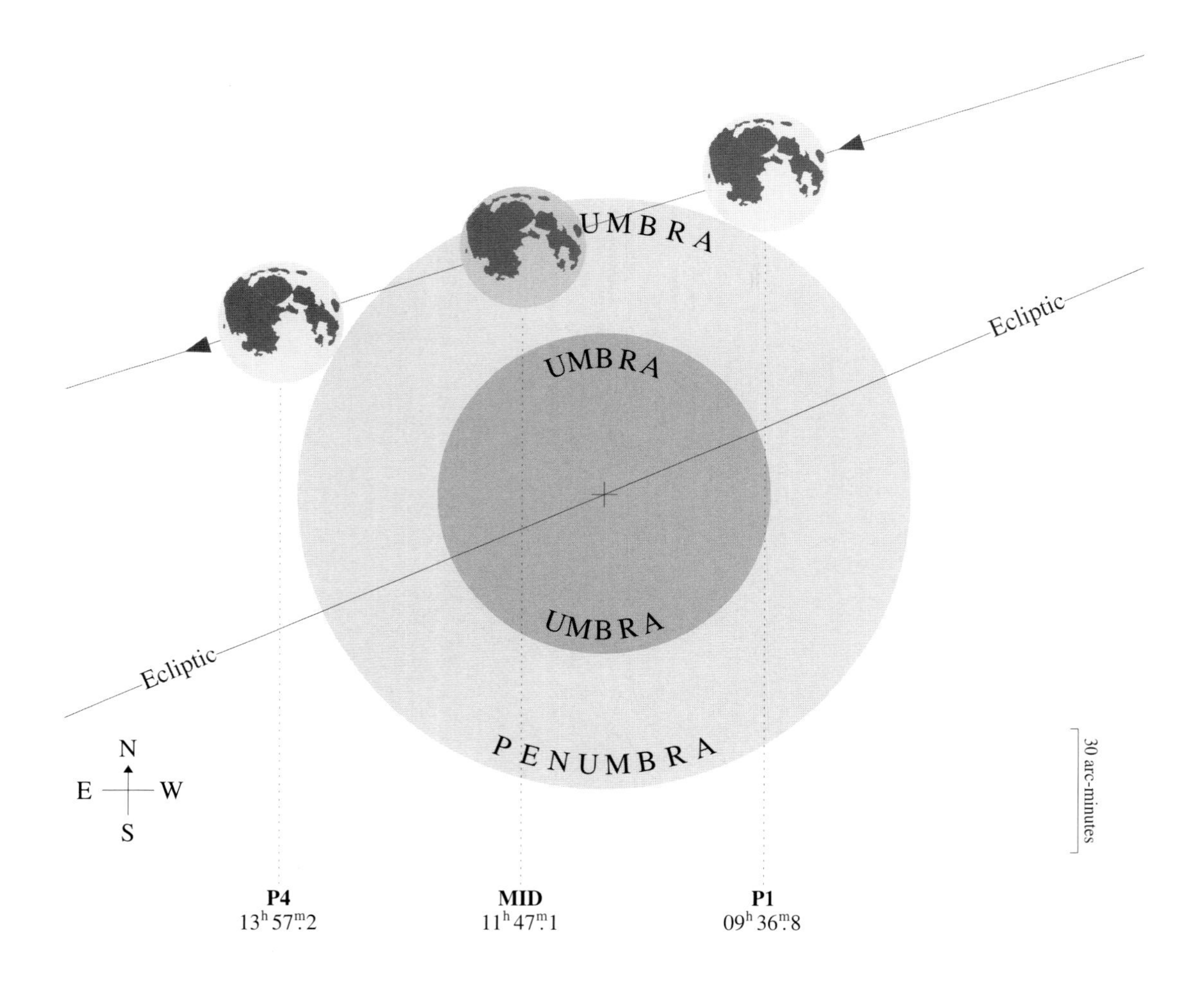

Delta T = $77^{s}.3$

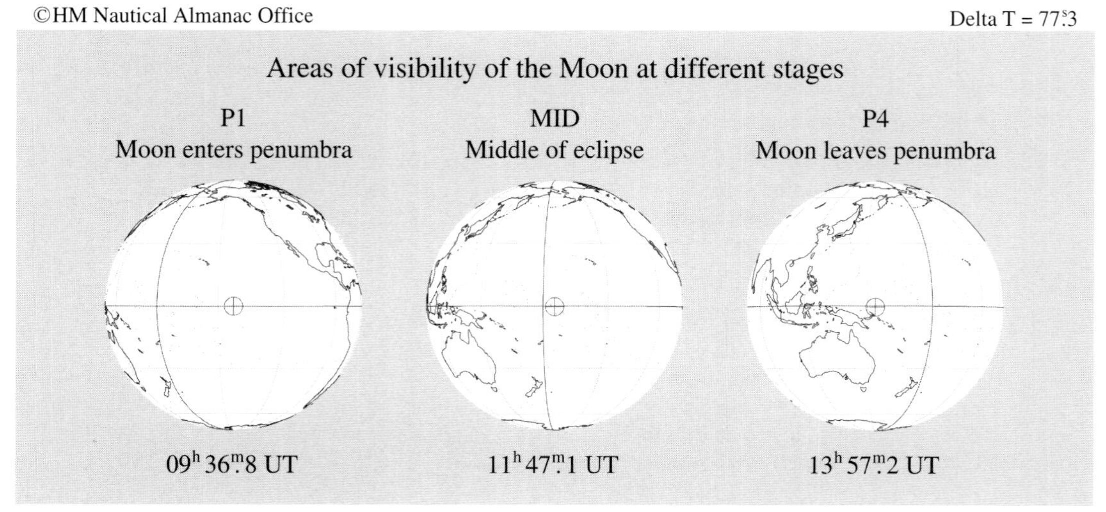

Penumbral Eclipse of the Moon 2016 August 18

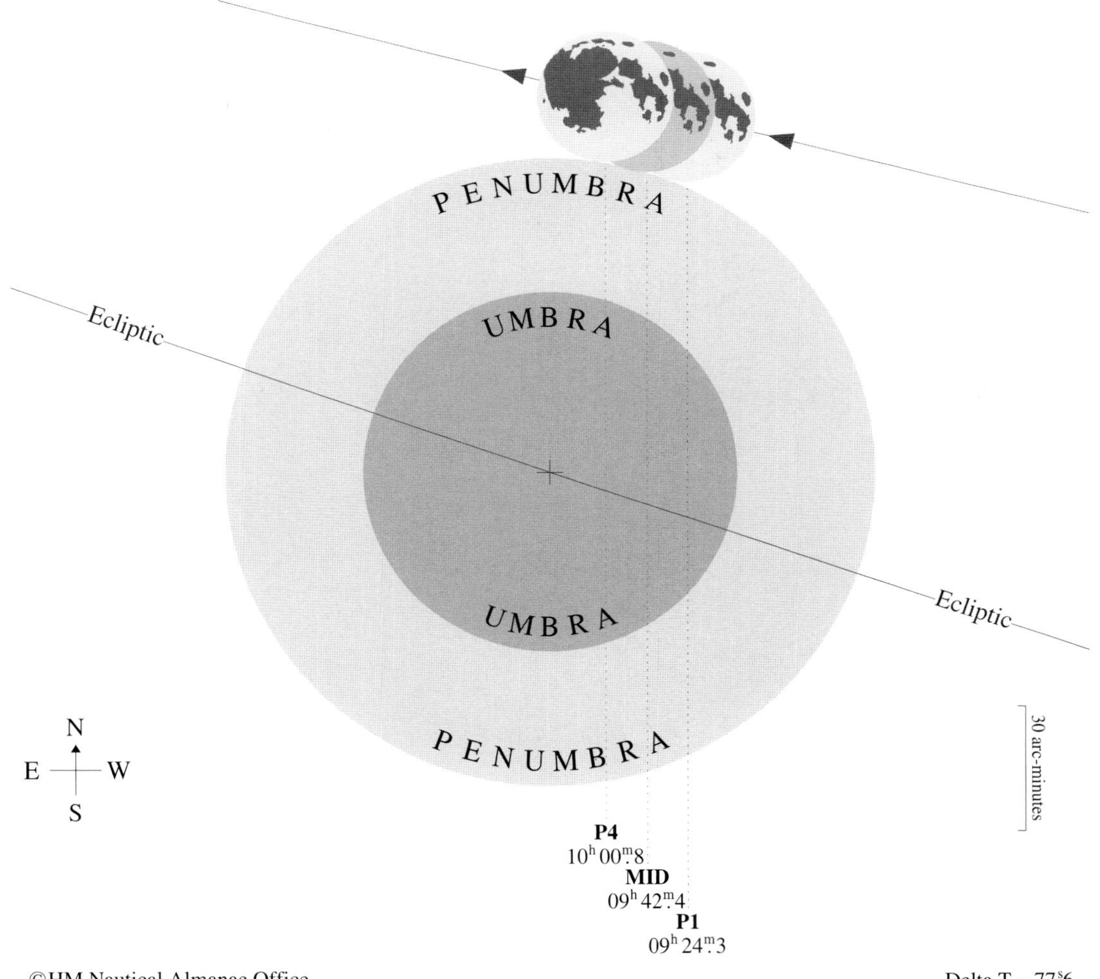

 Delta T = 77s.6

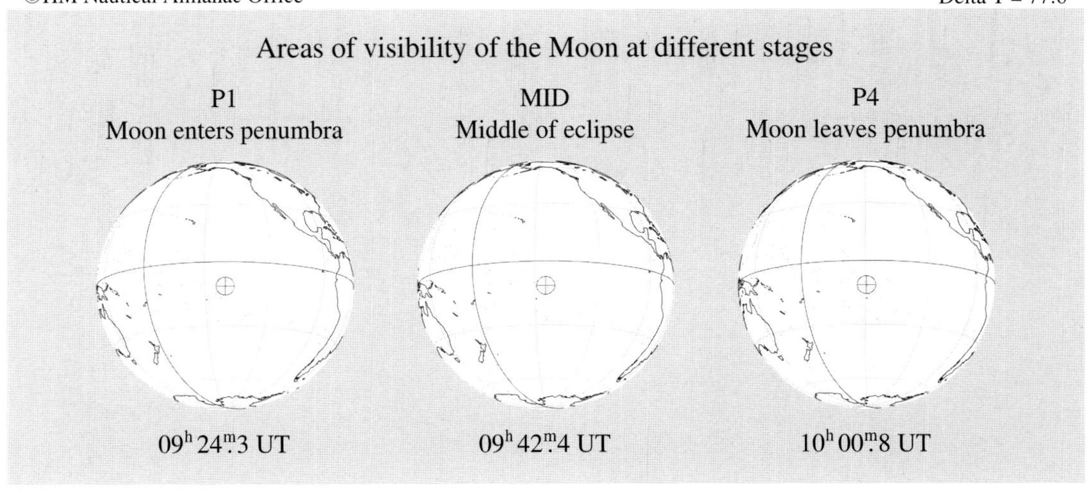

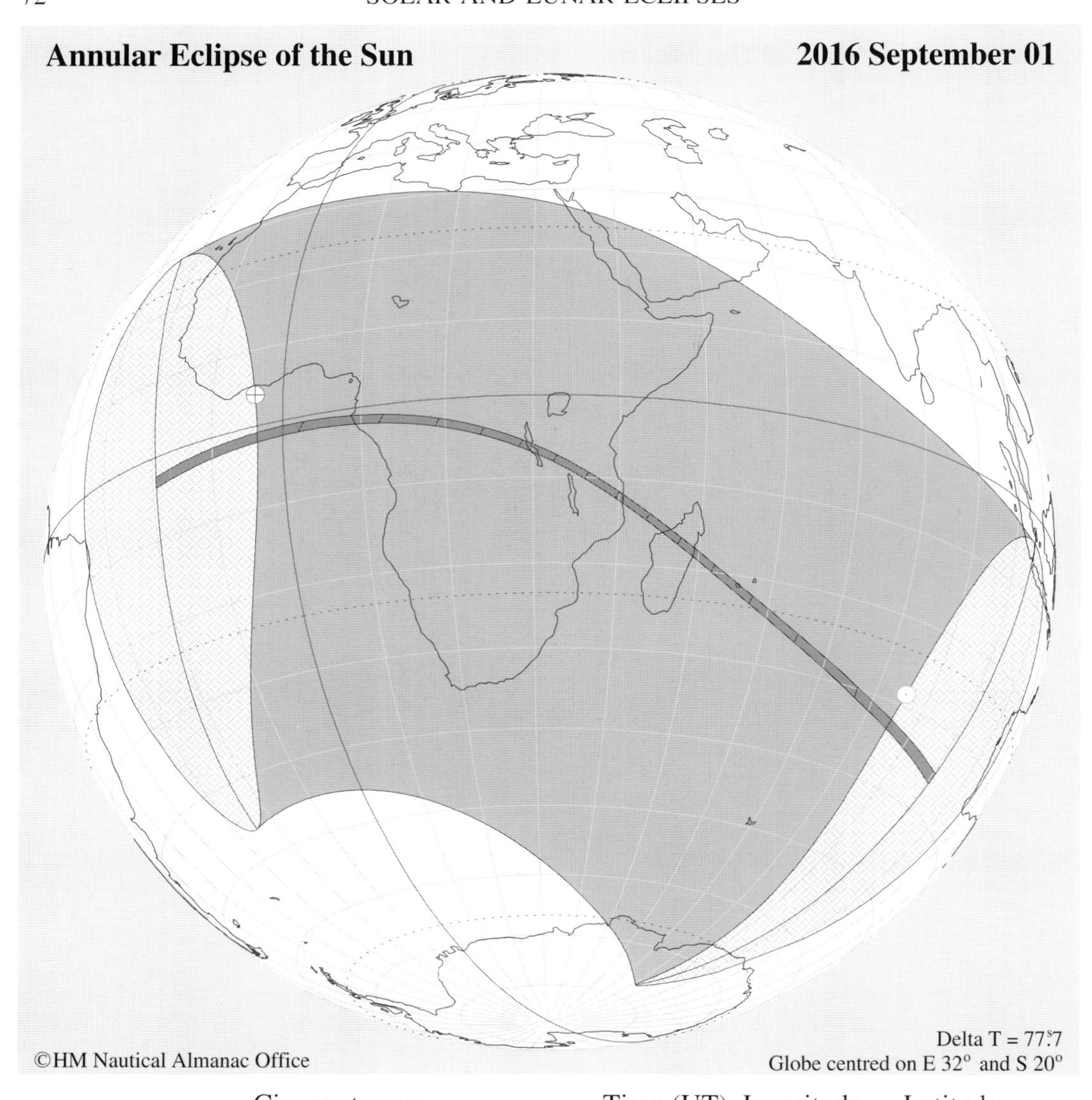

Circumstances	Time (UT)	Longitude	Latitude
	h m	° ′	° ′
⊕ Eclipse begins; first contact with Earth	6 13.0	W 3 49.8	N 4 00.4
Beginning of northern limit of umbra	7 18.7	W 19 21.2	S 2 23.5
Beginning of centre line; central eclipse begins	7 19.0	W 19 20.8	S 3 04.4
Beginning of southern limit of umbra	7 19.4	W 19 20.6	S 3 45.4
Beginning of northern limit of penumbra	7 19.7	W 24 29.1	N 29 08.1
Beginning of southern limit of penumbra	8 13.0	W 24 43.2	S 46 21.3
Central eclipse at local apparent noon	9 17.9	E 40 29.6	S 12 20.4
End of southern limit of penumbra	10 00.0	E 82 07.3	S 77 01.4
End of southern limit of umbra	10 53.9	E 100 31.9	S 36 20.3
End of northern limit of penumbra	10 53.9	E 105 59.7	S 3 26.3
End of centre line; central eclipse ends	10 54.3	E 100 35.1	S 35 38.2
End of northern limit of umbra	10 54.6	E 100 38.4	S 34 56.2
⊙ Eclipse ends; last contact with Earth	12 00.5	E 85 25.8	S 28 35.4

Penumbral Eclipse of the Moon 2016 September 16

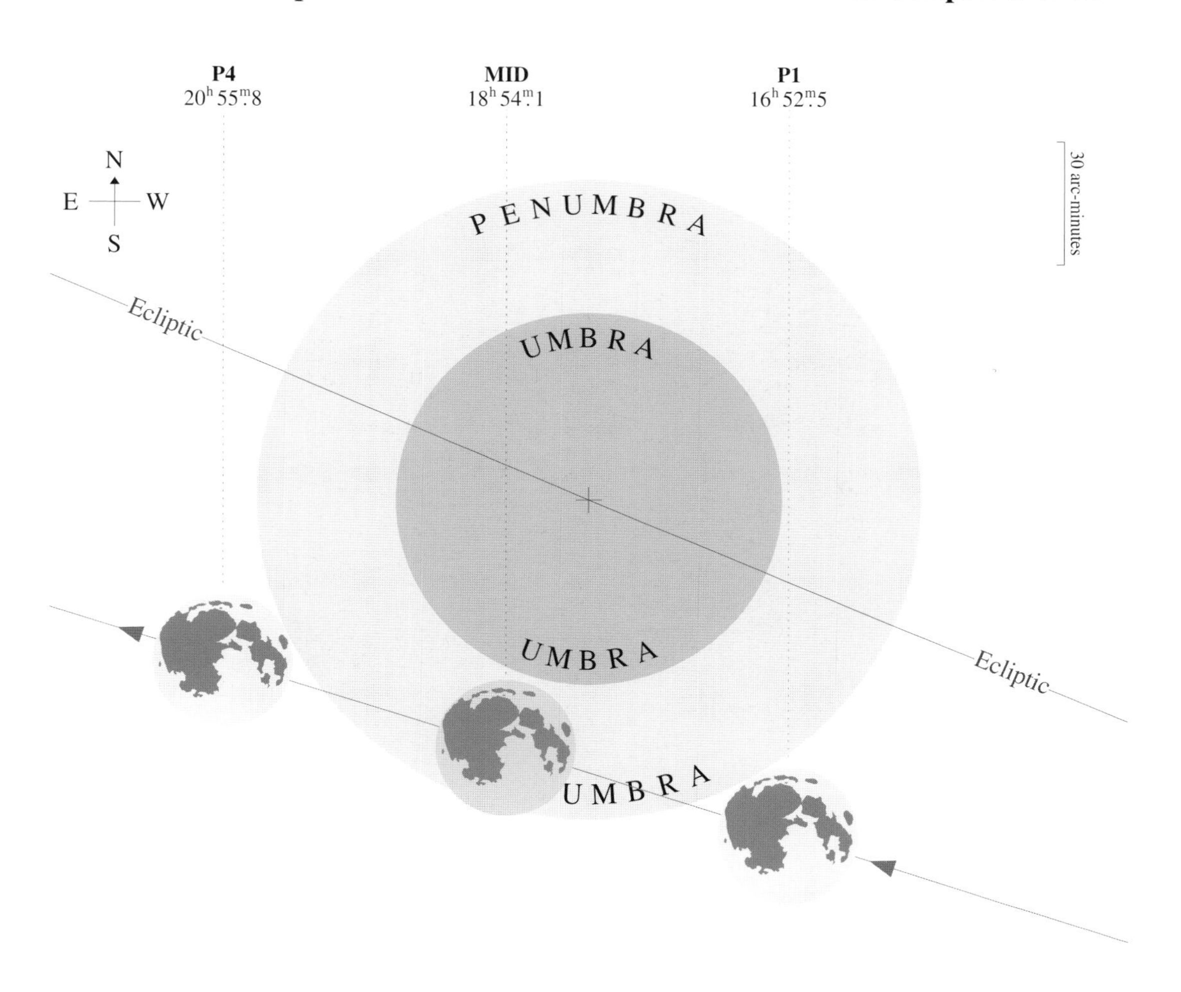

Delta T = $77^s.7$

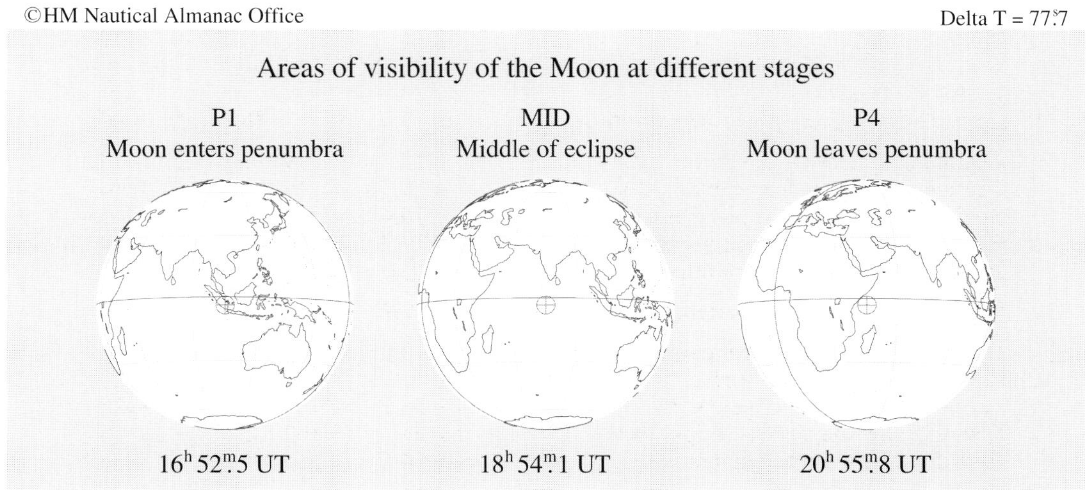

Penumbral Eclipse of the Moon 2017 February 10-11

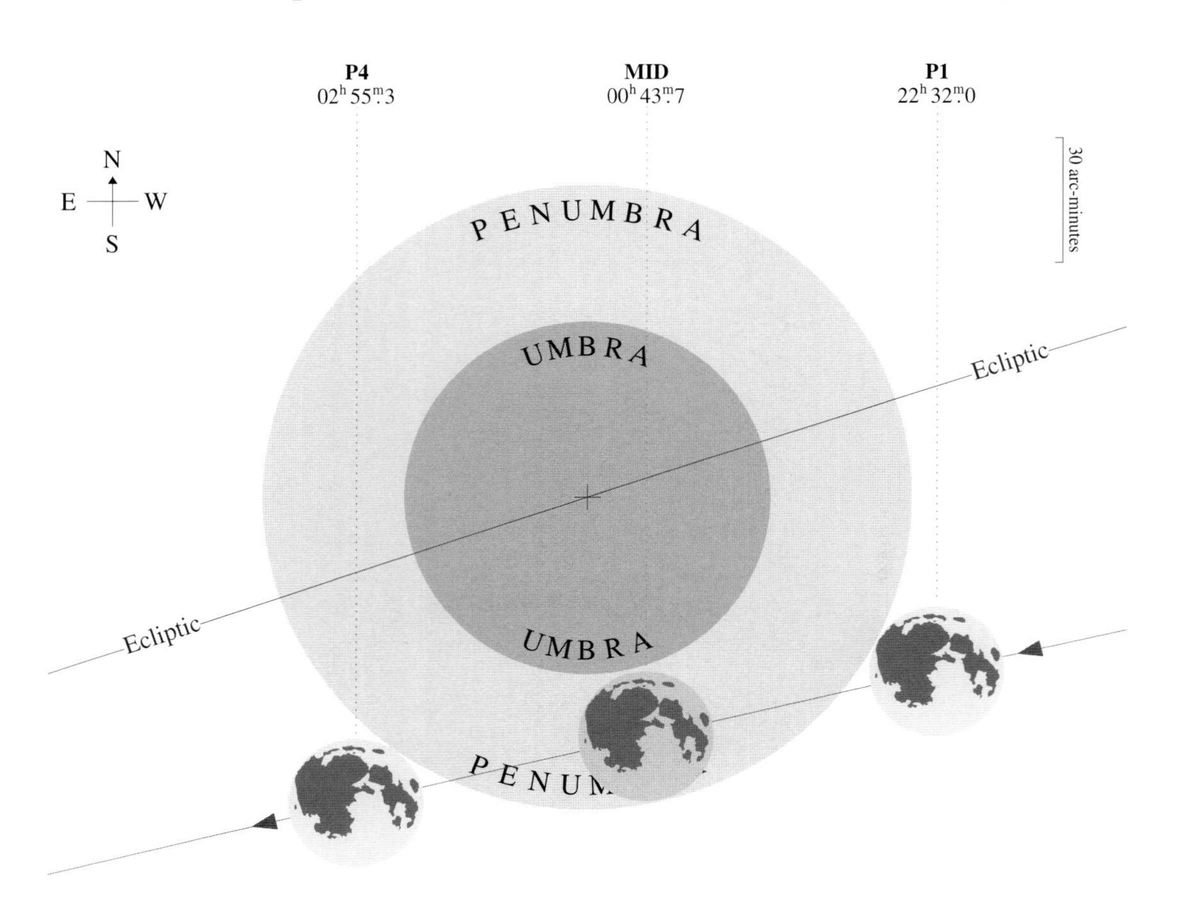

Delta T = 78s.0

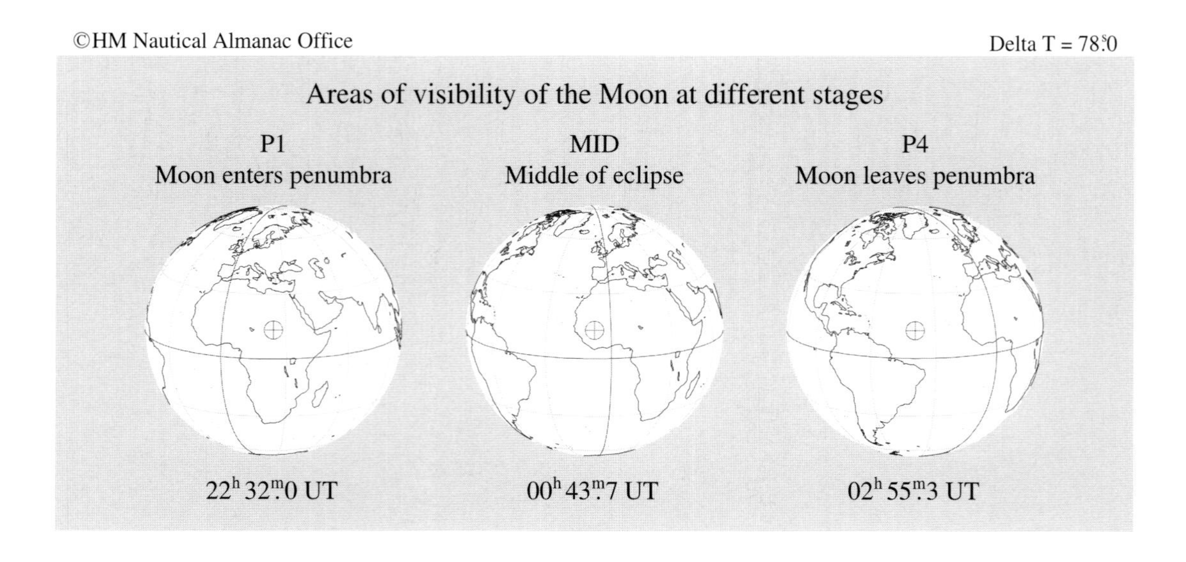

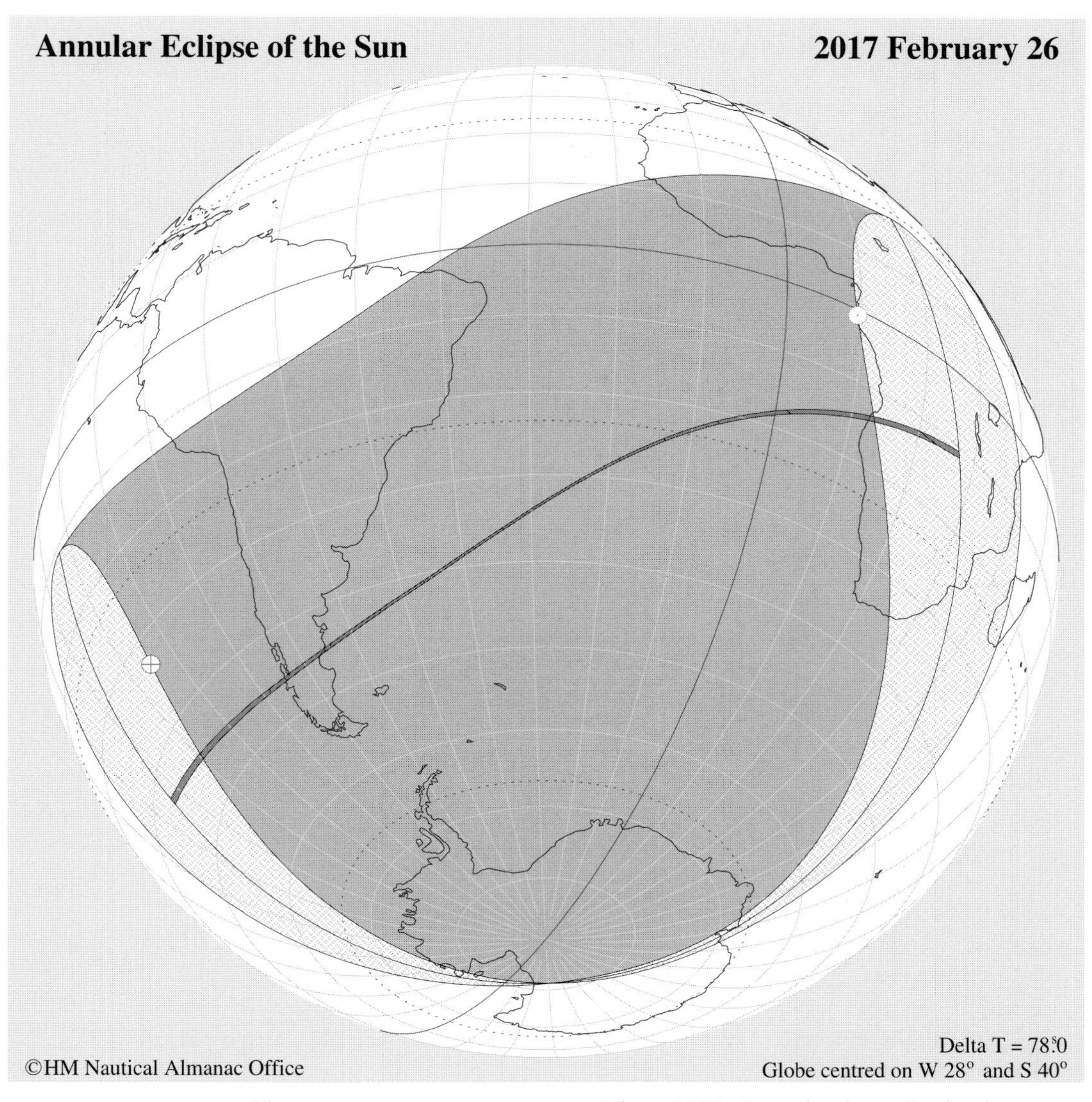

Circumstances	Time (UT)		Longitude		Latitude	
	h	m	°	′	°	′
⊕ Eclipse begins; first contact with Earth	12	10.6	W 95	04.1	S 33	09.2
Beginning of northern limit of penumbra	12	59.6	W103	20.1	S 10	50.2
Beginning of northern limit of umbra	13	15.5	W113	37.7	S 42	44.0
Beginning of centre line; central eclipse begins	13	15.9	W113	50.3	S 43	07.8
Beginning of southern limit of umbra	13	16.3	W114	03.1	S 43	31.5
Central eclipse at local apparent noon	14	38.6	W 36	26.6	S 37	12.3
End of southern limit of umbra	16	30.3	E 27	19.3	S 11	18.1
End of centre line; central eclipse ends	16	30.7	E 27	10.3	S 10	55.6
End of northern limit of umbra	16	31.1	E 27	01.3	S 10	33.1
End of northern limit of penumbra	16	46.7	E 18	09.2	N21	29.5
⊙ Eclipse ends; last contact with Earth	17	35.8	E 9	21.7	S 0	52.2

Partial Eclipse of the Moon 2017 August 07

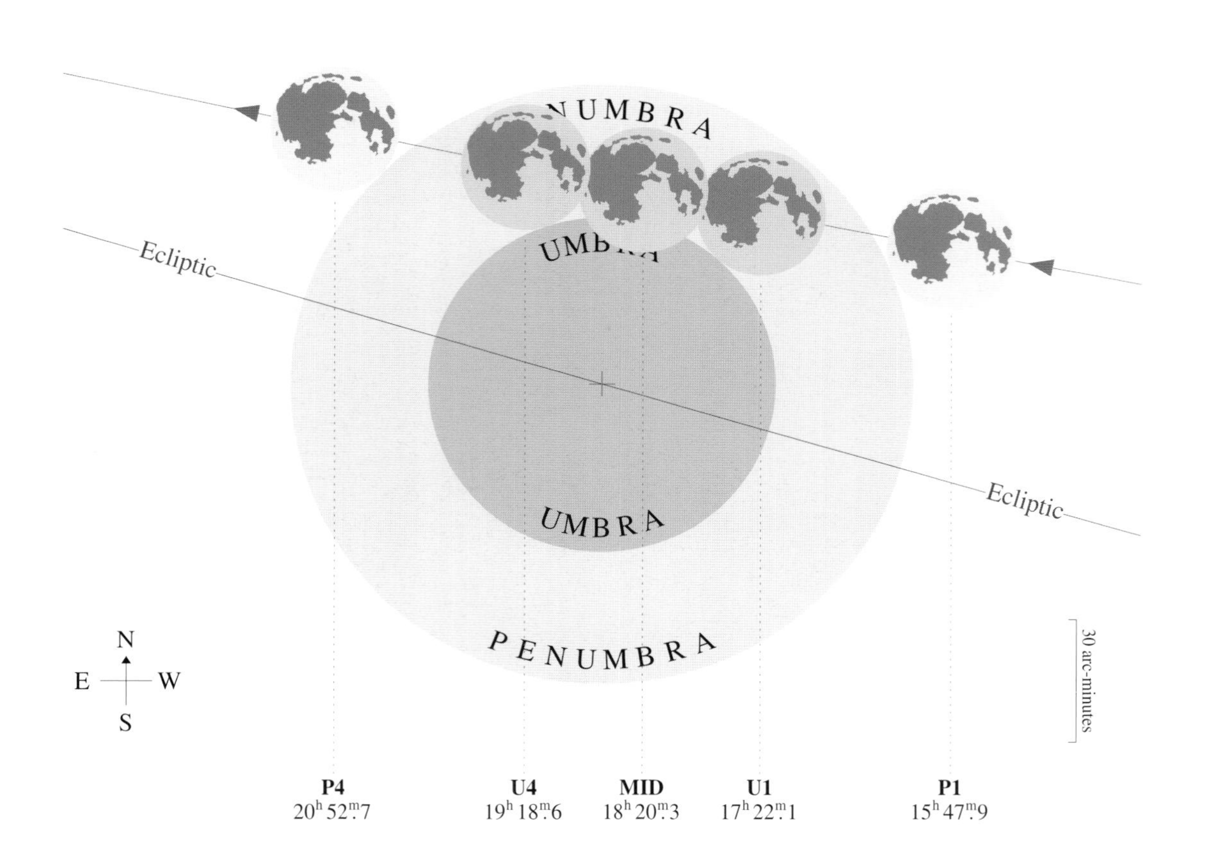

Delta T = $78^{s}.4$

Areas of visibility of the Moon at different stages

U1	MID	U4
Moon enters umbra	Middle of eclipse	Moon leaves umbra
$17^{h}22^{m}.1$ UT	$18^{h}20^{m}.3$ UT	$19^{h}18^{m}.6$ UT

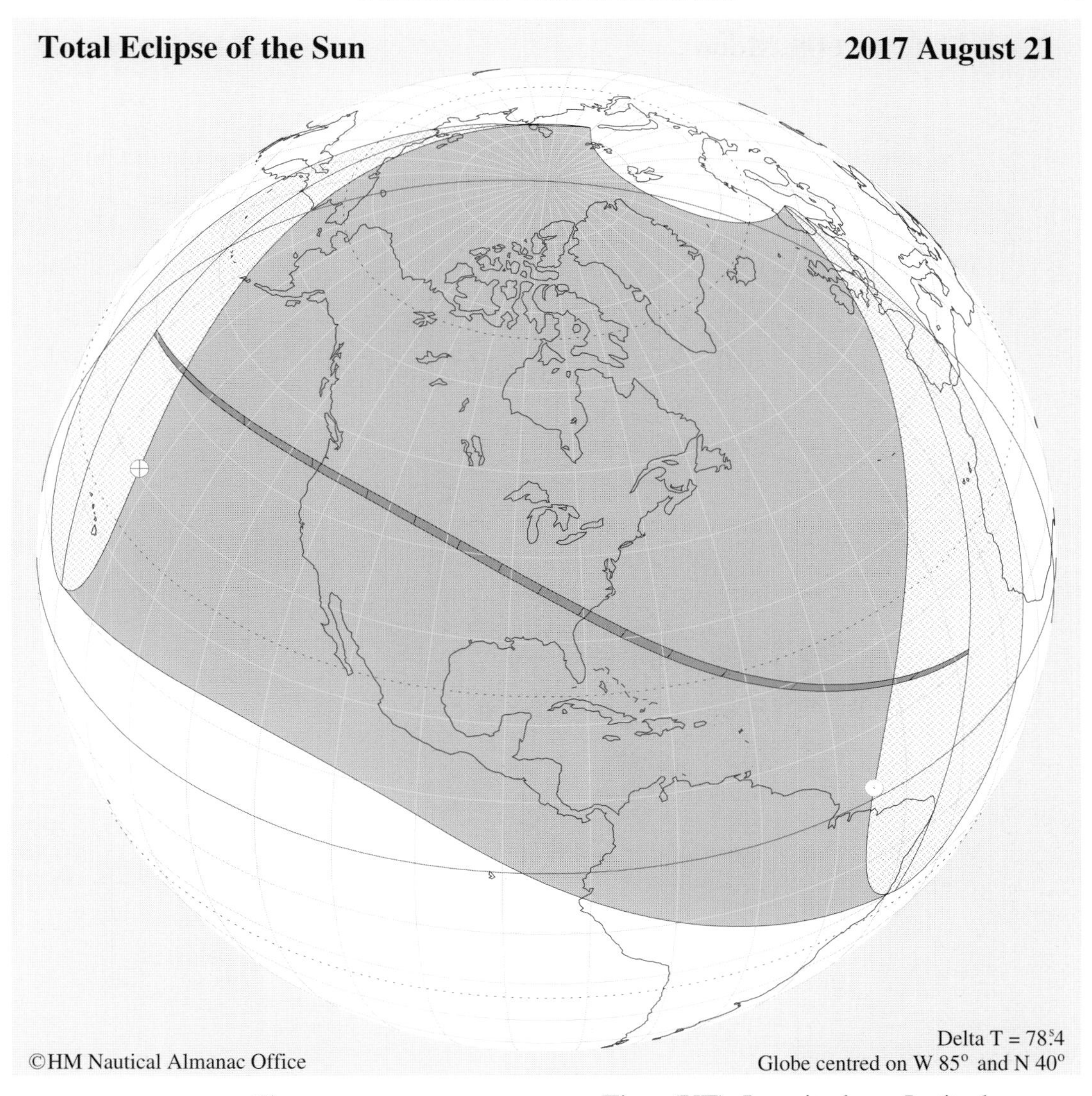

Circumstances	Time (UT)	Longitude	Latitude
	h m	° ′	° ′
⊕ Eclipse begins; first contact with Earth	15 46.7	W153 02.6	N 30 32.5
Beginning of southern limit of penumbra	16 32.8	W159 16.5	N 8 38.6
Beginning of southern limit of umbra	16 48.6	W171 22.5	N 39 27.7
Beginning of centre line; central eclipse begins	16 48.9	W171 32.4	N 39 43.9
Beginning of northern limit of umbra	16 49.2	W171 42.4	N 40 00.2
Beginning of northern limit of penumbra	18 06.0	E 71 42.2	N 77 34.4
Central eclipse at local apparent noon	18 13.1	W 92 30.6	N 38 55.3
End of northern limit of penumbra	18 45.0	E 12 40.3	N 61 55.2
End of northern limit of umbra	20 01.7	W 27 16.8	N 11 16.1
End of centre line; central eclipse ends	20 01.9	W 27 23.9	N 11 00.9
End of southern limit of umbra	20 02.2	W 27 30.8	N 10 45.8
End of southern limit of penumbra	20 17.9	W 38 09.5	S 20 17.2
⊙ Eclipse ends; last contact with Earth	21 04.2	W 44 57.1	N 1 42.2

Total Eclipse of the Moon 2018 January 31

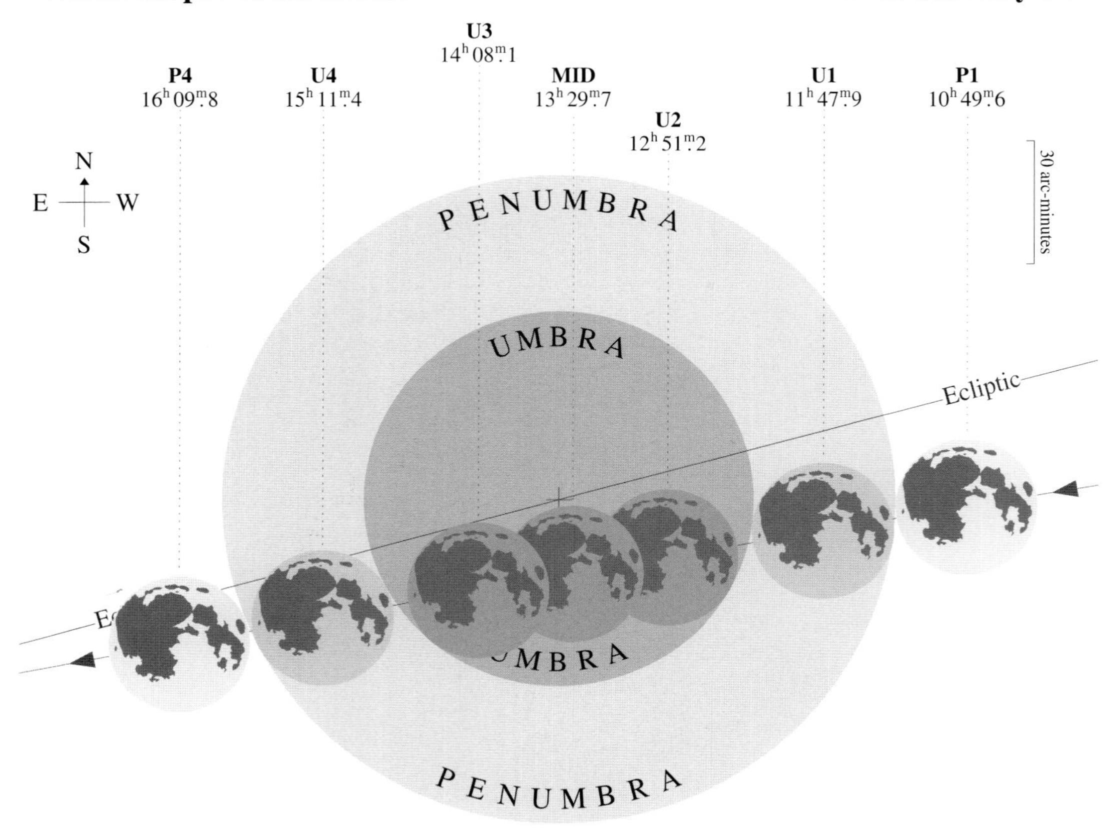

 Delta T = 78s.7

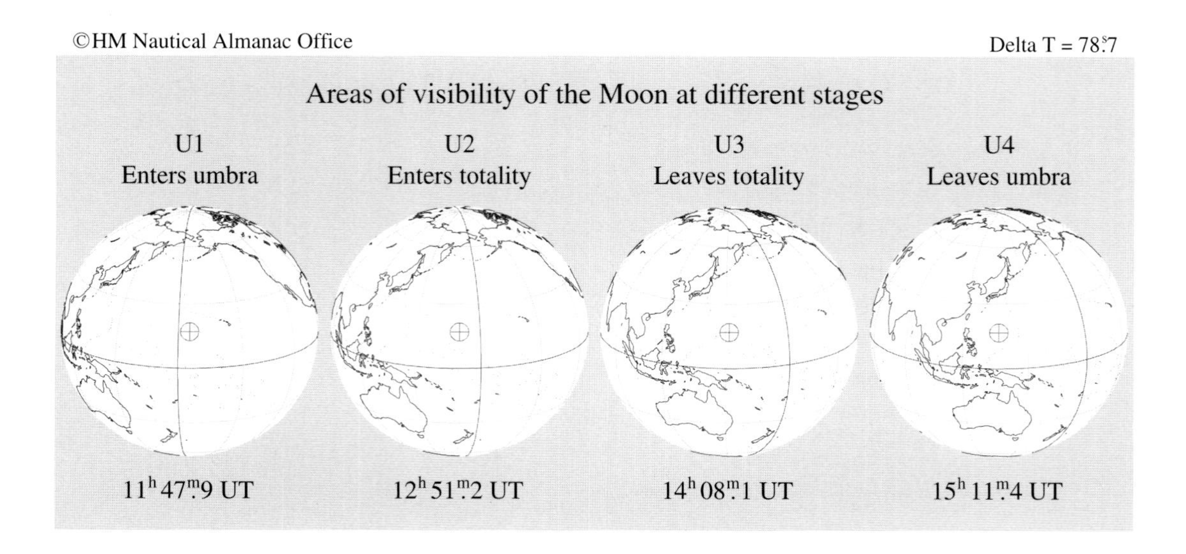

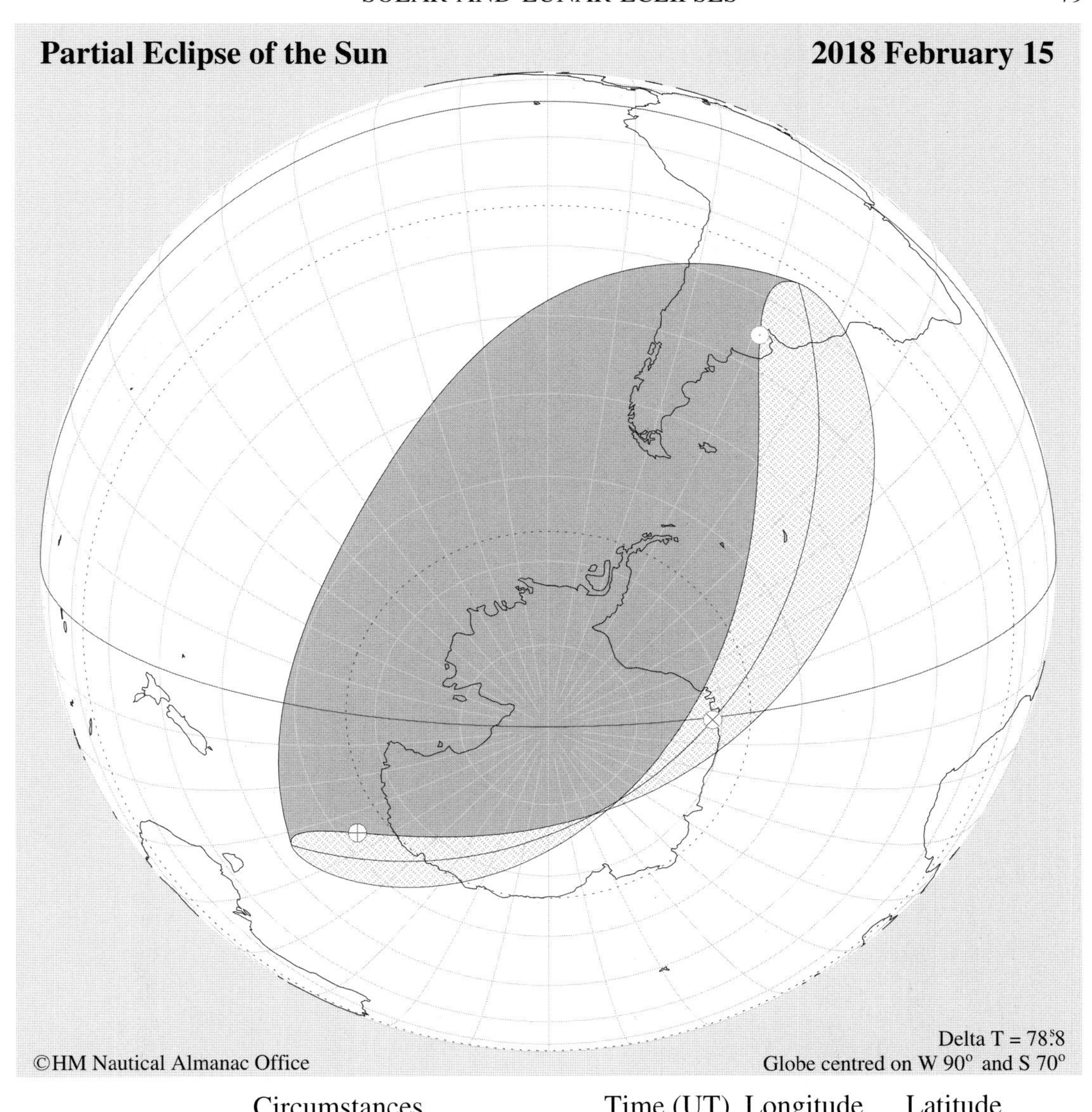

Circumstances	Time (UT)	Longitude	Latitude
	h m	° ′	° ′
⊕ Eclipse begins; first contact with Earth	18 55.7	E 144 29.4	S 62 26.3
Beginning of northern limit of penumbra	19 17.2	E 147 08.7	S 52 57.8
⊗ Greatest eclipse (magnitude = 0.5992)	20 51.2	E 0 54.9	S 71 05.6
End of northern limit of penumbra	22 25.4	W 56 51.3	S 25 16.5
⊙ Eclipse ends; last contact with Earth	22 47.0	W 59 12.6	S 35 23.6

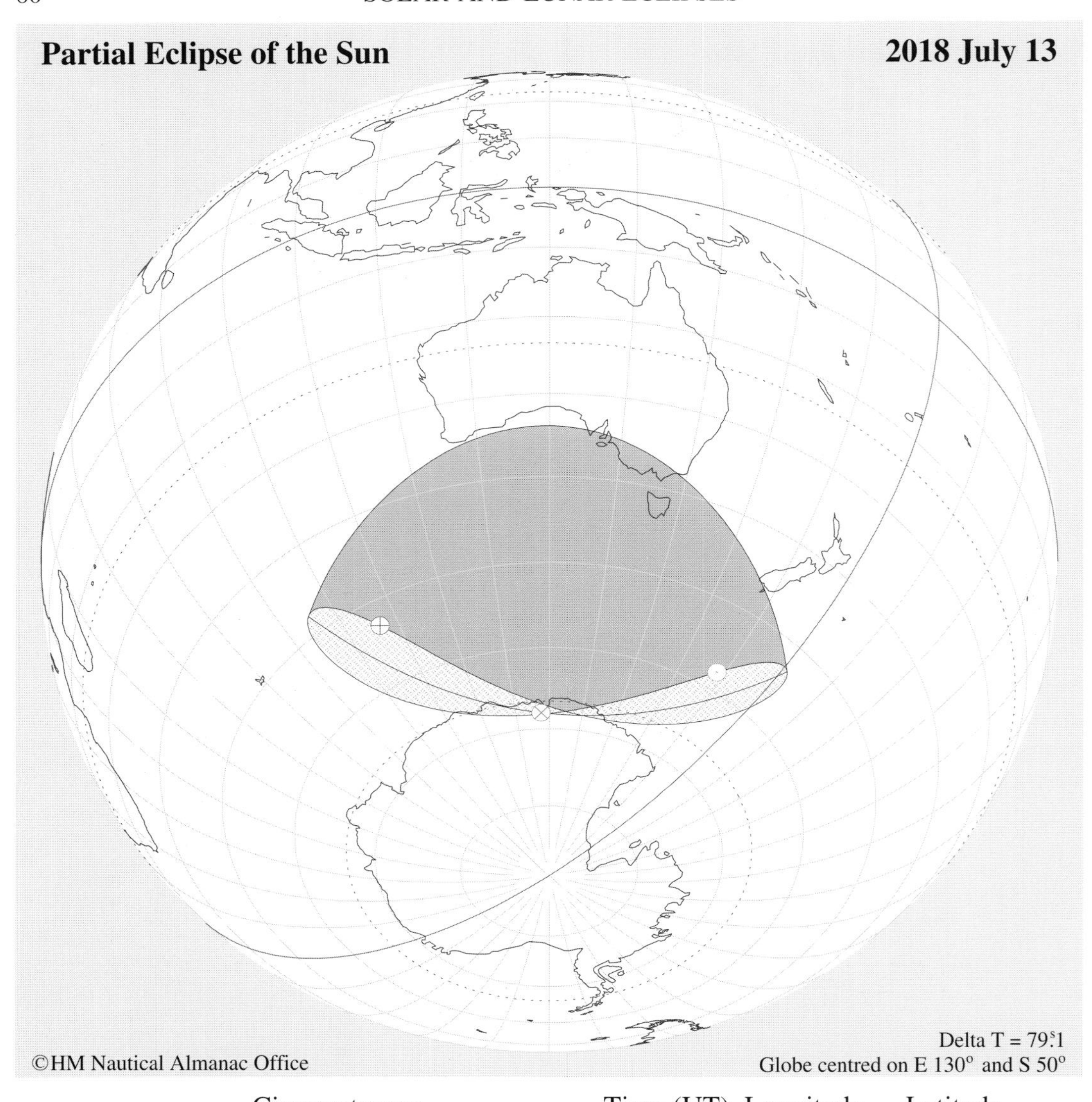

	Circumstances	Time (UT)	Longitude	Latitude
		h m	° ′	° ′
⊕	Eclipse begins; first contact with Earth	1 48.2	E 96 27.3	S 52 56.6
	Beginning of northern limit of penumbra	2 09.3	E 85 16.1	S 47 43.2
⊗	Greatest eclipse (magnitude = 0.3365)	3 00.9	E 127 31.0	S 67 55.6
	End of northern limit of penumbra	3 52.5	W 178 52.4	S 53 02.4
⊙	Eclipse ends; last contact with Earth	4 13.6	E 168 22.3	S 57 52.6

Total Eclipse of the Moon 2018 July 27

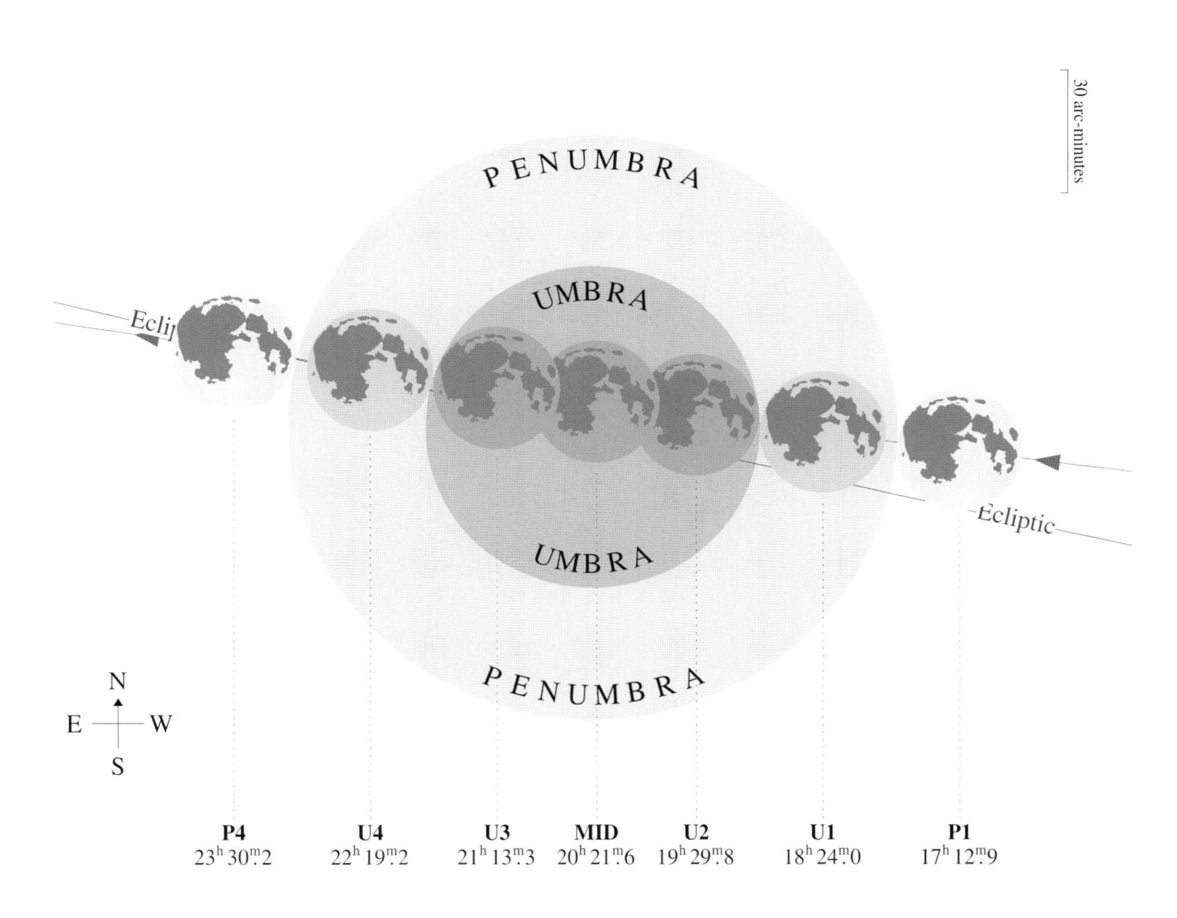

Delta T = $79^s.1$

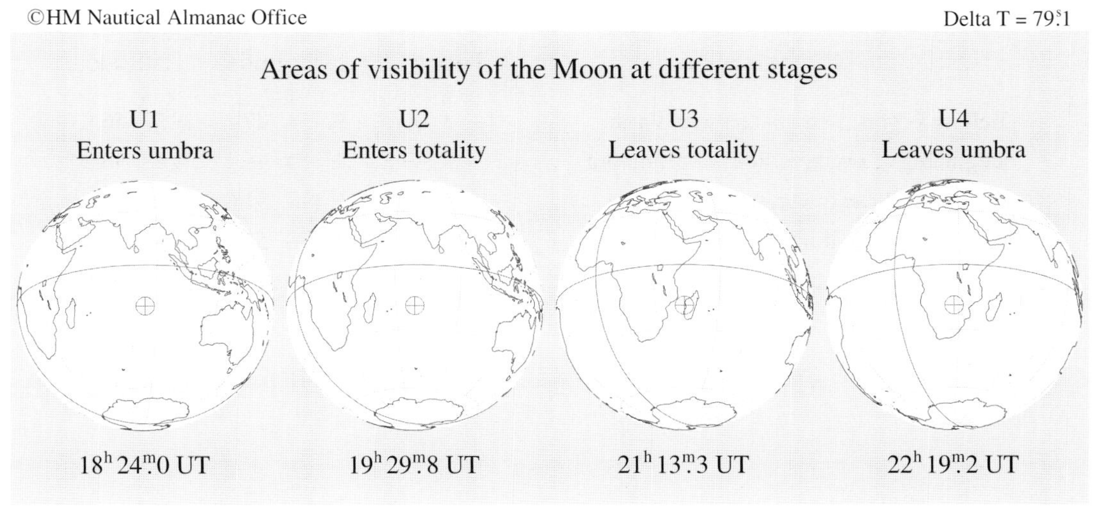

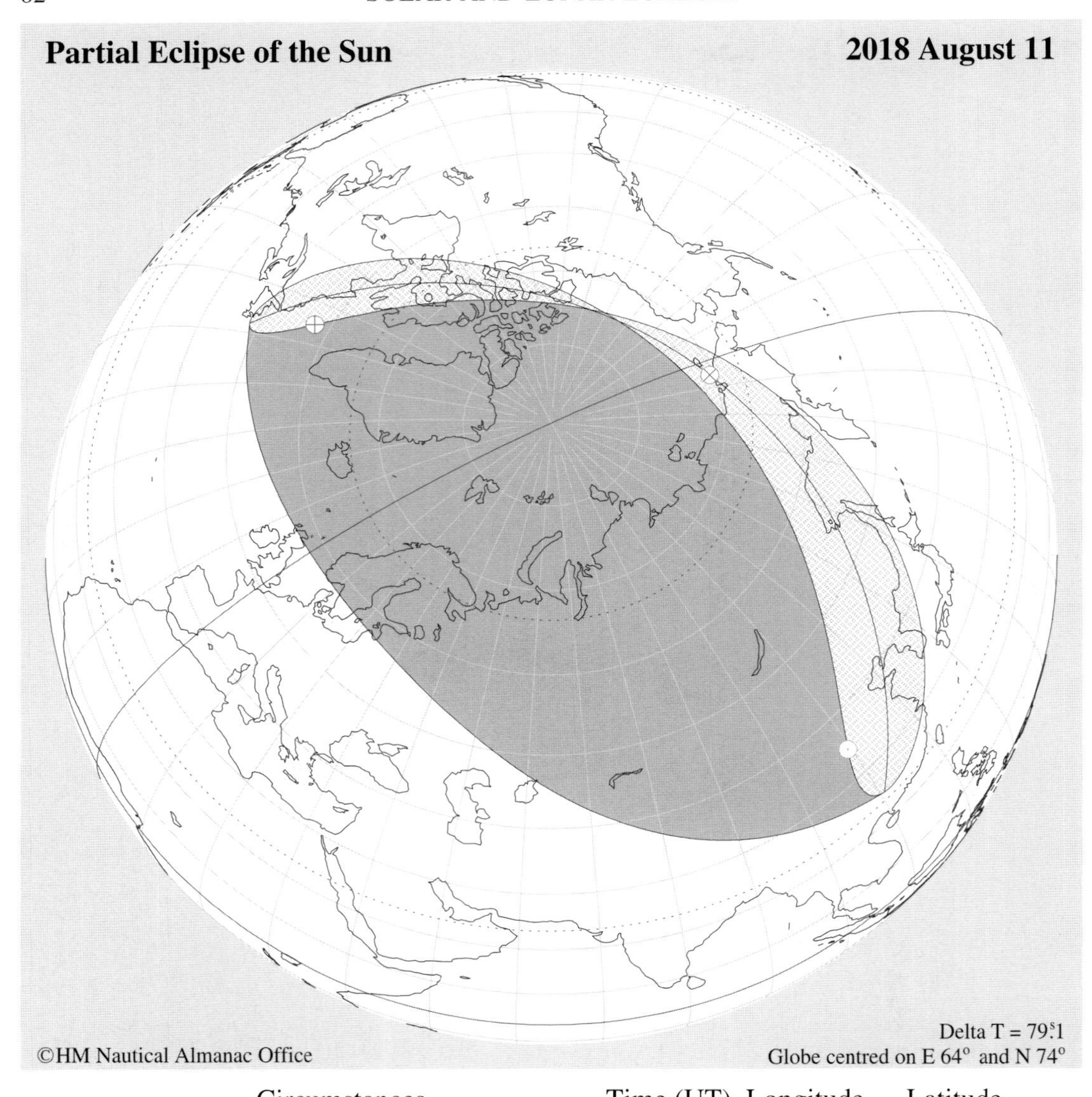

	Circumstances	Time (UT)	Longitude	Latitude
		h m	° ′	° ′
⊕	Eclipse begins; first contact with Earth	8 01.9	W 54 47.1	N 57 46.8
	Beginning of southern limit of penumbra	8 20.5	W 51 45.2	N 48 30.0
⊗	Greatest eclipse (magnitude = 0.7372)	9 46.1	E 174 43.6	N 70 26.1
	End of southern limit of penumbra	11 11.9	E 110 32.0	N 24 49.1
○	Eclipse ends; last contact with Earth	11 30.6	E 109 30.9	N 34 44.0

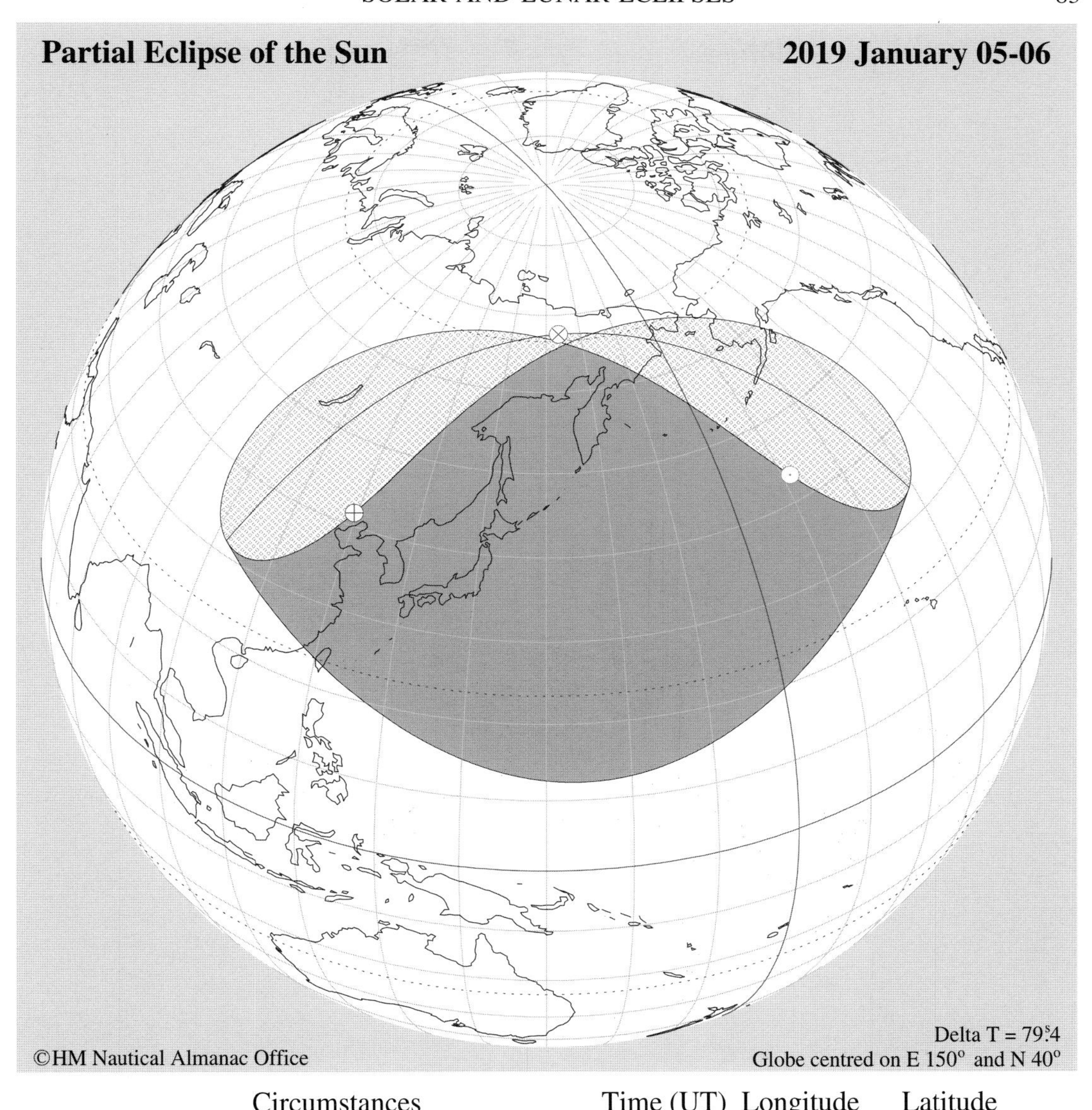

Circumstances	Time (UT)		Longitude		Latitude	
	h	m	°	′	°	′
⊕ Eclipse begins; first contact with Earth	23	33.9	E 119	27.4	N 41	30.4
Beginning of southern limit of penumbra	0	17.1	E 101	54.8	N 31	36.9
⊗ Greatest eclipse (magnitude = 0.7149)	1	41.3	E 153	37.3	N 67	26.1
End of southern limit of penumbra	3	05.5	W 150	47.6	N 33	17.5
⊙ Eclipse ends; last contact with Earth	3	48.7	W 168	38.3	N 43	07.2

Total Eclipse of the Moon — 2019 January 21

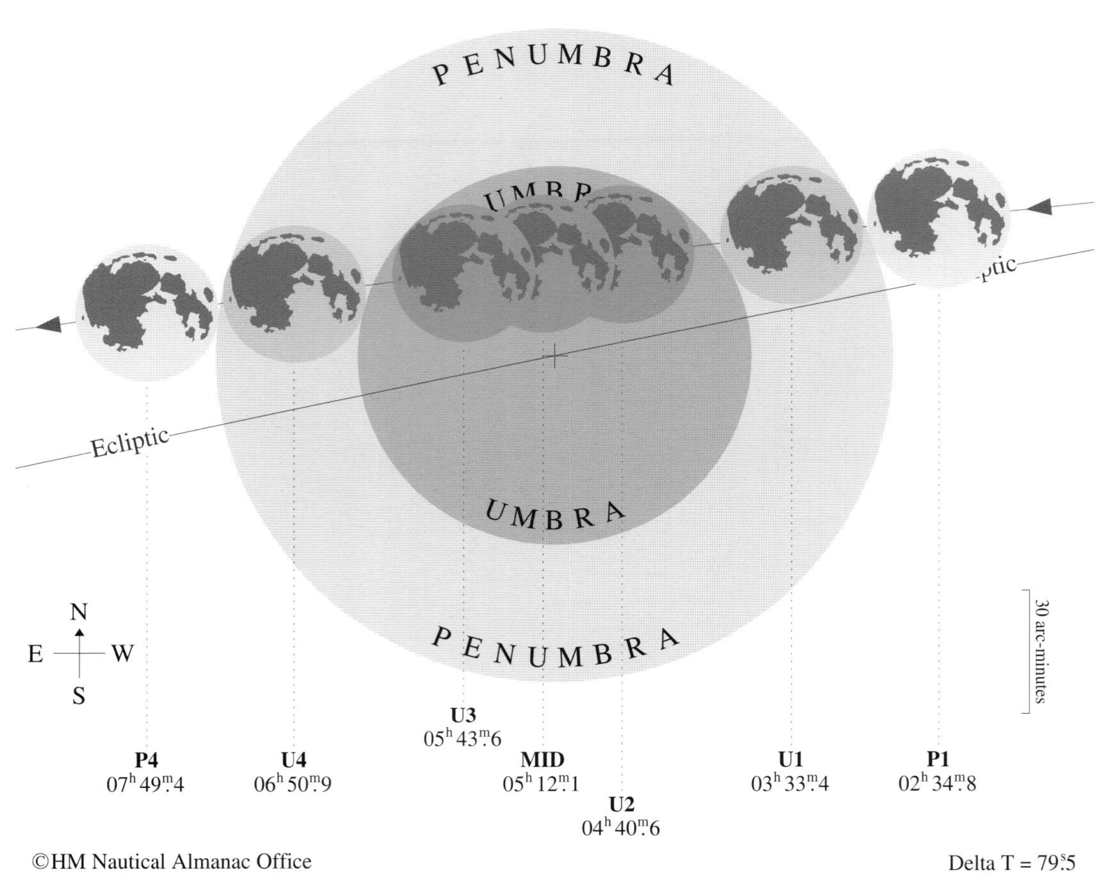

Areas of visibility of the Moon at different stages

U1 Enters umbra	U2 Enters totality	U3 Leaves totality	U4 Leaves umbra
$03^{h}\,33^{m}.4$ UT	$04^{h}\,40^{m}.6$ UT	$05^{h}\,43^{m}.6$ UT	$06^{h}\,50^{m}.9$ UT

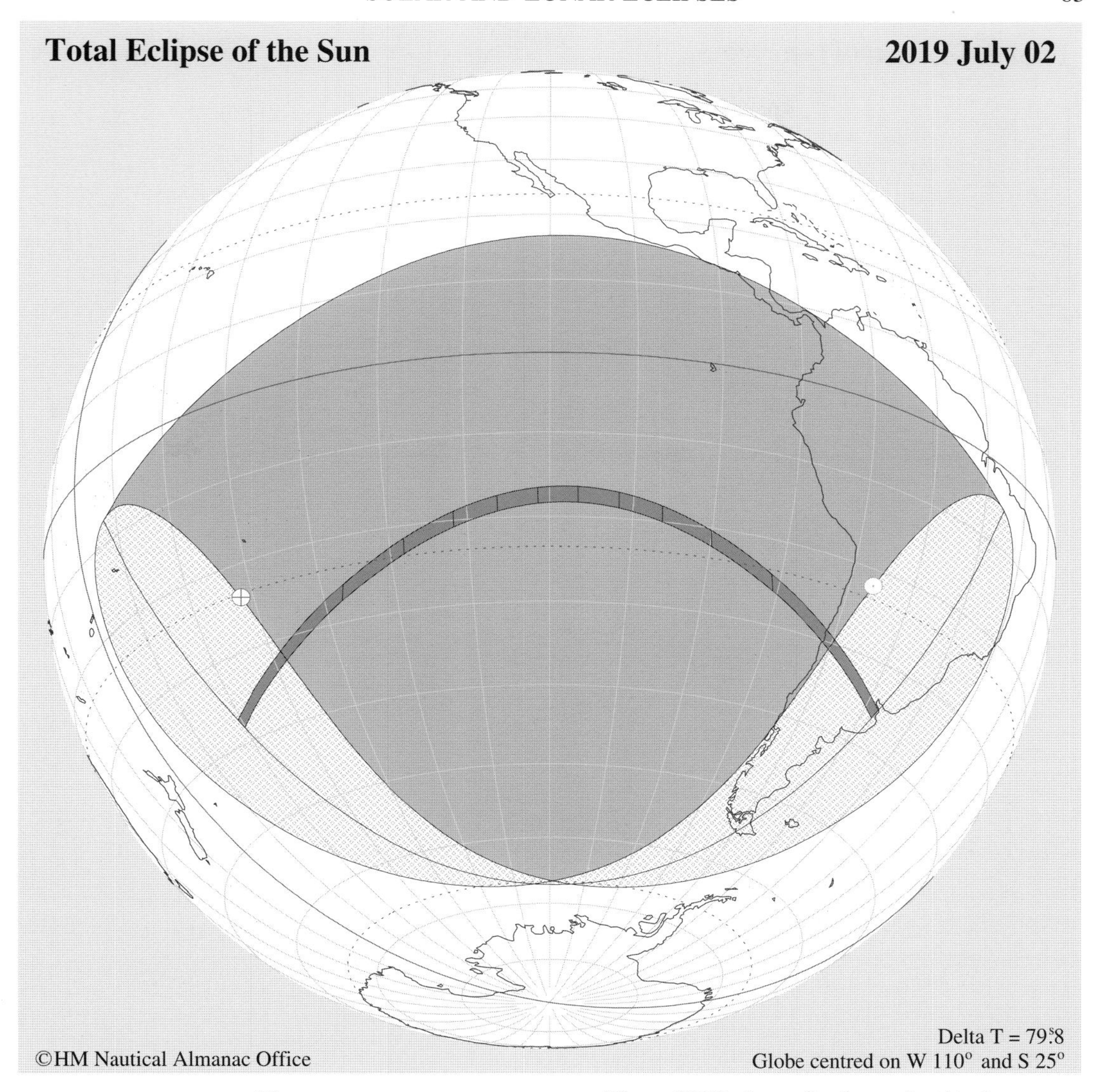

	Circumstances	Time (UT)		Longitude		Latitude	
		h	m	°	′	°	′
⊕	Eclipse begins; first contact with Earth	16	55.0	W151	53.6	S 23	53.3
	Beginning of northern limit of penumbra	17	47.8	W172	49.4	S 7	15.1
	Beginning of northern limit of umbra	18	01.5	W160	37.0	S 37	06.4
	Beginning of centre line; central eclipse begins	18	02.1	W160	22.5	S 37	39.8
	Beginning of southern limit of umbra	18	02.8	W160	07.8	S 38	13.5
	Central eclipse at local apparent noon	19	21.5	W109	21.2	S 17	24.6
	End of southern limit of umbra	20	42.7	W 57	53.8	S 36	22.9
	End of centre line; central eclipse ends	20	43.4	W 57	39.8	S 35	47.9
	End of northern limit of umbra	20	44.0	W 57	26.0	S 35	13.2
	End of northern limit of penumbra	20	57.8	W 45	40.5	S 5	17.9
⊙	Eclipse ends; last contact with Earth	21	50.4	W 66	26.7	S 21	57.6

Partial Eclipse of the Moon — 2019 July 16-17

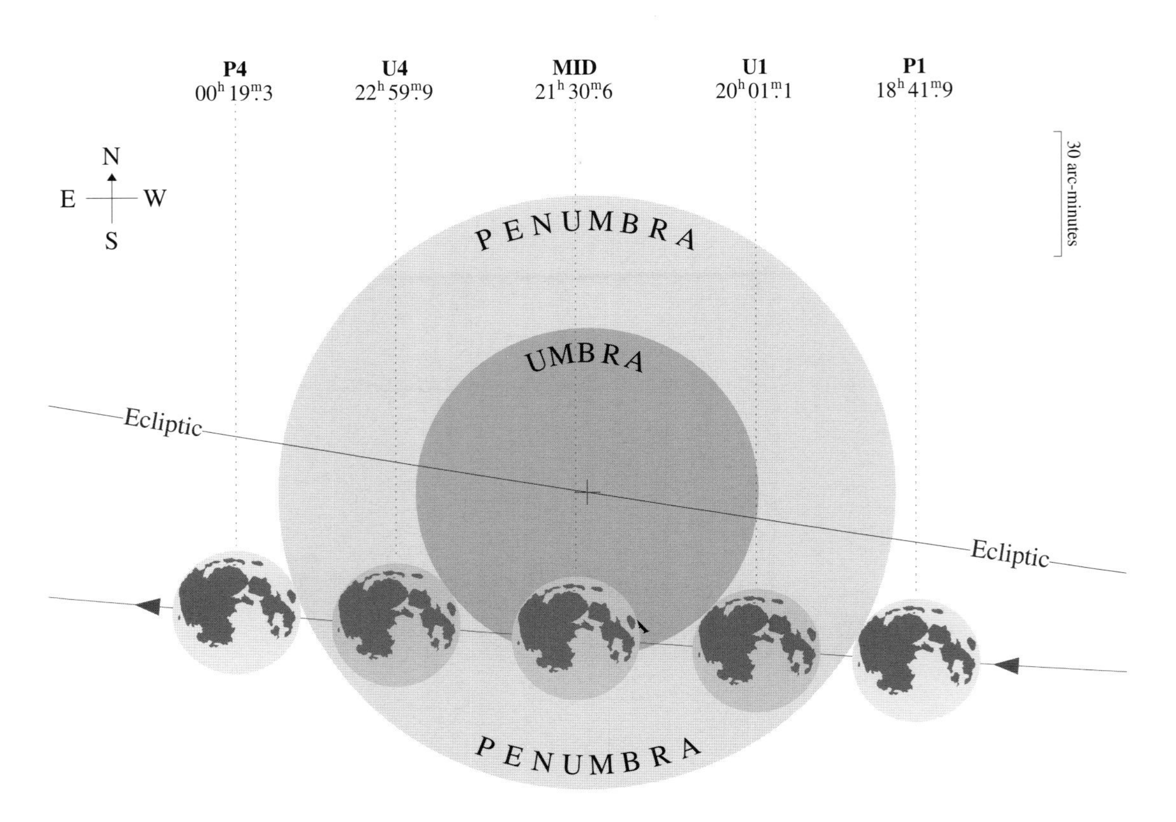

 Delta T = 79ˢ.8

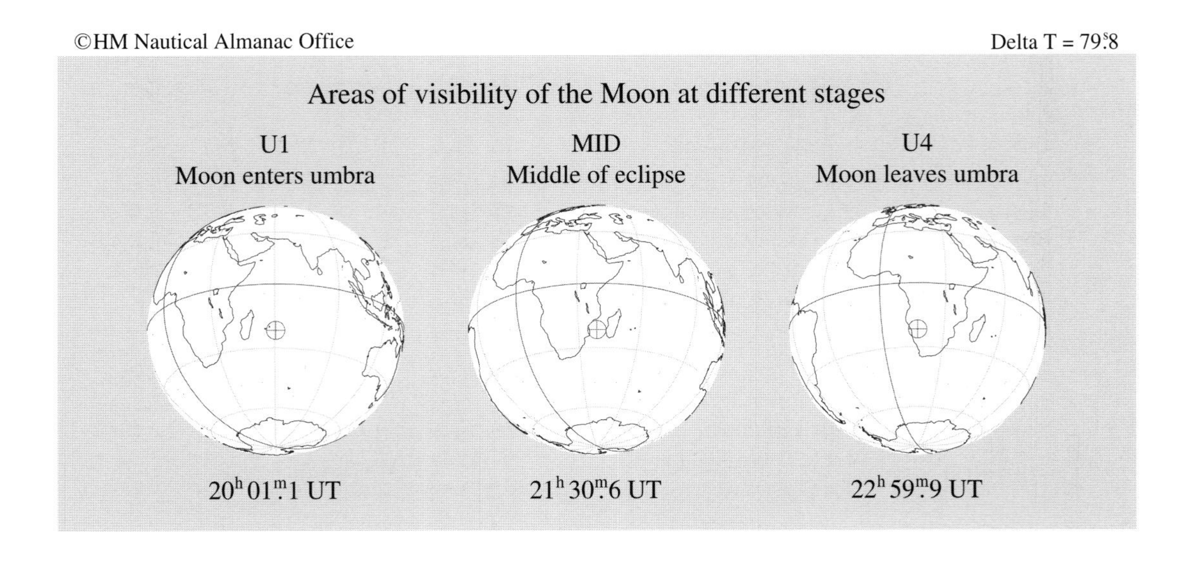

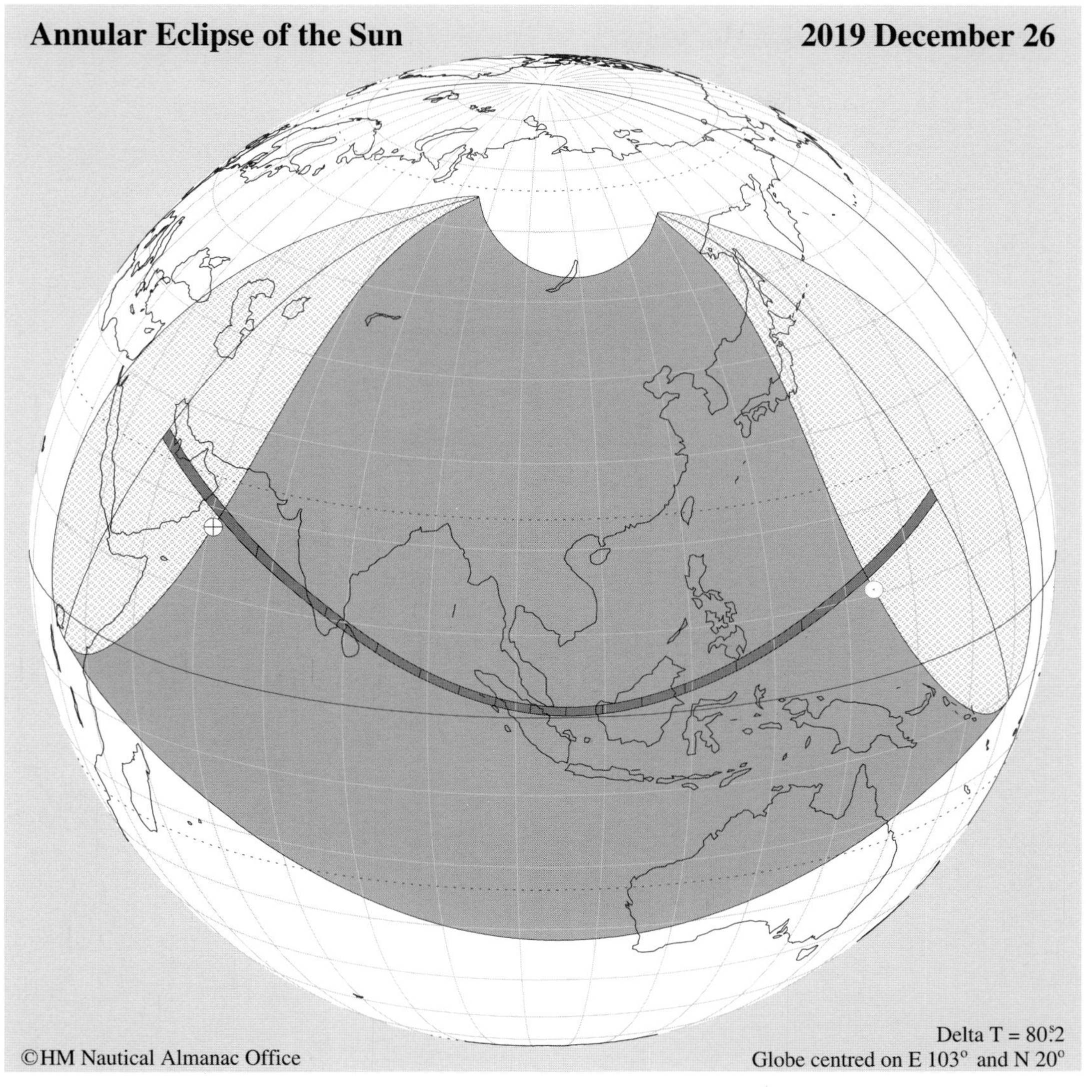

	Circumstances	Time (UT)	Longitude	Latitude
		h m	° ′	° ′
⊕	Eclipse begins; first contact with Earth	2 29.7	E 60 36.8	N 17 47.1
	Beginning of southern limit of umbra	3 35.5	E 47 58.1	N 25 18.0
	Beginning of centre line; central eclipse begins	3 35.9	E 48 15.2	N 25 58.9
	Beginning of northern limit of umbra	3 36.2	E 48 32.3	N 26 40.0
	Beginning of southern limit of penumbra	3 38.2	E 33 57.5	S 3 34.9
	Beginning of northern limit of penumbra	4 49.0	E 85 52.6	N 65 01.1
	Central eclipse at local apparent noon	5 14.4	E 101 28.4	N 1 07.0
	End of northern limit of penumbra	5 46.1	E 130 55.4	N 61 27.4
	End of southern limit of penumbra	6 56.9	E 170 35.2	S 10 48.2
	End of northern limit of umbra	6 58.9	E 156 30.0	N 19 36.8
	End of centre line; central eclipse ends	6 59.2	E 156 45.4	N 18 53.8
	End of southern limit of umbra	6 59.6	E 157 00.8	N 18 11.0
⊙	Eclipse ends; last contact with Earth	8 05.5	E 144 02.8	N 10 37.1

Penumbral Eclipse of the Moon 2020 January 10

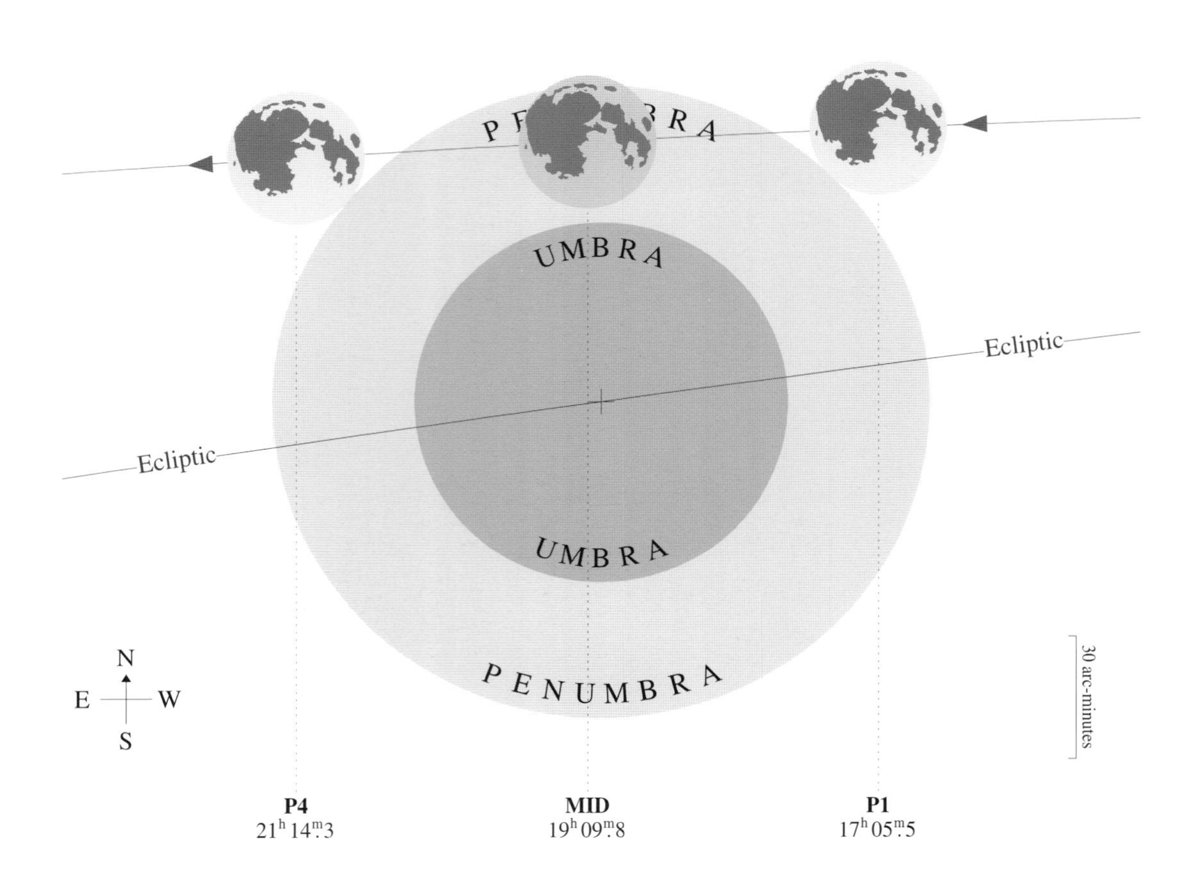

 Delta T = 80s.2

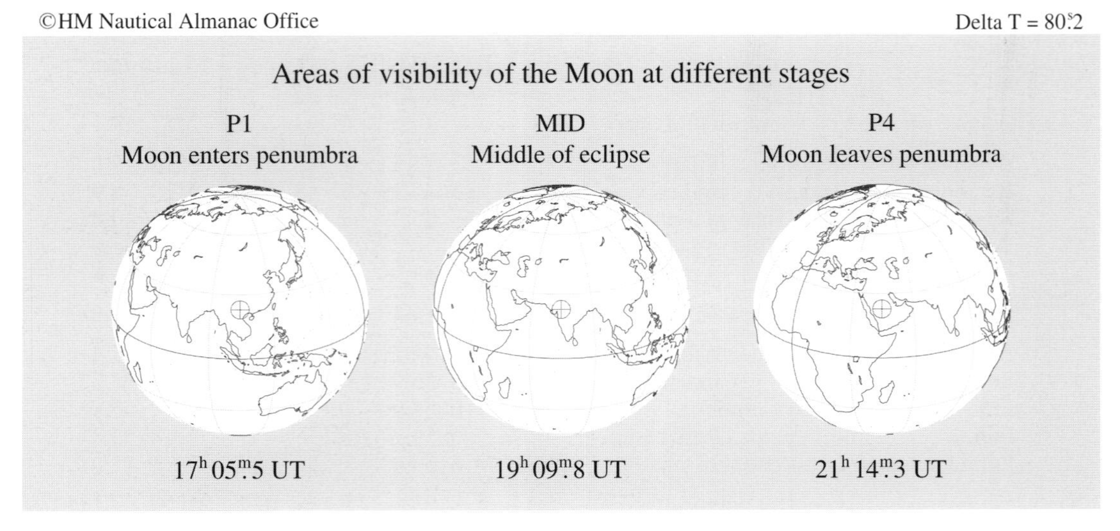

Penumbral Eclipse of the Moon **2020 June 05**

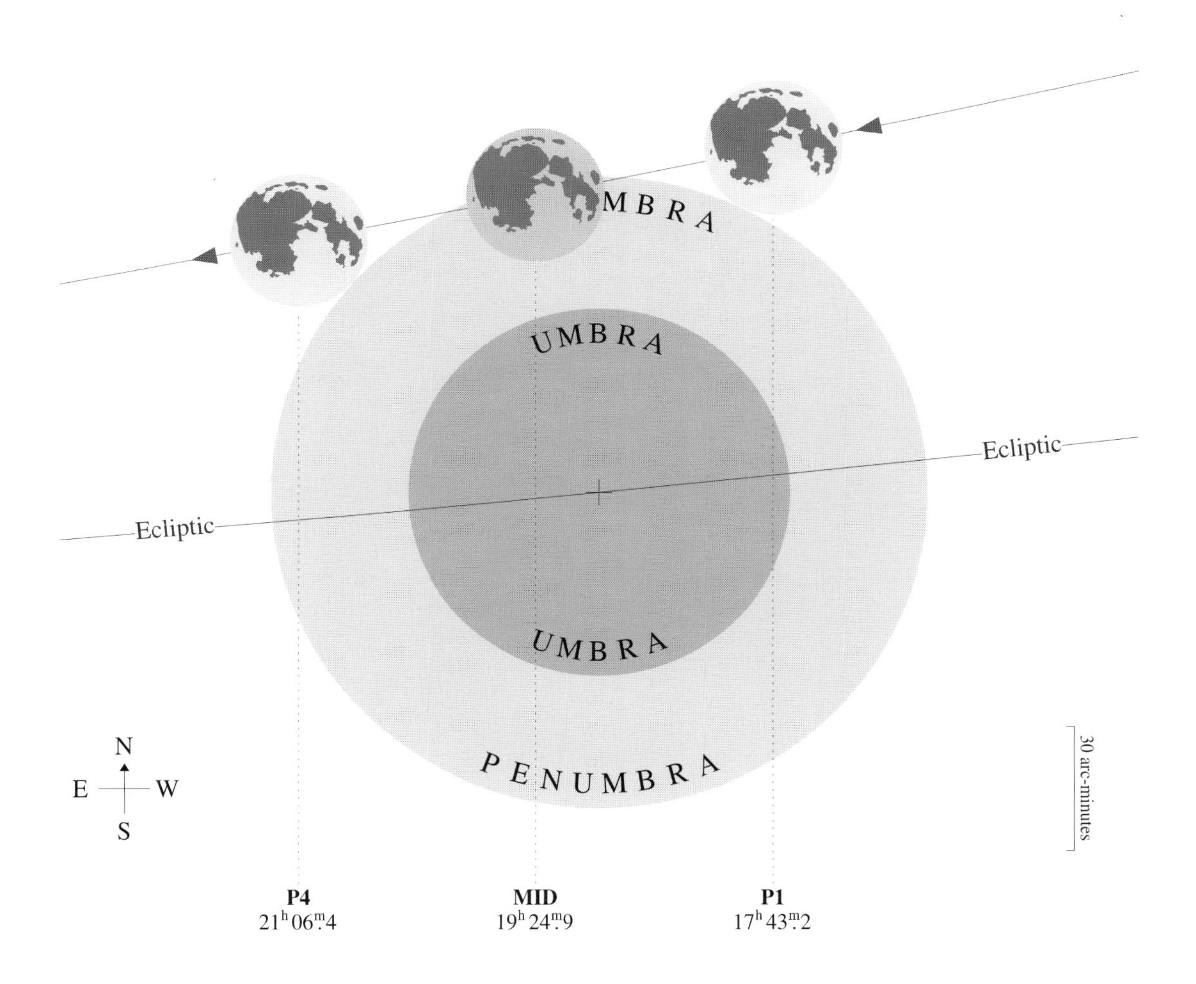

Delta T = $80^{s}.5$

Areas of visibility of the Moon at different stages

P1	MID	P4
Moon enters penumbra	Middle of eclipse	Moon leaves penumbra
$17^{h} 43^{m}.2$ UT	$19^{h} 24^{m}.9$ UT	$21^{h} 06^{m}.4$ UT

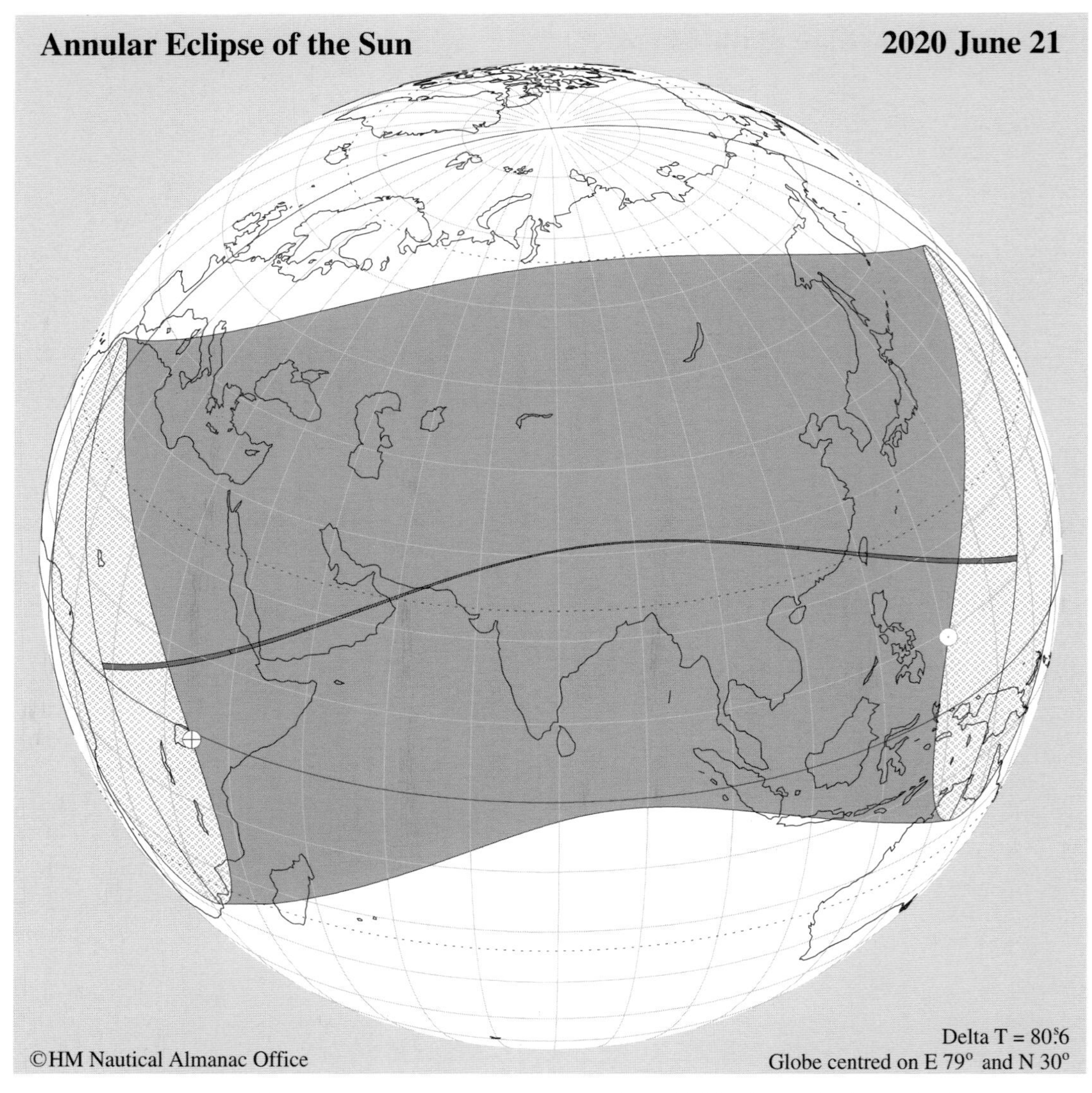

Circumstances	Time (UT) h m	Longitude ° ′	Latitude ° ′
⊕ Eclipse begins; first contact with Earth	3 45.8	E 34 27.7	S 1 02.3
Beginning of southern limit of penumbra	4 46.1	E 32 12.6	S 27 53.8
Beginning of southern limit of umbra	4 48.0	E 18 02.6	N 0 56.6
Beginning of centre line; central eclipse begins	4 48.3	E 17 51.1	N 1 15.8
Beginning of northern limit of umbra	4 48.5	E 17 39.5	N 1 35.1
Beginning of northern limit of penumbra	5 24.3	W 7 04.6	N 33 11.5
Central eclipse at local apparent noon	6 41.2	E 80 09.8	N 30 34.9
End of northern limit of penumbra	7 55.4	E 175 13.7	N 42 44.9
End of northern limit of umbra	8 31.3	E 147 49.2	N 11 45.8
End of centre line; central eclipse ends	8 31.5	E 147 38.2	N 11 27.9
End of southern limit of umbra	8 31.7	E 147 27.1	N 11 10.1
End of southern limit of penumbra	8 33.7	E 134 00.7	S 17 50.6
⊙ Eclipse ends; last contact with Earth	9 33.9	E 131 00.9	N 9 10.4

Penumbral Eclipse of the Moon 2020 July 05

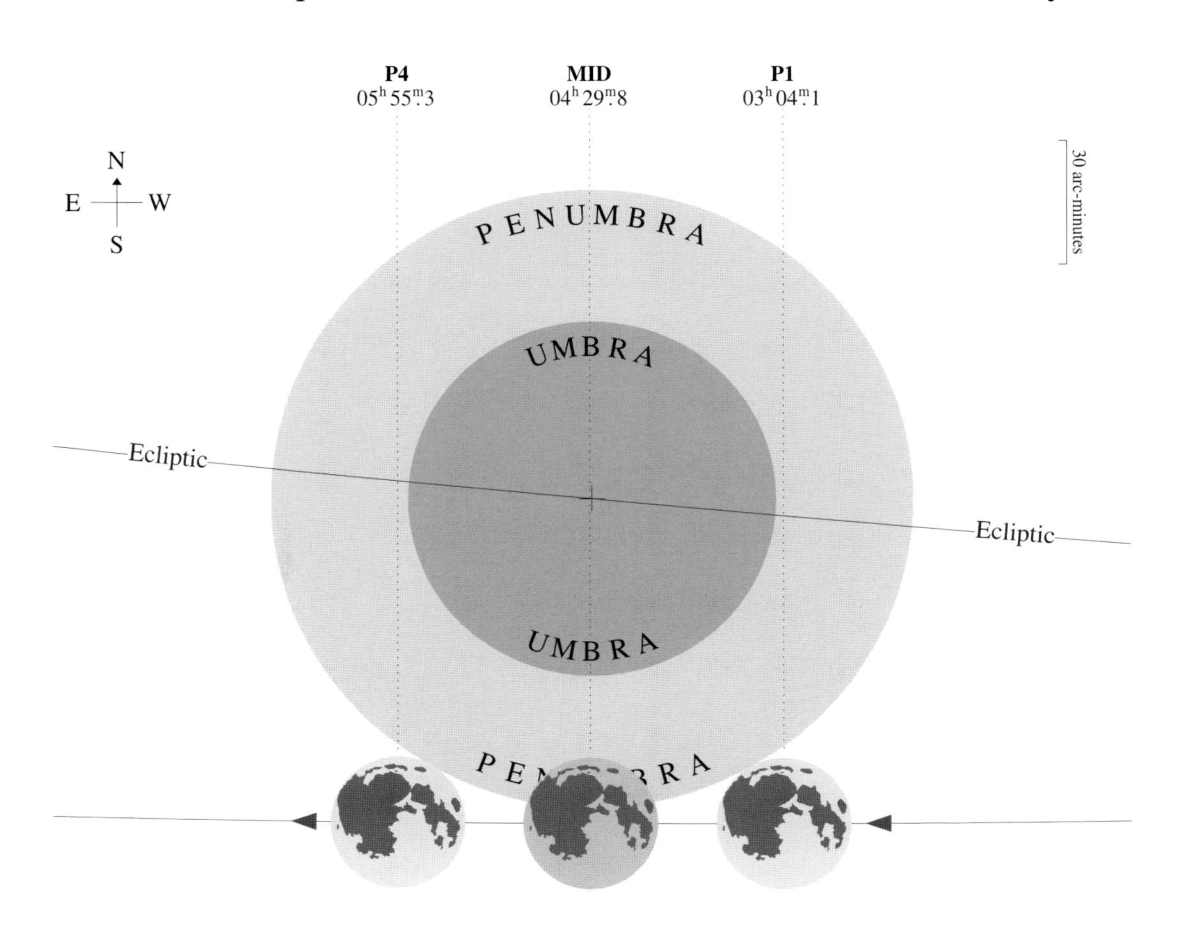

Delta T = $80^{s}\!.6$

Areas of visibility of the Moon at different stages

P1	MID	P4
Moon enters penumbra	Middle of eclipse	Moon leaves penumbra
$03^{h}\,04^{m}\!.1$ UT	$04^{h}\,29^{m}\!.8$ UT	$05^{h}\,55^{m}\!.3$ UT

Penumbral Eclipse of the Moon 2020 November 30

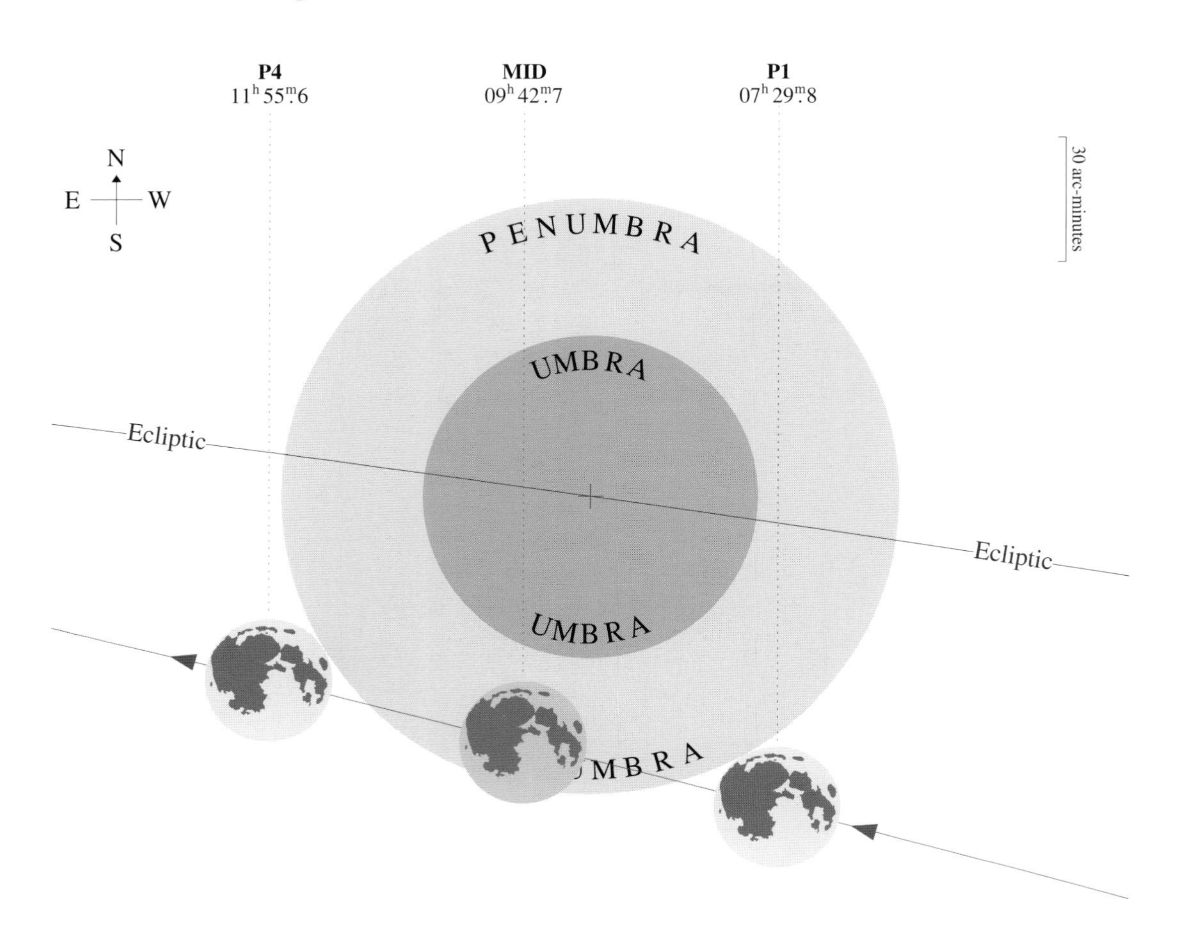

 Delta T = 80.s9

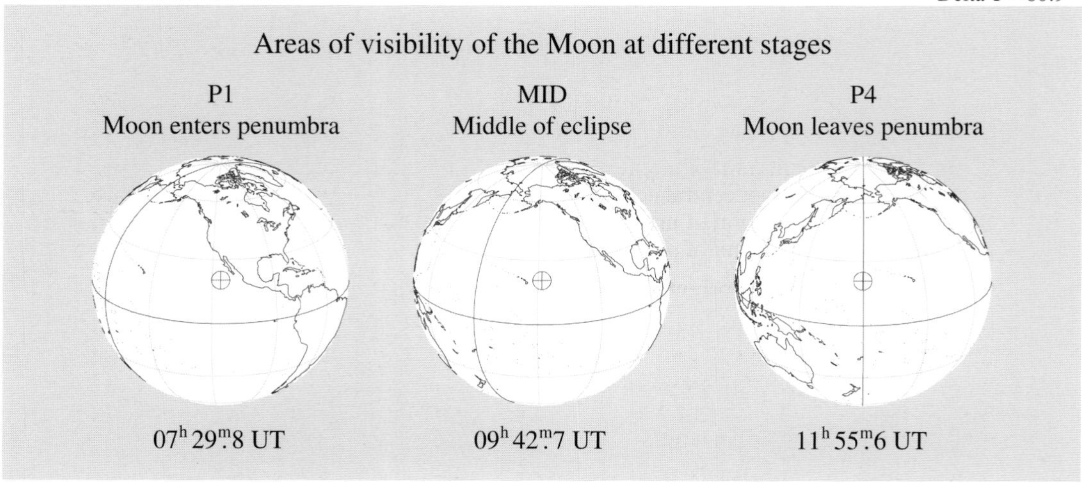

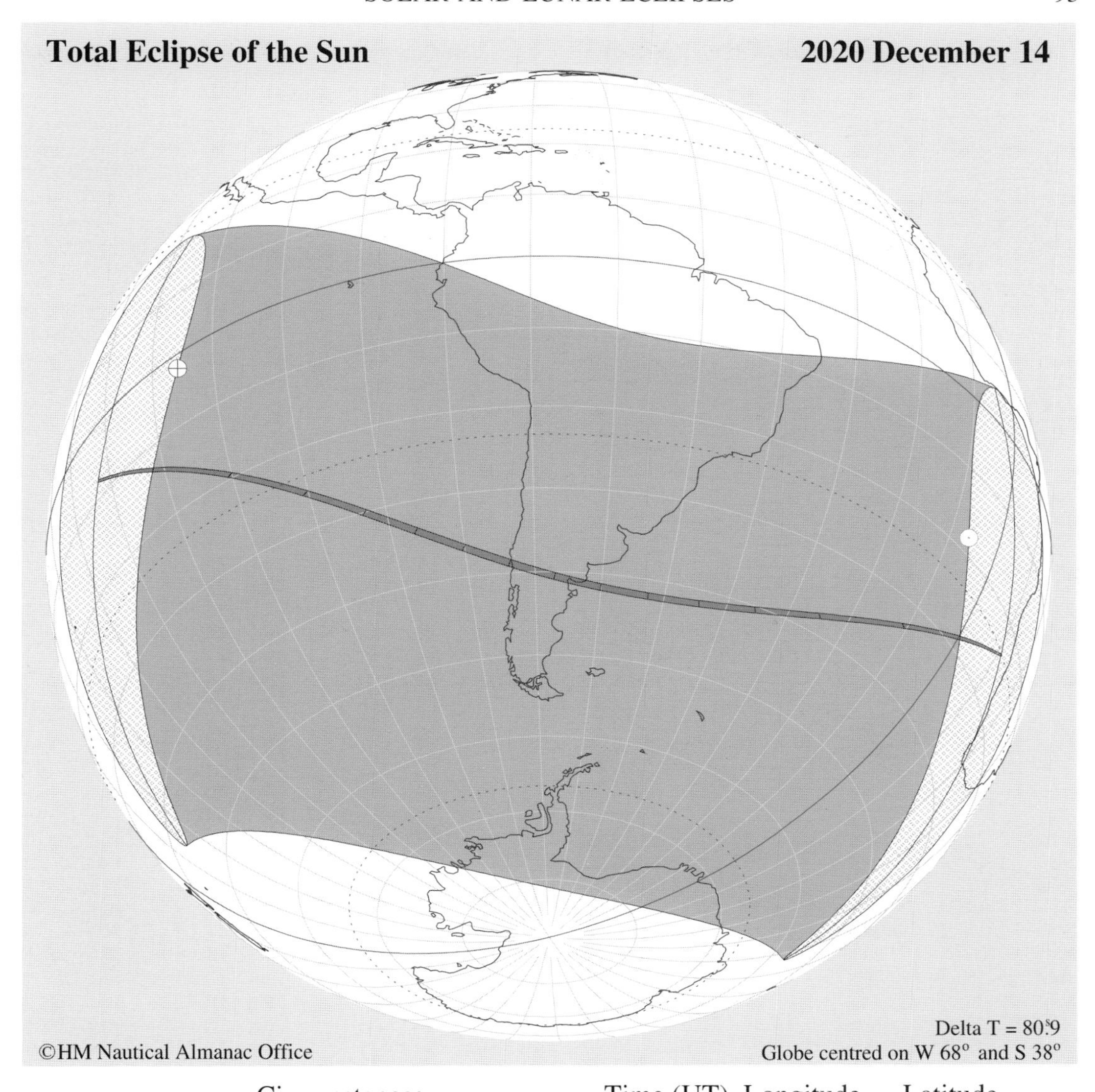

Circumstances	Time (UT)	Longitude	Latitude
	h m	° ′	° ′
⊕ Eclipse begins; first contact with Earth	13 33.7	W115 36.2	S 2 06.3
Beginning of northern limit of penumbra	14 20.6	W117 02.2	N20 47.3
Beginning of northern limit of umbra	14 32.5	W132 41.7	S 7 37.5
Beginning of centre line; central eclipse begins	14 32.6	W132 47.4	S 7 46.2
Beginning of southern limit of umbra	14 32.8	W132 53.2	S 7 54.9
Beginning of southern limit of penumbra	15 20.6	W165 21.3	S 43 20.5
Central eclipse at local apparent noon	16 18.0	W 65 45.8	S 40 46.4
End of southern limit of penumbra	17 05.8	E 53 27.1	S 56 50.8
End of southern limit of umbra	17 53.8	E 11 11.9	S 23 44.6
End of centre line; central eclipse ends	17 53.9	E 11 06.2	S 23 37.0
End of northern limit of umbra	17 54.0	E 11 00.5	S 23 29.5
End of northern limit of penumbra	18 06.0	W 4 53.0	N 4 54.7
⊙ Eclipse ends; last contact with Earth	18 52.9	W 6 26.4	S 18 01.4

New Moon	First Quarter	Full Moon	Last Quarter	Perigee	Apogee
		2001			
d h m	d h m	d h m	d h m	d h	d h
	Jan. 2 22 31	Jan. 9 20 24	Jan. 16 12 35	Jan. 10 09	Jan. 24 19
Jan. 24 13 07	Feb. 1 14 02	Feb. 8 07 12	Feb. 15 03 23	Feb. 7 22	Feb. 20 22
Feb. 23 08 21	Mar. 3 02 03	Mar. 9 17 23	Mar. 16 20 45	Mar. 8 09	Mar. 20 11
Mar. 25 01 21	Apr. 1 10 49	Apr. 8 03 22	Apr. 15 15 31	Apr. 5 10	Apr. 17 06
Apr. 23 15 26	Apr. 30 17 08	May 7 13 52	May 15 10 11	May 2 04	May 15 01
May 23 02 46	May 29 22 09	June 6 01 39	June 14 03 28	May 27 07	June 11 20
June 21 11 58	June 28 03 19	July 5 15 04	July 13 18 45	June 23 17	July 9 11
July 20 19 44	July 27 10 08	Aug. 4 05 56	Aug. 12 07 53	July 21 21	Aug. 5 21
Aug. 19 02 55	Aug. 25 19 55	Sept. 2 21 43	Sept. 10 18 59	Aug. 19 06	Sept. 1 23
Sept. 17 10 27	Sept. 24 09 31	Oct. 2 13 49	Oct. 10 04 20	Sept. 16 16	Sept. 29 06
Oct. 16 19 23	Oct. 24 02 58	Nov. 1 05 41	Nov. 8 12 21	Oct. 14 23	Oct. 26 20
Nov. 15 06 40	Nov. 22 23 21	Nov. 30 20 49	Dec. 7 19 52	Nov. 11 17	Nov. 23 16
Dec. 14 20 47	Dec. 22 20 56	Dec. 30 10 41		Dec. 6 23	Dec. 21 13
		2002			
			Jan. 6 03 55	Jan. 2 07	Jan. 18 09
Jan. 13 13 29	Jan. 21 17 46	Jan. 28 22 50	Feb. 4 13 33	Jan. 30 09	Feb. 14 22
Feb. 12 07 41	Feb. 20 12 02	Feb. 27 09 17	Mar. 6 01 24	Feb. 27 20	Mar. 14 01
Mar. 14 02 02	Mar. 22 02 28	Mar. 28 18 25	Apr. 4 15 29	Mar. 28 08	Apr. 10 05
Apr. 12 19 21	Apr. 20 12 48	Apr. 27 03 00	May 4 07 16	Apr. 25 16	May 7 19
May 12 10 45	May 19 19 42	May 26 11 51	June 3 00 05	May 23 16	June 4 13
June 10 23 46	June 18 00 29	June 24 21 42	July 2 17 19	June 19 07	July 2 08
July 10 10 26	July 17 04 47	July 24 09 07	Aug. 1 10 22	July 14 13	July 30 02
Aug. 8 19 15	Aug. 15 10 12	Aug. 22 22 29	Aug. 31 02 31	Aug. 10 23	Aug. 26 18
Sept. 7 03 10	Sept. 13 18 08	Sept. 21 13 59	Sept. 29 17 03	Sept. 8 03	Sept. 23 03
Oct. 6 11 18	Oct. 13 05 33	Oct. 21 07 20	Oct. 29 05 28	Oct. 6 13	Oct. 20 05
Nov. 4 20 34	Nov. 11 20 52	Nov. 20 01 34	Nov. 27 15 46	Nov. 4 01	Nov. 16 11
Dec. 4 07 34	Dec. 11 15 49	Dec. 19 19 10	Dec. 27 00 31	Dec. 2 09	Dec. 14 04
				Dec. 30 01	
		2003			
Jan. 2 20 23	Jan. 10 13 15	Jan. 18 10 48	Jan. 25 08 33		Jan. 11 01
Feb. 1 10 48	Feb. 9 11 11	Feb. 16 23 51	Feb. 23 16 46	Jan. 23 22	Feb. 7 22
Mar. 3 02 35	Mar. 11 07 15	Mar. 18 10 34	Mar. 25 01 51	Feb. 19 16	Mar. 7 17
Apr. 1 19 19	Apr. 9 23 40	Apr. 16 19 36	Apr. 23 12 18	Mar. 19 19	Apr. 4 04
May 1 12 15	May 9 11 53	May 16 03 36	May 23 00 31	Apr. 17 05	May 1 08
May 31 04 20	June 7 20 28	June 14 11 16	June 21 14 45	May 15 16	May 28 13
June 29 18 39	July 7 02 32	July 13 19 21	July 21 07 01	June 12 23	June 25 02
July 29 06 53	Aug. 5 07 28	Aug. 12 04 48	Aug. 20 00 48	July 10 22	July 22 20
Aug. 27 17 26	Sept. 3 12 34	Sept. 10 16 36	Sept. 18 19 03	Aug. 6 14	Aug. 19 14
Sept. 26 03 09	Oct. 2 19 09	Oct. 10 07 27	Oct. 18 12 31	Aug. 31 19	Sept. 16 09
Oct. 25 12 50	Nov. 1 04 25	Nov. 9 01 13	Nov. 17 04 15	Sept. 28 06	Oct. 14 02
Nov. 23 22 59	Nov. 30 17 16	Dec. 8 20 37	Dec. 16 17 42	Oct. 26 12	Nov. 10 12
Dec. 23 09 43	Dec. 30 10 03			Nov. 23 23	Dec. 7 12
				Dec. 22 12	
		2004			
		Jan. 7 15 40	Jan. 15 04 46		Jan. 3 20
Jan. 21 21 05	Jan. 29 06 03	Feb. 6 08 47	Feb. 13 13 39	Jan. 19 19	Jan. 31 14
Feb. 20 09 18	Feb. 28 03 24	Mar. 6 23 14	Mar. 13 21 01	Feb. 16 08	Feb. 28 11
Mar. 20 22 41	Mar. 28 23 48	Apr. 5 11 03	Apr. 12 03 46	Mar. 12 04	Mar. 27 07
Apr. 19 13 21	Apr. 27 17 32	May 4 20 33	May 11 11 04	Apr. 8 02	Apr. 24 00
May 19 04 52	May 27 07 57	June 3 04 19	June 9 20 02	May 6 05	May 21 12
June 17 20 27	June 25 19 08	July 2 11 09	July 9 07 33	June 3 13	June 17 16
July 17 11 24	July 25 03 37	July 31 18 05	Aug. 7 22 01	July 1 23	July 14 21
Aug. 16 01 24	Aug. 23 10 12	Aug. 30 02 22	Sept. 6 15 10	July 30 06	Aug. 11 10
Sept. 14 14 29	Sept. 21 15 53	Sept. 28 13 09	Oct. 6 10 12	Aug. 27 06	Sept. 8 03
Oct. 14 02 48	Oct. 20 21 59	Oct. 28 03 07	Nov. 5 05 53	Sept. 22 21	Oct. 5 22
Nov. 12 14 27	Nov. 19 05 50	Nov. 26 20 07	Dec. 5 00 53	Oct. 18 00	Nov. 2 18
Dec. 12 01 29	Dec. 18 16 40	Dec. 26 15 06		Nov. 14 14	Nov. 30 11
				Dec. 12 21	Dec. 27 19

New Moon	First Quarter	Full Moon	Last Quarter	Perigee	Apogee
		2005			
d h m	d h m	d h m	d h m	d h	d h
			Jan. 3 17 46	Jan. 10 10	Jan. 23 19
Jan. 10 12 03	Jan. 17 06 57	Jan. 25 10 32	Feb. 2 07 27	Feb. 7 22	Feb. 20 05
Feb. 8 22 28	Feb. 16 00 16	Feb. 24 04 54	Mar. 3 17 36	Mar. 8 04	Mar. 19 23
Mar. 10 09 10	Mar. 17 19 19	Mar. 25 20 58	Apr. 2 00 50	Apr. 4 11	Apr. 16 19
Apr. 8 20 32	Apr. 16 14 37	Apr. 24 10 06	May 1 06 24	Apr. 29 10	May 14 14
May 8 08 45	May 16 08 56	May 23 20 18	May 30 11 47	May 26 11	June 11 06
June 6 21 55	June 15 01 22	June 22 04 14	June 28 18 23	June 23 12	July 8 18
July 6 12 02	July 14 15 20	July 21 11 00	July 28 03 19	July 21 20	Aug. 4 22
Aug. 5 03 05	Aug. 13 02 38	Aug. 19 17 53	Aug. 26 15 18	Aug. 19 06	Sept. 1 03
Sept. 3 18 45	Sept. 11 11 37	Sept. 18 02 01	Sept. 25 06 41	Sept. 16 14	Sept. 28 15
Oct. 3 10 28	Oct. 10 19 01	Oct. 17 12 14	Oct. 25 01 17	Oct. 14 14	Oct. 26 10
Nov. 2 01 24	Nov. 9 01 57	Nov. 16 00 57	Nov. 23 22 11	Nov. 10 00	Nov. 23 06
Dec. 1 15 01	Dec. 8 09 36	Dec. 15 16 15	Dec. 23 19 36	Dec. 5 05	Dec. 21 03
Dec. 31 03 12					
		2006			
	Jan. 6 18 56	Jan. 14 09 48	Jan. 22 15 14	Jan. 1 23	Jan. 17 19
Jan. 29 14 14	Feb. 5 06 29	Feb. 13 04 44	Feb. 21 07 17	Jan. 30 08	Feb. 14 01
Feb. 28 00 31	Mar. 6 20 16	Mar. 14 23 35	Mar. 22 19 10	Feb. 27 20	Mar. 13 02
Mar. 29 10 15	Apr. 5 12 01	Apr. 13 16 40	Apr. 21 03 28	Mar. 28 07	Apr. 9 13
Apr. 27 19 44	May 5 05 13	May 13 06 51	May 20 09 20	Apr. 25 11	May 7 07
May 27 05 25	June 3 23 06	June 11 18 03	June 18 14 08	May 22 15	June 4 02
June 25 16 05	July 3 16 37	July 11 03 02	July 17 19 12	June 16 17	July 1 20
July 25 04 31	Aug. 2 08 46	Aug. 9 10 54	Aug. 16 01 51	July 13 18	July 29 13
Aug. 23 19 10	Aug. 31 22 56	Sept. 7 18 42	Sept. 14 11 15	Aug. 10 18	Aug. 26 01
Sept. 22 11 45	Sept. 30 11 04	Oct. 7 03 13	Oct. 14 00 25	Sept. 8 03	Sept. 22 05
Oct. 22 05 14	Oct. 29 21 25	Nov. 5 12 58	Nov. 12 17 45	Oct. 6 14	Oct. 19 10
Nov. 20 22 18	Nov. 28 06 29	Dec. 5 00 25	Dec. 12 14 32	Nov. 4 00	Nov. 15 23
Dec. 20 14 01	Dec. 27 14 48			Dec. 2 00	Dec. 13 19
				Dec. 28 02	
		2007			
		Jan. 3 13 57	Jan. 11 12 44		Jan. 10 16
Jan. 19 04 01	Jan. 25 23 01	Feb. 2 05 45	Feb. 10 09 51	Jan. 22 13	Feb. 7 13
Feb. 17 16 14	Feb. 24 07 56	Mar. 3 23 17	Mar. 12 03 54	Feb. 19 10	Mar. 7 04
Mar. 19 02 42	Mar. 25 18 16	Apr. 2 17 15	Apr. 10 18 04	Mar. 19 19	Apr. 3 09
Apr. 17 11 36	Apr. 24 06 35	May 2 10 09	May 10 04 27	Apr. 17 06	Apr. 30 11
May 16 19 27	May 23 21 02	June 1 01 04	June 8 11 43	May 15 15	May 27 22
June 15 03 13	June 22 13 15	June 30 13 49	July 7 16 53	June 12 17	June 24 14
July 14 12 04	July 22 06 29	July 30 00 48	Aug. 5 21 19	July 9 22	July 22 09
Aug. 12 23 02	Aug. 20 23 54	Aug. 28 10 35	Sept. 4 02 32	Aug. 4 00	Aug. 19 03
Sept. 11 12 44	Sept. 19 16 48	Sept. 26 19 45	Oct. 3 10 06	Aug. 31 00	Sept. 15 21
Oct. 11 05 01	Oct. 19 08 33	Oct. 26 04 51	Nov. 1 21 18	Sept. 28 02	Oct. 13 10
Nov. 9 23 03	Nov. 17 22 32	Nov. 24 14 30	Dec. 1 12 44	Oct. 26 12	Nov. 9 13
Dec. 9 17 40	Dec. 17 10 17	Dec. 24 01 15	Dec. 31 07 51	Nov. 24 00	Dec. 6 17
				Dec. 22 10	
		2008			
Jan. 8 11 37	Jan. 15 19 46	Jan. 22 13 35	Jan. 30 05 03		Jan. 3 08
Feb. 7 03 44	Feb. 14 03 33	Feb. 21 03 30	Feb. 29 02 18	Jan. 19 09	Jan. 31 04
Mar. 7 17 14	Mar. 14 10 45	Mar. 21 18 40	Mar. 29 21 47	Feb. 14 01	Feb. 28 01
Apr. 6 03 55	Apr. 12 18 32	Apr. 20 10 25	Apr. 28 14 12	Mar. 10 22	Mar. 26 20
May 5 12 18	May 12 03 47	May 20 02 11	May 28 02 56	Apr. 7 19	Apr. 23 10
June 3 19 22	June 10 15 03	June 18 17 30	June 26 12 10	May 6 03	May 20 14
July 3 02 18	July 10 04 35	July 18 07 59	July 25 18 41	June 3 13	June 16 18
Aug. 1 10 12	Aug. 8 20 20	Aug. 16 21 16	Aug. 23 23 49	July 1 21	July 14 04
Aug. 30 19 58	Sept. 7 14 04	Sept. 15 09 13	Sept. 22 05 04	July 29 23	Aug. 10 20
Sept. 29 08 12	Oct. 7 09 04	Oct. 14 20 02	Oct. 21 11 54	Aug. 26 04	Sept. 7 15
Oct. 28 23 14	Nov. 6 04 03	Nov. 13 06 17	Nov. 19 21 31	Sept. 20 03	Oct. 5 11
Nov. 27 16 54	Dec. 5 21 25	Dec. 12 16 37	Dec. 19 10 29	Oct. 17 06	Nov. 2 05
Dec. 27 12 22				Nov. 14 10	Nov. 29 17
				Dec. 12 22	Dec. 26 18

2009

New Moon (d h m)	First Quarter (d h m)	Full Moon (d h m)	Last Quarter (d h m)	Perigee (d h)	Apogee (d h)
	Jan. 4 11 56	Jan. 11 03 27	Jan. 18 02 46	Jan. 10 11	Jan. 23 00
Jan. 26 07 55	Feb. 2 23 13	Feb. 9 14 49	Feb. 16 21 37	Feb. 7 20	Feb. 19 17
Feb. 25 01 35	Mar. 4 07 46	Mar. 11 02 38	Mar. 18 17 47	Mar. 7 15	Mar. 19 13
Mar. 26 16 06	Apr. 2 14 34	Apr. 9 14 56	Apr. 17 13 36	Apr. 2 02	Apr. 16 09
Apr. 25 03 22	May 1 20 44	May 9 04 01	May 17 07 26	Apr. 28 06	May 14 03
May 24 12 11	May 31 03 22	June 7 18 12	June 15 22 14	May 26 04	June 10 16
June 22 19 35	June 29 11 28	July 7 09 21	July 15 09 53	June 23 11	July 7 22
July 22 02 34	July 28 22 00	Aug. 6 00 55	Aug. 13 18 55	July 21 20	Aug. 4 01
Aug. 20 10 01	Aug. 27 11 42	Sept. 4 16 02	Sept. 12 02 16	Aug. 19 05	Aug. 31 11
Sept. 18 18 44	Sept. 26 04 50	Oct. 4 06 10	Oct. 11 08 56	Sept. 16 08	Sept. 28 04
Oct. 18 05 33	Oct. 26 00 42	Nov. 2 19 14	Nov. 9 15 56	Oct. 13 12	Oct. 25 23
Nov. 16 19 14	Nov. 24 21 39	Dec. 2 07 30	Dec. 9 00 13	Nov. 7 07	Nov. 22 20
Dec. 16 12 02	Dec. 24 17 36	Dec. 31 19 13		Dec. 4 14	Dec. 20 15

2010

New Moon (d h m)	First Quarter (d h m)	Full Moon (d h m)	Last Quarter (d h m)	Perigee (d h)	Apogee (d h)
			Jan. 7 10 39	Jan. 1 21	Jan. 17 02
Jan. 15 07 11	Jan. 23 10 53	Jan. 30 06 17	Feb. 5 23 48	Jan. 30 09	Feb. 13 02
Feb. 14 02 51	Feb. 22 00 42	Feb. 28 16 38	Mar. 7 15 42	Feb. 27 22	Mar. 12 10
Mar. 15 21 01	Mar. 23 11 00	Mar. 30 02 25	Apr. 6 09 37	Mar. 28 05	Apr. 9 03
Apr. 14 12 29	Apr. 21 18 20	Apr. 28 12 18	May 6 04 15	Apr. 24 21	May 6 22
May 14 01 04	May 20 23 42	May 27 23 07	June 4 22 13	May 20 09	June 3 17
June 12 11 14	June 19 04 29	June 26 11 30	July 4 14 35	June 15 15	July 1 10
July 11 19 40	July 18 10 10	July 26 01 36	Aug. 3 04 58	July 13 11	July 29 00
Aug. 10 03 08	Aug. 16 18 14	Aug. 24 17 04	Sept. 1 17 22	Aug. 10 18	Aug. 25 06
Sept. 8 10 30	Sept. 15 05 50	Sept. 23 09 17	Oct. 1 03 52	Sept. 8 04	Sept. 21 08
Oct. 7 18 44	Oct. 14 21 27	Oct. 23 01 36	Oct. 30 12 46	Oct. 6 14	Oct. 18 18
Nov. 6 04 52	Nov. 13 16 38	Nov. 21 17 27	Nov. 28 20 36	Nov. 3 17	Nov. 15 12
Dec. 5 17 36	Dec. 13 13 59	Dec. 21 08 13	Dec. 28 04 18	Nov. 30 19	Dec. 13 09
				Dec. 25 12	

2011

New Moon (d h m)	First Quarter (d h m)	Full Moon (d h m)	Last Quarter (d h m)	Perigee (d h)	Apogee (d h)
Jan. 4 09 02	Jan. 12 11 31	Jan. 19 21 21	Jan. 26 12 57		Jan. 10 06
Feb. 3 02 30	Feb. 11 07 18	Feb. 18 08 36	Feb. 24 23 26	Jan. 22 00	Feb. 6 23
Mar. 4 20 46	Mar. 12 23 45	Mar. 19 18 10	Mar. 26 12 07	Feb. 19 07	Mar. 6 08
Apr. 3 14 32	Apr. 11 12 05	Apr. 18 02 44	Apr. 25 02 47	Mar. 19 19	Apr. 2 09
May 3 06 51	May 10 20 33	May 17 11 08	May 24 18 52	Apr. 17 06	Apr. 29 18
June 1 21 02	June 9 02 10	June 15 20 13	June 23 11 48	May 15 11	May 27 10
July 1 08 54	July 8 06 29	July 15 06 39	July 23 05 02	June 12 02	June 24 04
July 30 18 40	Aug. 6 11 08	Aug. 13 18 57	Aug. 21 21 54	July 7 14	July 21 23
Aug. 29 03 04	Sept. 4 17 39	Sept. 12 09 26	Sept. 20 13 38	Aug. 2 21	Aug. 18 16
Sept. 27 11 09	Oct. 4 03 15	Oct. 12 02 06	Oct. 20 03 30	Aug. 30 18	Sept. 15 06
Oct. 26 19 56	Nov. 2 16 38	Nov. 10 20 16	Nov. 18 15 09	Sept. 28 01	Oct. 12 12
Nov. 25 06 10	Dec. 2 09 52	Dec. 10 14 36	Dec. 18 00 47	Oct. 26 12	Nov. 8 13
Dec. 24 18 06				Nov. 23 23	Dec. 6 01
				Dec. 22 03	

2012

New Moon (d h m)	First Quarter (d h m)	Full Moon (d h m)	Last Quarter (d h m)	Perigee (d h)	Apogee (d h)
	Jan. 1 06 14	Jan. 9 07 30	Jan. 16 09 08		Jan. 2 20
Jan. 23 07 39	Jan. 31 04 09	Feb. 7 21 54	Feb. 14 17 04	Jan. 17 21	Jan. 30 18
Feb. 21 22 34	Mar. 1 01 21	Mar. 8 09 39	Mar. 15 01 25	Feb. 11 19	Feb. 27 14
Mar. 22 14 37	Mar. 30 19 40	Apr. 6 19 18	Apr. 13 10 49	Mar. 10 10	Mar. 26 06
Apr. 21 07 18	Apr. 29 09 57	May 6 03 35	May 12 21 47	Apr. 7 17	Apr. 22 14
May 20 23 47	May 28 20 16	June 4 11 11	June 11 10 41	May 6 04	May 19 16
June 19 15 02	June 27 03 30	July 3 18 52	July 11 01 48	June 3 13	June 16 01
July 19 04 24	July 26 08 56	Aug. 2 03 27	Aug. 9 18 55	July 1 18	July 13 17
Aug. 17 15 54	Aug. 24 13 53	Aug. 31 13 58	Sept. 8 13 15	July 29 08	Aug. 10 11
Sept. 16 02 10	Sept. 22 19 41	Sept. 30 03 18	Oct. 8 07 33	Aug. 23 19	Sept. 7 06
Oct. 15 12 02	Oct. 22 03 32	Oct. 29 19 49	Nov. 7 00 35	Sept. 19 03	Oct. 5 01
Nov. 13 22 08	Nov. 20 14 31	Nov. 28 14 46	Dec. 6 15 31	Oct. 17 01	Nov. 1 15
Dec. 13 08 41	Dec. 20 05 19	Dec. 28 10 21		Nov. 14 10	Nov. 28 20
				Dec. 12 23	Dec. 25 21

2013

New Moon	First Quarter	Full Moon	Last Quarter	Perigee	Apogee
d h m	d h m	d h m	d h m	d h	d h
			Jan. 5 03 58	Jan. 10 10	Jan. 22 11
Jan. 11 19 43	Jan. 18 23 45	Jan. 27 04 38	Feb. 3 13 56	Feb. 7 12	Feb. 19 06
Feb. 10 07 20	Feb. 17 20 30	Feb. 25 20 26	Mar. 4 21 53	Mar. 5 23	Mar. 19 03
Mar. 11 19 51	Mar. 19 17 26	Mar. 27 09 27	Apr. 3 04 36	Mar. 31 04	Apr. 15 22
Apr. 10 09 35	Apr. 18 12 31	Apr. 25 19 57	May 2 11 14	Apr. 27 20	May 13 14
May 10 00 28	May 18 04 34	May 25 04 25	May 31 18 58	May 26 02	June 9 22
June 8 15 56	June 16 17 24	June 23 11 32	June 30 04 53	June 23 11	July 7 01
July 8 07 14	July 16 03 18	July 22 18 15	July 29 17 43	July 21 20	Aug. 3 09
Aug. 6 21 51	Aug. 14 10 56	Aug. 21 01 44	Aug. 28 09 35	Aug. 19 01	Aug. 31 00
Sept. 5 11 36	Sept. 12 17 08	Sept. 19 11 13	Sept. 27 03 55	Sept. 15 17	Sept. 27 18
Oct. 5 00 34	Oct. 11 23 02	Oct. 18 23 37	Oct. 26 23 40	Oct. 10 23	Oct. 25 14
Nov. 3 12 50	Nov. 10 05 57	Nov. 17 15 16	Nov. 25 19 28	Nov. 6 09	Nov. 22 10
Dec. 3 00 22	Dec. 9 15 12	Dec. 17 09 28	Dec. 25 13 47	Dec. 4 10	Dec. 20 00

2014

New Moon	First Quarter	Full Moon	Last Quarter	Perigee	Apogee
Jan. 1 11 14	Jan. 8 03 39	Jan. 16 04 52	Jan. 24 05 19	Jan. 1 21	Jan. 16 02
Jan. 30 21 38	Feb. 6 19 22	Feb. 14 23 53	Feb. 22 17 15	Jan. 30 10	Feb. 12 05
Mar. 1 07 59	Mar. 8 13 27	Mar. 16 17 08	Mar. 24 01 46	Feb. 27 20	Mar. 11 20
Mar. 30 18 44	Apr. 7 08 30	Apr. 15 07 42	Apr. 22 07 51	Mar. 27 19	Apr. 8 15
Apr. 29 06 14	May 7 03 15	May 14 19 16	May 21 12 59	Apr. 23 00	May 6 10
May 28 18 40	June 5 20 39	June 13 04 11	June 19 18 39	May 18 12	June 3 04
June 27 08 08	July 5 11 59	July 12 11 25	July 19 02 08	June 15 03	June 30 19
July 26 22 42	Aug. 4 00 49	Aug. 10 18 09	Aug. 17 12 26	July 13 08	July 28 03
Aug. 25 14 13	Sept. 2 11 11	Sept. 9 01 38	Sept. 16 02 05	Aug. 10 18	Aug. 24 06
Sept. 24 06 14	Oct. 1 19 32	Oct. 8 10 50	Oct. 15 19 12	Sept. 8 04	Sept. 20 14
Oct. 23 21 56	Oct. 31 02 48	Nov. 6 22 23	Nov. 14 15 15	Oct. 6 10	Oct. 18 06
Nov. 22 12 32	Nov. 29 10 06	Dec. 6 12 27	Dec. 14 12 51	Nov. 3 00	Nov. 15 02
Dec. 22 01 36	Dec. 28 18 31			Nov. 27 23	Dec. 12 23
				Dec. 24 17	

2015

New Moon	First Quarter	Full Moon	Last Quarter	Perigee	Apogee
		Jan. 5 04 53	Jan. 13 09 46		Jan. 9 18
Jan. 20 13 13	Jan. 27 04 48	Feb. 3 23 09	Feb. 12 03 50	Jan. 21 20	Feb. 6 06
Feb. 18 23 47	Feb. 25 17 14	Mar. 5 18 05	Mar. 13 17 48	Feb. 19 07	Mar. 5 08
Mar. 20 09 36	Mar. 27 07 42	Apr. 4 12 05	Apr. 12 03 44	Mar. 19 20	Apr. 1 13
Apr. 18 18 57	Apr. 25 23 55	May 4 03 42	May 11 10 36	Apr. 17 04	Apr. 29 04
May 18 04 13	May 25 17 19	June 2 16 19	June 9 15 42	May 15 00	May 26 22
June 16 14 05	June 24 11 02	July 2 02 19	July 8 20 24	June 10 05	June 23 17
July 16 01 24	July 24 04 04	July 31 10 43	Aug. 7 02 02	July 5 19	July 21 11
Aug. 14 14 53	Aug. 22 19 31	Aug. 29 18 35	Sept. 5 09 54	Aug. 2 10	Aug. 18 03
Sept. 13 06 41	Sept. 21 08 59	Sept. 28 02 50	Oct. 4 21 06	Aug. 30 15	Sept. 14 11
Oct. 13 00 05	Oct. 20 20 31	Oct. 27 12 05	Nov. 3 12 24	Sept. 28 02	Oct. 11 13
Nov. 11 17 47	Nov. 19 06 27	Nov. 25 22 44	Dec. 3 07 40	Oct. 26 13	Nov. 7 22
Dec. 11 10 29	Dec. 18 15 14	Dec. 25 11 11		Nov. 23 20	Dec. 5 15
				Dec. 21 09	

2016

New Moon	First Quarter	Full Moon	Last Quarter	Perigee	Apogee
			Jan. 2 05 30		Jan. 2 12
Jan. 10 01 30	Jan. 16 23 26	Jan. 24 01 45	Feb. 1 03 28	Jan. 15 02	Jan. 30 09
Feb. 8 14 39	Feb. 15 07 46	Feb. 22 18 20	Mar. 1 23 10	Feb. 11 03	Feb. 27 03
Mar. 9 01 54	Mar. 15 17 03	Mar. 23 12 01	Mar. 31 15 17	Mar. 10 07	Mar. 25 14
Apr. 7 11 23	Apr. 14 03 59	Apr. 22 05 23	Apr. 30 03 28	Apr. 7 18	Apr. 21 16
May 6 19 29	May 13 17 02	May 21 21 14	May 29 12 12	May 6 04	May 18 22
June 5 02 59	June 12 08 10	June 20 11 02	June 27 18 18	June 3 11	June 15 12
July 4 11 01	July 12 00 52	July 19 22 56	July 26 22 59	July 1 07	July 13 05
Aug. 2 20 44	Aug. 10 18 21	Aug. 18 09 26	Aug. 25 03 41	July 27 12	Aug. 10 00
Sept. 1 09 03	Sept. 9 11 49	Sept. 16 19 05	Sept. 23 09 56	Aug. 22 01	Sept. 6 19
Oct. 1 00 11	Oct. 9 04 33	Oct. 16 04 23	Oct. 22 19 14	Sept. 18 17	Oct. 4 11
Oct. 30 17 38	Nov. 7 19 51	Nov. 14 13 52	Nov. 21 08 33	Oct. 17 00	Oct. 31 19
Nov. 29 12 18	Dec. 7 09 03	Dec. 14 00 05	Dec. 21 01 55	Nov. 14 11	Nov. 27 20
Dec. 29 06 53				Dec. 12 23	Dec. 25 06

New Moon	First Quarter	Full Moon	Last Quarter	Perigee	Apogee
		2017			
d h m	d h m	d h m	d h m	d h	d h
	Jan. 5 19 47	Jan. 12 11 34	Jan. 19 22 13	Jan. 10 06	Jan. 22 00
Jan. 28 00 07	Feb. 4 04 19	Feb. 11 00 33	Feb. 18 19 33	Feb. 6 14	Feb. 18 21
Feb. 26 14 58	Mar. 5 11 32	Mar. 12 14 54	Mar. 20 15 58	Mar. 3 08	Mar. 18 17
Mar. 28 02 57	Apr. 3 18 39	Apr. 11 06 08	Apr. 19 09 56	Mar. 30 13	Apr. 15 10
Apr. 26 12 16	May 3 02 47	May 10 21 42	May 19 00 33	Apr. 27 16	May 12 20
May 25 19 44	June 1 12 42	June 9 13 09	June 17 11 32	May 26 01	June 8 22
June 24 02 30	July 1 00 51	July 9 04 06	July 16 19 25	June 23 11	July 6 04
July 23 09 45	July 30 15 23	Aug. 7 18 10	Aug. 15 01 15	July 21 17	Aug. 2 18
Aug. 21 18 30	Aug. 29 08 13	Sept. 6 07 03	Sept. 13 06 25	Aug. 18 13	Aug. 30 11
Sept. 20 05 30	Sept. 28 02 53	Oct. 5 18 40	Oct. 12 12 25	Sept. 13 16	Sept. 27 07
Oct. 19 19 12	Oct. 27 22 22	Nov. 4 05 23	Nov. 10 20 36	Oct. 9 06	Oct. 25 02
Nov. 18 11 42	Nov. 26 17 03	Dec. 3 15 47	Dec. 10 07 51	Nov. 6 00	Nov. 21 19
Dec. 18 06 30	Dec. 26 09 20			Dec. 4 09	Dec. 19 01
		2018			
		Jan. 2 02 24	Jan. 8 22 25	Jan. 1 22	Jan. 15 02
Jan. 17 02 17	Jan. 24 22 20	Jan. 31 13 26	Feb. 7 15 54	Jan. 30 10	Feb. 11 14
Feb. 15 21 05	Feb. 23 08 09	Mar. 2 00 51	Mar. 9 11 20	Feb. 27 15	Mar. 11 09
Mar. 17 13 11	Mar. 24 15 35	Mar. 31 12 37	Apr. 8 07 17	Mar. 26 17	Apr. 8 06
Apr. 16 01 57	Apr. 22 21 45	Apr. 30 00 58	May 8 02 08	Apr. 20 15	May 6 01
May 15 11 48	May 22 03 49	May 29 14 19	June 6 18 31	May 17 21	June 2 17
June 13 19 43	June 20 10 51	June 28 04 53	July 6 07 50	June 15 00	June 30 03
July 13 02 48	July 19 19 52	July 27 20 20	Aug. 4 18 18	July 13 08	July 27 06
Aug. 11 09 57	Aug. 18 07 48	Aug. 26 11 56	Sept. 3 02 37	Aug. 10 18	Aug. 23 11
Sept. 9 18 01	Sept. 16 23 15	Sept. 25 02 52	Oct. 2 09 45	Sept. 8 01	Sept. 20 01
Oct. 9 03 47	Oct. 16 18 01	Oct. 24 16 45	Oct. 31 16 40	Oct. 5 22	Oct. 17 19
Nov. 7 16 02	Nov. 15 14 54	Nov. 23 05 39	Nov. 30 00 19	Oct. 31 20	Nov. 14 16
Dec. 7 07 20	Dec. 15 11 49	Dec. 22 17 48	Dec. 29 09 34	Nov. 26 12	Dec. 12 12
				Dec. 24 10	
		2019			
Jan. 6 01 28	Jan. 14 06 45	Jan. 21 05 16	Jan. 27 21 10		Jan. 9 04
Feb. 4 21 03	Feb. 12 22 26	Feb. 19 15 53	Feb. 26 11 27	Jan. 21 20	Feb. 5 09
Mar. 6 16 04	Mar. 14 10 27	Mar. 21 01 43	Mar. 28 04 09	Feb. 19 09	Mar. 4 11
Apr. 5 08 50	Apr. 12 19 06	Apr. 19 11 12	Apr. 26 22 18	Mar. 19 20	Apr. 1 00
May 4 22 45	May 12 01 12	May 18 21 11	May 26 16 33	Apr. 16 22	Apr. 28 18
June 3 10 02	June 10 05 59	June 17 08 30	June 25 09 46	May 13 22	May 26 13
July 2 19 16	July 9 10 55	July 16 21 38	July 25 01 18	June 7 23	June 23 08
Aug. 1 03 12	Aug. 7 17 31	Aug. 15 12 29	Aug. 23 14 56	July 5 05	July 21 00
Aug. 30 10 37	Sept. 6 03 10	Sept. 14 04 32	Sept. 22 02 41	Aug. 2 07	Aug. 17 11
Sept. 28 18 26	Oct. 5 16 47	Oct. 13 21 08	Oct. 21 12 39	Aug. 30 16	Sept. 13 14
Oct. 28 03 38	Nov. 4 10 23	Nov. 12 13 34	Nov. 19 21 11	Sept. 28 02	Oct. 10 18
Nov. 26 15 05	Dec. 4 06 58	Dec. 12 05 12	Dec. 19 04 57	Oct. 26 11	Nov. 7 09
Dec. 26 05 13				Nov. 23 08	Dec. 5 04
				Dec. 18 20	
		2020			
	Jan. 3 04 45	Jan. 10 19 21	Jan. 17 12 58		Jan. 2 01
Jan. 24 21 42	Feb. 2 01 41	Feb. 9 07 33	Feb. 15 22 17	Jan. 13 20	Jan. 29 21
Feb. 23 15 32	Mar. 2 19 57	Mar. 9 17 47	Mar. 16 09 34	Feb. 10 20	Feb. 26 12
Mar. 24 09 28	Apr. 1 10 21	Apr. 8 02 35	Apr. 14 22 56	Mar. 10 06	Mar. 24 15
Apr. 23 02 26	Apr. 30 20 38	May 7 10 45	May 14 14 02	Apr. 7 18	Apr. 20 19
May 22 17 39	May 30 03 30	June 5 19 12	June 13 06 23	May 6 03	May 18 08
June 21 06 41	June 28 08 15	July 5 04 44	July 12 23 29	June 3 04	June 15 01
July 20 17 33	July 27 12 32	Aug. 3 15 58	Aug. 11 16 44	June 30 02	July 12 19
Aug. 19 02 41	Aug. 25 17 57	Sept. 2 05 22	Sept. 10 09 25	July 25 05	Aug. 9 14
Sept. 17 11 00	Sept. 24 01 55	Oct. 1 21 05	Oct. 10 00 39	Aug. 21 11	Sept. 6 06
Oct. 16 19 31	Oct. 23 13 23	Oct. 31 14 49	Nov. 8 13 46	Sept. 18 14	Oct. 3 17
Nov. 15 05 07	Nov. 22 04 45	Nov. 30 09 29	Dec. 8 00 36	Oct. 17 00	Oct. 30 19
Dec. 14 16 16	Dec. 21 23 41	Dec. 30 03 28		Nov. 14 12	Nov. 27 00
				Dec. 12 21	Dec. 24 17

SEASONS AND APSIDES

	Perihelion	Equinox	Solstice	Aphelion	Equinox	Solstice
	d h	d h m	d h m	d h	d h m	d h m
2001	Jan. 4 09	Mar. 20 13 31	June 21 07 38	July 4 14	Sept. 22 23 04	Dec. 21 19 21
2002	2 14	20 19 16	21 13 24	6 04	23 04 55	22 01 14
2003	4 05	21 01 00	21 19 10	4 06	23 10 47	22 07 04
2004	4 18	20 06 49	21 00 57	5 11	22 16 30	21 12 42
2005	Jan. 2 01	Mar. 20 12 33	June 21 06 46	July 5 05	Sept. 22 22 23	Dec. 21 18 35
2006	4 15	20 18 26	21 12 26	3 23	23 04 03	22 00 22
2007	3 20	21 00 07	21 18 06	7 00	23 09 51	22 06 08
2008	3 00	20 05 48	20 23 59	4 08	22 15 44	21 12 04
2009	Jan. 4 15	Mar. 20 11 44	June 21 05 45	July 4 02	Sept. 22 21 18	Dec. 21 17 47
2010	3 00	20 17 32	21 11 28	6 11	23 03 09	21 23 38
2011	3 19	20 23 21	21 17 16	4 15	23 09 04	22 05 30
2012	5 00	20 05 14	20 23 09	5 03	22 14 49	21 11 11
2013	Jan. 2 05	Mar. 20 11 02	June 21 05 04	July 5 15	Sept. 22 20 44	Dec. 21 17 11
2014	4 12	20 16 57	21 10 51	4 00	23 02 29	21 23 03
2015	4 07	20 22 45	21 16 38	6 19	23 08 20	22 04 48
2016	2 23	20 04 30	20 22 34	4 16	22 14 21	21 10 44
2017	Jan. 4 14	Mar. 20 10 28	June 21 04 24	July 3 20	Sept. 22 20 02	Dec. 21 16 28
2018	3 06	20 16 15	21 10 07	6 17	23 01 54	21 22 22
2019	3 05	20 21 58	21 15 54	4 22	23 07 50	22 04 19
2020	5 08	20 03 49	20 21 43	4 12	22 13 30	21 10 02

GREATEST ELONGATIONS OF MERCURY

	East	West	East	West	East	West	East	West
	d h	d h	d h	d h	d h	d h	d h	d h
2001	Jan 28 13	Mar 11 06	May 22 04	Jul 9 17	Sep 18 22	Oct 29 17	...	...
2002	Jan 11 23	Feb 21 16	May 4 04	Jun 21 15	Sep 1 10	Oct 13 08	Dec 26 05	...
2003	...	Feb 4 01	Apr 16 15	Jun 3 06	Aug 14 21	Sep 27 00	Dec 9 06	...
2004	...	Jan 17 10	Mar 29 12	May 14 21	Jul 27 03	Sep 9 14	Nov 21 01	Dec 29 21
2005	Mar 12 18	Apr 26 17	Jul 9 03	Aug 23 23	Nov 3 16	Dec 12 13	...	...
2006	Feb 24 05	Apr 8 19	Jun 20 20	Aug 7 01	Oct 17 04	Nov 25 13	...	...
2007	Feb 7 17	Mar 22 02	Jun 2 10	Jul 20 15	Sep 29 16	Nov 8 21	...	...
2008	Jan 22 05	Mar 3 11	May 14 04	Jul 1 18	Sep 11 04	Oct 22 10	...	...
2009	Jan 4 14	Feb 13 21	Apr 26 08	Jun 13 12	Aug 24 16	Oct 6 02	Dec 18 17	...
2010	...	Jan 27 06	Apr 8 23	May 26 02	Aug 7 01	Sep 19 18	Dec 1 15	...
2011	...	Jan 9 15	Mar 23 01	May 7 19	Jul 20 05	Sep 3 06	Nov 14 08	Dec 23 04
2012	Mar 5 09	Apr 18 18	Jul 1 02	Aug 16 12	Oct 26 22	Dec 4 23	...	...
2013	Feb 16 21	Mar 31 22	Jun 12 17	Jul 30 09	Oct 9 10	Nov 18 03	...	...
2014	Jan 31 10	Mar 14 07	May 25 07	Jul 12 18	Sep 21 22	Nov 1 13	...	...
2015	Jan 14 20	Feb 24 17	May 7 05	Jun 24 17	Sep 4 10	Oct 16 04	Dec 29 03	...
2016	...	Feb 7 02	Apr 18 14	Jun 5 09	Aug 16 21	Sep 28 20	Dec 11 04	...
2017	...	Jan 19 10	Apr 1 10	May 18 00	Jul 30 04	Sep 12 10	Nov 24 00	...
2018	...	Jan 1 20	Mar 15 15	Apr 29 19	Jul 12 05	Aug 26 21	Nov 6 15	Dec 15 12
2019	Feb 27 01	Apr 11 20	Jun 23 23	Aug 9 23	Oct 20 04	Nov 28 11	...	...
2020	Feb 10 14	Mar 24 02	Jun 4 13	Jul 22 15	Oct 1 16	Nov 10 17	...	...

GREATEST ELONGATIONS OF VENUS

	East	West		East	West		East	West
	d h	d h		d h	d h		d h	d h
2001	Jan 17 06	Jun 8 05	**2008**	...	...	**2015**	Jun 6 18	Oct 26 07
2002	Aug 22 13	...	**2009**	Jan 14 21	Jun 5 21	**2016**	...	...
2003	...	Jan 11 03	**2010**	Aug 20 04	...	**2017**	Jan 12 13	Jun 3 13
2004	Mar 29 17	Aug 17 19	**2011**	...	Jan 8 16	**2018**	Aug 17 17	...
2005	Nov 3 19	...	**2012**	Mar 27 08	Aug 15 09	**2019**	...	Jan 6 05
2006	...	Mar 25 07	**2013**	Nov 1 08	...	**2020**	Mar 24 22	Aug 13 00
2007	Jun 9 03	Oct 28 15	**2014**	...	Mar 22 20			

SUPERIOR PLANETS

		Mars	**Jupiter**	**Saturn**	**Uranus**	**Neptune**	**Pluto**
		d h	d h	d h	d h	d h	d h
2001	Stationary	May 11 15	Nov 2 17	Sep 27 03	May 29 22	May 11 02	Mar 18 22
	Opposition	Jun 13 18	...	Dec 3 14	Aug 15 15	Jul 30 12	Jun 4 12
	Stationary	Jul 19 23	Jan 25 15	Jan 25 16	Oct 31 04	Oct 17 23	Aug 25 07
	Conjunction	...	Jun 14 13	May 25 13	Feb 9 12	Jan 26 04	Dec 7 04
2002	Stationary	...	Dec 4 21	Oct 11 13	Jun 3 07	May 13 14	Mar 21 06
	Opposition	...	Jan 1 06	Dec 17 17	Aug 20 01	Aug 2 01	Jun 7 05
	Stationary	...	Mar 1 15	Feb 8 10	Nov 4 12	Oct 20 11	Aug 27 20
	Conjunction	Aug 10 22	Jul 20 01	Jun 9 11	Feb 13 17	Jan 28 14	Dec 9 17
2003	Stationary	Jul 30 22	...	Oct 26 00	Jun 7 15	May 16 03	Mar 23 17
	Opposition	Aug 28 18	Feb 2 09	Dec 31 21	Aug 24 10	Aug 4 14	Jun 9 21
	Stationary	Sep 29 14	Apr 4 05	Feb 22 10	Nov 8 19	Oct 23 00	Aug 30 06
	Conjunction	...	Aug 22 10	Jun 24 14	Feb 17 22	Jan 31 00	Dec 12 05
2004	Stationary	...	Jan 4 15	Nov 8 11	Jun 11 00	May 17 15	Mar 24 23
	Opposition	...	Mar 4 05	...	Aug 27 19	Aug 6 03	Jun 11 12
	Stationary	...	May 5 13	Mar 7 15	Nov 12 02	Oct 24 10	Aug 31 17
	Conjunction	Sep 15 13	Sep 22 00	Jul 8 17	Feb 22 02	Feb 2 09	Dec 13 17
2005	Stationary	Oct 1 10	Feb 2 16	Nov 22 18	Jun 15 07	May 20 03	Mar 27 08
	Opposition	Nov 7 08	Apr 3 16	Jan 13 23	Sep 1 03	Aug 8 16	Jun 14 03
	Stationary	Dec 10 23	Jun 5 22	Mar 22 00	Nov 16 07	Oct 26 22	Sep 3 03
	Conjunction	...	Oct 22 13	Jul 23 17	Feb 25 07	Feb 3 19	Dec 16 04
2006	Stationary	...	Mar 5 00	Dec 6 20	Jun 19 16	May 22 17	Mar 29 15
	Opposition	...	May 4 15	Jan 27 23	Sep 5 11	Aug 11 05	Jun 16 17
	Stationary	...	Jul 6 19	Apr 5 12	Nov 20 14	Oct 29 07	Sep 5 11
	Conjunction	Oct 23 07	Nov 21 23	Aug 7 12	Mar 1 11	Feb 6 06	Dec 18 15
2007	Stationary	Nov 15 16	Apr 6 02	Dec 20 12	Jun 23 23	May 25 06	Apr 1 00
	Opposition	Dec 24 20	Jun 5 23	Feb 10 19	Sep 9 19	Aug 13 18	Jun 19 07
	Stationary	...	Aug 7 06	Apr 20 01	Nov 24 18	Oct 31 20	Sep 7 22
	Conjunction	...	Dec 23 06	Aug 21 23	Mar 5 16	Feb 8 16	Dec 21 00
2008	Stationary	...	May 9 15	...	Jun 27 08	May 26 22	Apr 2 09
	Opposition	...	Jul 9 08	Feb 24 10	Sep 13 02	Aug 15 08	Jun 20 20
	Stationary	Jan 30 21	Sep 8 03	May 3 13	Nov 28 00	Nov 2 07	Sep 9 06
	Conjunction	Dec 5 22	...	Sep 4 02	Mar 8 20	Feb 11 02	Dec 22 09
2009	Stationary	Dec 21 16	Jun 15 20	Jan 1 20	Jul 1 16	May 29 11	Apr 4 16
	Opposition	...	Aug 14 18	Mar 8 20	Sep 17 10	Aug 17 21	Jun 23 08
	Stationary	...	Oct 13 09	May 17 19	Dec 2 05	Nov 4 19	Sep 11 16
	Conjunction	...	Jan 24 06	Sep 17 18	Mar 13 01	Feb 12 13	Dec 24 18
2010	Stationary	...	Jul 24 04	Jan 14 19	Jul 6 01	Jun 1 02	Apr 7 01
	Opposition	Jan 29 20	Sep 21 12	Mar 22 01	Sep 21 17	Aug 20 10	Jun 25 19
	Stationary	Mar 11 09	Nov 19 06	May 31 16	Dec 6 10	Nov 7 08	Sep 14 01
	Conjunction	...	Feb 28 11	Oct 1 01	Mar 17 07	Feb 14 23	Dec 27 01

SUPERIOR PLANETS

		Mars	Jupiter	Saturn	Uranus	Neptune	Pluto
		d h	d h	d h	d h	d h	d h
2011	Stationary	...	Aug 30 18	Jan 27 08	Jul 10 08	Jun 3 15	Apr 9 07
	Opposition	...	Oct 29 02	Apr 4 00	Sep 26 00	Aug 22 23	Jun 28 05
	Stationary	...	Dec 26 11	Jun 14 05	Dec 10 15	Nov 9 21	Sep 16 12
	Conjunction	Feb 4 17	Apr 6 15	Oct 13 21	Mar 21 12	Feb 17 10	Dec 29 08
2012	Stationary	Jan 25 01	Oct 4 14	Feb 8 12	Jul 13 17	Jun 5 06	Apr 10 15
	Opposition	Mar 3 20	Dec 3 02	Apr 15 18	Sep 29 07	Aug 24 13	Jun 29 15
	Stationary	Apr 15 12	...	Jun 26 09	Dec 13 20	Nov 11 11	Sep 17 21
	Conjunction	...	May 13 13	Oct 25 09	Mar 24 18	Feb 19 21	Dec 30 14
2013	Stationary	...	Nov 7 07	Feb 19 11	Jul 18 00	Jun 7 18	Apr 12 19
	Opposition	...	...	Apr 28 08	Oct 3 14	Aug 27 02	Jul 2 00
	Stationary	...	Jan 30 16	Jul 9 04	Dec 18 02	Nov 13 22	Sep 20 05
	Conjunction	Apr 18 00	Jun 19 16	Nov 6 12	Mar 29 01	Feb 21 07	...
2014	Stationary	Mar 1 21	Dec 9 07	Mar 3 04	Jul 22 09	Jun 10 06	Apr 15 01
	Opposition	Apr 8 21	Jan 5 21	May 10 18	Oct 7 21	Aug 29 15	Jul 4 08
	Stationary	May 21 09	Mar 6 10	Jul 21 15	Dec 22 06	Nov 16 11	Sep 22 13
	Conjunction	...	Jul 24 21	Nov 18 09	Apr 2 07	Feb 23 18	Jan 1 19
2015	Stationary	...	...	Mar 14 22	Jul 26 16	Jun 12 20	Apr 17 07
	Opposition	...	Feb 6 18	May 23 02	Oct 12 04	Sep 1 04	Jul 6 16
	Stationary	...	Apr 8 20	Aug 2 20	Dec 26 11	Nov 18 21	Sep 24 19
	Conjunction	Jun 14 16	Aug 26 22	Nov 30 00	Apr 6 14	Feb 26 05	Jan 4 00
2016	Stationary	Apr 17 02	Jan 8 20	Mar 25 13	Jul 30 02	Jun 14 08	Apr 18 13
	Opposition	May 22 11	Mar 8 11	Jun 3 07	Oct 15 11	Sep 2 17	Jul 7 22
	Stationary	Jun 30 08	May 9 23	Aug 13 18	Dec 29 16	Nov 20 10	Sep 26 03
	Conjunction	...	Sep 26 07	Dec 10 12	Apr 9 21	Feb 28 16	Jan 6 03
2017	Stationary	...	Feb 6 19	Apr 6 05	Aug 3 10	Jun 16 23	Apr 20 21
	Opposition	...	Apr 7 22	Jun 15 10	Oct 19 18	Sep 5 05	Jul 10 05
	Stationary	...	Jun 10 05	Aug 25 15	...	Nov 22 21	Sep 28 08
	Conjunction	Jul 27 01	Oct 26 18	Dec 21 21	Apr 14 06	Mar 2 03	Jan 7 07
2018	Stationary	Jun 28 14	Mar 9 10	Apr 18 02	Aug 7 20	Jun 19 12	Apr 23 02
	Opposition	Jul 27 05	May 9 01	Jun 27 13	Oct 24 01	Sep 7 18	Jul 12 10
	Stationary	Aug 28 10	Jul 11 04	Sep 6 10	Jan 2 21	Nov 25 08	Sep 30 16
	Conjunction	...	Nov 26 07	...	Apr 18 14	Mar 4 14	Jan 9 10
2019	Stationary	...	Apr 10 17	Apr 30 02	Aug 12 06	Jun 22 04	Apr 25 09
	Opposition	...	Jun 10 15	Jul 9 17	Oct 28 08	Sep 10 07	Jul 14 15
	Stationary	...	Aug 11 16	Sep 18 06	Jan 7 02	Nov 27 20	Oct 2 21
	Conjunction	Sep 2 11	Dec 27 18	Jan 2 06	Apr 22 23	Mar 7 01	Jan 11 12
2020	Stationary	Sep 9 18	May 14 18	May 11 09	Aug 15 17	Jun 23 18	Apr 26 13
	Opposition	Oct 13 23	Jul 14 08	Jul 20 22	Oct 31 16	Sep 11 20	Jul 15 19
	Stationary	Nov 15 19	Sep 13 00	Sep 29 03	Jan 11 07	Nov 29 09	Oct 4 06
	Conjunction	...	...	Jan 13 15	Apr 26 09	Mar 8 12	Jan 13 13

SPECIAL PHENOMENA

Passage of the Earth through the ring-plane of Saturn

2009 September 4 South to North

Transit of Mercury

2003	May	7
2006	November	8/9
2016	May	9
2019	November	11

Transit of Venus

2004	June	8
2012	June	6/7

Observability Data

		Mercury			Venus			Mars		
Date	JD	Transit	Elong.	Mag.	Transit	Elong.	Mag.	Transit	Elong.	Mag.
	245	h m	°		h m	°		h m	°	
00 Jun 25	**1720·5**	13 08	E 16	+ 2·5	12 20	E 4	− 3·9	12 11	E 2	+ 1·6
Jul 5	**1730·5**	12 09	E 5	4·9	12 34	7	3·9	12 00	W 1	1·6
15	**1740·5**	11 10	W 13	2·8	12 46	9	3·9	11 49	4	1·6
25	**1750·5**	10 43	20	0·6	12 57	12	3·9	11 38	7	1·7
00 Aug 4	**1760·5**	10 56	W 17	− 0·7	13 07	E 15	− 3·9	11 26	W 10	+ 1·7
14	**1770·5**	11 35	W 9	1·5	13 14	17	3·9	11 13	13	1·8
24	**1780·5**	12 14	E 3	1·7	13 20	20	3·9	10 59	16	1·8
Sep 3	**1790·5**	12 42	11	0·8	13 25	23	3·9	10 45	20	1·8
00 Sep 13	**1800·5**	13 01	E 18	− 0·3	13 30	E 25	− 3·9	10 31	W 23	+ 1·8
23	**1810·5**	13 14	23	− 0·1	13 36	28	3·9	10 15	27	1·8
Oct 3	**1820·5**	13 20	25	+ 0·0	13 43	30	3·9	10 00	30	1·8
13	**1830·5**	13 14	24	0·2	13 52	32	4·0	9 43	34	1·8
00 Oct 23	**1840·5**	12 34	E 15	+ 1·6	14 03	E 35	− 4·0	9 27	W 37	+ 1·8
Nov 2	**1850·5**	11 15	W 6	+ 3·4	14 16	37	4·0	9 10	41	1·7
12	**1860·5**	10 33	19	− 0·3	14 29	39	4·0	8 54	45	1·7
22	**1870·5**	10 37	18	0·7	14 43	41	4·1	8 37	49	1·7
00 Dec 2	**1880·5**	10 57	W 13	− 0·7	14 55	E 42	− 4·1	8 20	W 53	+ 1·6
12	**1890·5**	11 22	8	0·8	15 05	44	4·2	8 03	57	1·5
22	**1900·5**	11 51	W 3	1·1	15 12	45	4·2	7 46	61	1·4
01 Jan 1	**1910·5**	12 22	E 4	1·1	15 16	46	4·3	7 29	66	1·4
01 Jan 11	**1920·5**	12 53	E 10	− 1·0	15 17	E 47	− 4·4	7 12	W 70	+ 1·2
21	**1930·5**	13 19	16	0·9	15 14	47	4·4	6 55	75	1·1
31	**1940·5**	13 22	18	− 0·2	15 07	46	4·5	6 38	79	1·0
Feb 10	**1950·5**	12 30	7	+ 3·4	14 56	45	4·6	6 21	84	0·8
01 Feb 20	**1960·5**	11 12	W 14	+ 2·1	14 38	E 41	− 4·6	6 03	W 89	+ 0·7
Mar 2	**1970·5**	10 34	25	0·5	14 11	36	4·6	5 44	94	0·5
12	**1980·5**	10 28	27	0·2	13 32	27	4·5	5 25	99	0·3
22	**1990·5**	10 37	25	0·0	12 38	15	4·2	5 04	105	0·1
01 Apr 1	**2000·5**	10 54	W 20	− 0·3	11 38	W 8	− 4·0	4 41	W 111	− 0·2
11	**2010·5**	11 18	13	0·8	10 43	19	4·3	4 16	117	0·5
21	**2020·5**	11 51	W 3	1·9	10 01	30	4·5	3 48	124	0·7
May 1	**2030·5**	12 34	E 9	1·5	9 34	37	4·5	3 16	132	1·1
01 May 11	**2040·5**	13 12	E 18	− 0·6	9 16	W 42	− 4·5	2 40	W 141	− 1·4
21	**2050·5**	13 31	22	+ 0·4	9 04	44	4·4	1 58	151	1·7
31	**2060·5**	13 21	20	1·6	8 58	46	4·4	1 11	163	2·0
Jun 10	**2070·5**	12 38	10	3·8	8 55	46	4·3	0 19	W 174	2·3
01 Jun 20	**2080·5**	11 37	W 7	+ 4·5	8 54	W 45	− 4·2	23 20	E 171	− 2·3
30	**2090·5**	10 50	17	1·9	8 56	44	4·2	22 28	159	2·2
Jul 10	**2100·5**	10 35	21	+ 0·4	9 01	43	4·1	21 40	148	2·0
20	**2110·5**	10 54	17	− 0·7	9 08	42	4·1	20 58	138	1·8
01 Jul 30	**2120·5**	11 38	W 8	− 1·6	9 17	W 40	− 4·0	20 23	E 129	− 1·5
Aug 9	**2130·5**	12 23	E 4	1·7	9 27	38	4·0	19 53	121	1·3
19	**2140·5**	12 55	13	0·7	9 38	36	4·0	19 28	115	1·1
29	**2150·5**	13 15	20	− 0·2	9 49	34	4·0	19 06	109	0·9
01 Sep 8	**2160·5**	13 26	E 24	+ 0·0	9 59	W 31	− 4·0	18 49	E 104	− 0·7
18	**2170·5**	13 28	27	0·1	10 07	29	4·0	18 33	100	0·6
28	**2180·5**	13 14	24	0·5	10 15	27	3·9	18 20	96	0·4
Oct 8	**2190·5**	12 26	E 13	+ 2·2	10 22	W 24	− 3·9	18 07	E 92	− 0·3

Observability Data

Date	JD	Jupiter Transit	Jupiter Elong.	Jupiter Mag.	Saturn Transit	Saturn Elong.	Saturn Mag.
	245	h m	°		h m	°	
00 Jun 25	**1720·5**	9 32	W 35	− 2·1	9 22	W 38	+ 0·2
Jul 5	**1730·5**	9 02	42	2·1	8 47	46	0·2
15	**1740·5**	8 31	50	2·1	8 11	55	0·2
25	**1750·5**	7 59	58	2·2	7 35	63	0·2
00 Aug 4	**1760·5**	7 26	W 66	− 2·2	6 59	W 72	+ 0·2
14	**1770·5**	6 53	74	2·3	6 22	81	0·2
24	**1780·5**	6 19	82	2·3	5 45	90	0·1
Sep 3	**1790·5**	5 44	91	2·4	5 06	100	0·1
00 Sep 13	**1800·5**	5 07	W 100	− 2·5	4 27	W 110	+ 0·0
23	**1810·5**	4 30	109	2·5	3 48	119	− 0·0
Oct 3	**1820·5**	3 51	119	2·6	3 07	129	0·1
13	**1830·5**	3 10	129	2·7	2 26	140	0·1
00 Oct 23	**1840·5**	2 28	W 140	− 2·7	1 44	W 150	− 0·2
Nov 2	**1850·5**	1 45	150	2·8	1 02	161	0·3
12	**1860·5**	1 01	162	2·8	0 20	W 171	0·3
22	**1870·5**	0 16	W 173	2·8	23 33	E 176	0·4
00 Dec 2	**1880·5**	23 27	E 175	− 2·8	22 50	E 166	− 0·3
12	**1890·5**	22 42	164	2·8	22 08	155	0·2
22	**1900·5**	21 58	153	2·8	21 26	145	0·2
01 Jan 1	**1910·5**	21 14	142	2·7	20 45	134	0·1
01 Jan 11	**1920·5**	20 33	E 131	− 2·6	20 04	E 123	− 0·0
21	**1930·5**	19 52	120	2·6	19 24	113	+ 0·0
31	**1940·5**	19 13	110	2·5	18 45	103	0·1
Feb 10	**1950·5**	18 35	100	2·4	18 06	93	0·1
01 Feb 20	**1960·5**	17 59	E 91	− 2·3	17 29	E 83	+ 0·2
Mar 2	**1970·5**	17 23	82	2·3	16 52	74	0·2
12	**1980·5**	16 49	73	2·2	16 15	65	0·2
22	**1990·5**	16 16	65	2·2	15 39	55	0·2
01 Apr 1	**2000·5**	15 44	E 56	− 2·1	15 04	E 46	+ 0·2
11	**2010·5**	15 12	48	2·1	14 29	38	0·2
21	**2020·5**	14 41	40	2·0	13 54	29	0·2
May 1	**2030·5**	14 11	33	2·0	13 20	21	0·1
01 May 11	**2040·5**	13 41	E 25	− 2·0	12 46	E 12	+ 0·1
21	**2050·5**	13 11	18	1·9	12 12	E 4	0·1
31	**2060·5**	12 42	11	1·9	11 38	W 5	0·1
Jun 10	**2070·5**	12 12	3	1·9	11 04	13	0·1
01 Jun 20	**2080·5**	11 43	W 4	− 1·9	10 30	W 21	+ 0·1
30	**2090·5**	11 13	11	1·9	9 56	29	0·1
Jul 10	**2100·5**	10 44	18	1·9	9 21	38	0·2
20	**2110·5**	10 14	26	1·9	8 46	46	0·2
01 Jul 30	**2120·5**	9 44	W 33	− 2·0	8 11	W 55	+ 0·2
Aug 9	**2130·5**	9 13	41	2·0	7 35	64	0·1
19	**2140·5**	8 42	48	2·0	6 59	72	0·1
29	**2150·5**	8 11	56	2·1	6 22	82	0·1
01 Sep 8	**2160·5**	7 39	W 64	− 2·1	5 45	W 91	+ 0·1
18	**2170·5**	7 05	73	2·2	5 07	100	+ 0·0
28	**2180·5**	6 31	81	2·2	4 28	110	− 0·0
Oct 8	**2190·5**	5 56	W 90	− 2·3	3 48	W 120	− 0·1

Uranus

JD	Transit	Elong.	Mag.
245	h m	°	
1720·5	3 18	W 133	+ 5·7
1760·5	0 36	W 173	5·7
1800·5	21 48	E 147	5·7
1840·5	19 08	107	5·8
1880·5	16 33	E 67	+ 5·9
1920·5	14 03	E 28	5·9
1960·5	11 34	W 10	5·9
2000·5	9 05	48	5·9
2040·5	6 32	W 86	+ 5·8
2080·5	3 55	124	5·8
2120·5	1 13	W 163	5·7
2160·5	22 26	E 157	+ 5·7

Neptune

JD	Transit	Elong.	Mag.
1720·5	2 19	W 148	+ 7·9
1760·5	23 34	E 173	7·8
1800·5	20 53	134	7·9
1840·5	18 14	94	7·9
1880·5	15 40	E 54	+ 8·0
1920·5	13 08	E 15	8·0
1960·5	10 37	W 24	8·0
2000·5	8 04	63	7·9
2040·5	5 28	W 102	+ 7·9
2080·5	2 49	140	7·9
2120·5	0 08	W 179	7·8
2160·5	21 23	E 141	+ 7·9

Pluto

JD	Transit	Elong.	Mag.
1720·5	22 26	E 155	+ 13·7
1760·5	19 46	118	13·8
1800·5	17 09	80	13·8
1840·5	14 35	42	13·9
1880·5	12 04	E 11	+ 13·9
1920·5	9 32	W 38	13·9
1960·5	6 59	77	13·8
2000·5	4 22	116	13·8
2040·5	1 42	W 154	+ 13·8
2080·5	22 57	E 161	13·8
2120·5	20 16	125	13·8
2160·5	17 39	E 87	+ 13·8

Observability Data

Date	JD	Mercury			Venus			Mars		
		Transit	Elong.	Mag.	Transit	Elong.	Mag.	Transit	Elong.	Mag.
	245	h m	°		h m	°		h m	°	
01 Oct 18	**2200·5**	11 11	W 8	+2·9	10 28	W 22	−3·9	17 56	E 89	−0·1
28	**2210·5**	10 37	18	−0·4	10 34	19	3·9	17 45	86	+0·0
Nov 7	**2220·5**	10 45	16	0·8	10 42	17	3·9	17 34	82	0·1
17	**2230·5**	11 06	10	0·9	10 50	14	3·9	17 23	79	0·3
01 Nov 27	**2240·5**	11 30	W 4	−1·0	11 01	W 12	−3·9	17 12	E 77	+0·4
Dec 7	**2250·5**	11 57	E 2	1·1	11 13	9	3·9	17 01	74	0·5
17	**2260·5**	12 27	7	0·9	11 27	7	3·9	16 49	71	0·6
27	**2270·5**	12 58	13	0·8	11 42	4	3·9	16 36	68	0·7
02 Jan 6	**2280·5**	13 22	E 18	−0·8	11 58	W 2	−3·9	16 24	E 65	+0·8
16	**2290·5**	13 22	18	−0·1	12 13	E 1	3·9	16 11	62	0·9
26	**2300·5**	12 21	E 5	+4·0	12 26	3	3·9	15 58	59	1·0
Feb 5	**2310·5**	11 02	W 17	1·4	12 37	5	3·9	15 45	56	1·1
02 Feb 15	**2320·5**	10 31	W 25	+0·2	12 46	E 8	−3·9	15 32	E 53	+1·2
25	**2330·5**	10 31	26	+0·0	12 53	10	3·9	15 19	51	1·2
Mar 7	**2340·5**	10 45	24	−0·1	12 59	13	3·9	15 07	48	1·3
17	**2350·5**	11 05	18	0·3	13 05	15	3·9	14 55	45	1·4
02 Mar 27	**2360·5**	11 30	W 11	−0·9	13 11	E 17	−3·9	14 43	E 42	+1·4
Apr 6	**2370·5**	12 01	W 2	1·9	13 19	20	3·9	14 32	39	1·5
16	**2380·5**	12 38	E 10	1·5	13 27	22	3·9	14 21	36	1·6
26	**2390·5**	13 10	18	−0·6	13 37	25	3·9	14 10	33	1·6
02 May 6	**2400·5**	13 19	E 21	+0·6	13 49	E 27	−3·9	13 59	E 30	+1·6
16	**2410·5**	12 56	15	2·4	14 03	30	3·9	13 49	27	1·7
26	**2420·5**	12 03	E 2	5·6	14 16	32	4·0	13 39	24	1·7
Jun 5	**2430·5**	11 07	W 13	3·1	14 29	34	4·0	13 28	21	1·7
02 Jun 15	**2440·5**	10 33	W 21	+1·3	14 39	E 36	−4·0	13 17	E 18	+1·7
25	**2450·5**	10 28	22	+0·2	14 48	39	4·0	13 06	15	1·8
Jul 5	**2460·5**	10 51	17	−0·7	14 54	41	4·1	12 54	12	1·8
15	**2470·5**	11 38	W 7	1·6	14 57	42	4·1	12 42	9	1·8
02 Jul 25	**2480·5**	12 29	E 5	−1·6	14 58	E 44	−4·1	12 29	E 6	+1·7
Aug 4	**2490·5**	13 06	14	0·7	14 57	45	4·2	12 16	E 2	1·7
14	**2500·5**	13 28	21	−0·2	14 54	46	4·3	12 02	W 2	1·7
24	**2510·5**	13 37	26	+0·1	14 49	46	4·3	11 47	4	1·8
02 Sep 3	**2520·5**	13 35	E 27	+0·3	14 42	E 45	−4·4	11 32	W 8	+1·8
13	**2530·5**	13 13	23	0·8	14 32	44	4·5	11 17	11	1·8
23	**2540·5**	12 19	E 10	3·0	14 16	41	4·6	11 01	14	1·8
Oct 3	**2550·5**	11 08	W 10	2·5	13 52	35	4·6	10 45	18	1·8
02 Oct 13	**2560·5**	10 42	W 18	−0·5	13 15	E 27	−4·5	10 29	W 21	+1·8
23	**2570·5**	10 55	14	0·9	12 24	E 15	4·3	10 13	25	1·8
Nov 2	**2580·5**	11 17	8	1·1	11 24	W 6	4·0	9 57	28	1·8
12	**2590·5**	11 40	W 1	1·3	10 28	18	4·4	9 41	32	1·8
02 Nov 22	**2600·5**	12 05	E 5	−0·9	9 46	W 29	−4·6	9 26	W 36	+1·7
Dec 2	**2610·5**	12 32	10	0·7	9 18	37	4·7	9 11	39	1·7
12	**2620·5**	13 01	15	0·6	9 01	42	4·6	8 56	43	1·7
22	**2630·5**	13 23	19	−0·6	8 52	45	4·6	8 42	47	1·6
03 Jan 1	**2640·5**	13 19	E 18	+0·1	8 48	W 47	−4·5	8 28	W 51	+1·5
11	**2650·5**	12 09	E 3	4·5	8 48	47	4·4	8 15	54	1·5
21	**2660·5**	10 51	W 18	0·9	8 52	47	4·4	8 02	58	1·4
31	**2670·5**	10 28	W 25	+0·0	8 59	W 46	−4·3	7 49	W 62	+1·3

Observability Data

Date	JD	Jupiter Transit	Jupiter Elong.	Jupiter Mag.	Saturn Transit	Saturn Elong.	Saturn Mag.
	245	h m	°		h m	°	
01 Oct 18	**2200·5**	5 19	W 99	− 2·3	3 07	W 130	− 0·2
28	**2210·5**	4 41	109	2·4	2 26	140	0·2
Nov 7	**2220·5**	4 02	119	2·5	1 44	151	0·3
17	**2230·5**	3 22	129	2·5	1 02	162	0·4
01 Nov 27	**2240·5**	2 39	W 140	− 2·6	0 19	W 173	− 0·4
Dec 7	**2250·5**	1 56	151	2·7	23 32	E 176	0·4
17	**2260·5**	1 12	162	2·7	22 49	165	0·4
27	**2270·5**	0 27	W 174	2·7	22 07	154	0·3
02 Jan 6	**2280·5**	23 37	E 175	− 2·7	21 25	E 144	− 0·2
16	**2290·5**	22 53	163	2·7	20 44	133	0·2
26	**2300·5**	22 08	152	2·6	20 03	122	0·1
Feb 5	**2310·5**	21 25	141	2·6	19 23	112	− 0·1
02 Feb 15	**2320·5**	20 43	E 130	− 2·5	18 44	E 102	+ 0·0
25	**2330·5**	20 03	119	2·4	18 06	92	0·0
Mar 7	**2340·5**	19 23	109	2·4	17 28	82	0·1
17	**2350·5**	18 46	100	2·3	16 51	73	0·1
02 Mar 27	**2360·5**	18 09	E 91	− 2·2	16 15	E 64	+ 0·1
Apr 6	**2370·5**	17 34	82	2·2	15 39	55	0·1
16	**2380·5**	17 00	73	2·1	15 04	46	0·1
26	**2390·5**	16 27	65	2·0	14 29	37	0·1
02 May 6	**2400·5**	15 54	E 56	− 2·0	13 55	E 29	+ 0·1
16	**2410·5**	15 23	49	1·9	13 21	20	0·1
26	**2420·5**	14 51	41	1·9	12 47	12	0·1
Jun 5	**2430·5**	14 21	33	1·9	12 13	4	0·0
02 Jun 15	**2440·5**	13 50	E 26	− 1·8	11 39	W 5	+ 0·0
25	**2450·5**	13 20	18	1·8	11 05	13	0·1
Jul 5	**2460·5**	12 50	11	1·8	10 32	21	0·1
15	**2470·5**	12 20	4	1·8	9 57	29	0·1
02 Jul 25	**2480·5**	11 50	W 4	− 1·8	9 23	W 38	+ 0·1
Aug 4	**2490·5**	11 20	11	1·8	8 48	46	0·1
14	**2500·5**	10 50	18	1·8	8 13	55	0·1
24	**2510·5**	10 19	26	1·8	7 38	64	0·1
02 Sep 3	**2520·5**	9 49	W 33	− 1·8	7 01	W 73	+ 0·1
13	**2530·5**	9 17	41	1·9	6 25	82	0·1
23	**2540·5**	8 45	49	1·9	5 47	91	+ 0·0
Oct 3	**2550·5**	8 13	57	1·9	5 09	101	− 0·0
02 Oct 13	**2560·5**	7 40	W 65	− 2·0	4 30	W 110	− 0·1
23	**2570·5**	7 06	74	2·0	3 50	120	0·1
Nov 2	**2580·5**	6 31	83	2·1	3 09	131	0·2
12	**2590·5**	5 55	92	2·2	2 28	141	0·3
02 Nov 22	**2600·5**	5 18	W 102	− 2·2	1 46	W 152	− 0·3
Dec 2	**2610·5**	4 40	112	2·3	1 04	163	0·4
12	**2620·5**	4 00	122	2·4	0 21	W 174	0·5
22	**2630·5**	3 19	132	2·4	23 34	E 175	0·5
03 Jan 1	**2640·5**	2 37	W 143	− 2·5	22 51	E 164	− 0·4
11	**2650·5**	1 54	154	2·5	22 09	153	0·3
21	**2660·5**	1 10	166	2·6	21 27	143	0·3
31	**2670·5**	0 25	W 177	− 2·6	20 45	E 132	− 0·2

Uranus

JD	Transit	Elong.	Mag.
245	h m	°	
2200·5	19 45	E 116	+ 5·8
2240·5	17 09	76	5·8
2280·5	14 37	E 37	5·9
2320·5	12 08	W 1	5·9
2360·5	9 40	W 39	+ 5·9
2400·5	7 08	77	5·8
2440·5	4 32	115	5·8
2480·5	1 51	W 154	5·7
2520·5	23 04	E 166	+ 5·7
2560·5	20 22	126	5·8
2600·5	17 44	86	5·8
2640·5	15 12	E 46	+ 5·9

Neptune

JD	Transit	Elong.	Mag.
2200·5	18 44	E 101	+ 7·9
2240·5	16 09	62	7·9
2280·5	13 36	E 22	8·0
2320·5	11 05	W 17	8·0
2360·5	8 33	W 56	+ 8·0
2400·5	5 58	94	7·9
2440·5	3 20	133	7·9
2480·5	0 39	W 172	7·8
2520·5	21 53	E 148	+ 7·9
2560·5	19 14	109	7·9
2600·5	16 38	69	7·9
2640·5	14 05	E 29	+ 8·0

Pluto

JD	Transit	Elong.	Mag.
2200·5	15 04	E 49	+ 13·9
2240·5	12 32	E 14	13·9
2280·5	10 01	W 31	13·9
2320·5	7 28	69	13·9
2360·5	4 52	W 108	+ 13·8
2400·5	2 13	W 147	13·8
2440·5	23 27	E 167	13·8
2480·5	20 47	132	13·8
2520·5	18 08	E 95	+ 13·8
2560·5	15 33	57	13·9
2600·5	13 01	E 20	13·9
2640·5	10 30	W 24	+ 13·9

Observability Data

Date	JD	Mercury Transit	Mercury Elong.	Mercury Mag.	Venus Transit	Venus Elong.	Venus Mag.	Mars Transit	Mars Elong.	Mars Mag.
	245	h m	°		h m	°		h m	°	
03 Feb 10	**2680·5**	10 35	W 25	− 0·1	9 08	W 45	− 4·2	7 37	W 65	+ 1·2
20	**2690·5**	10 53	21	0·2	9 18	43	4·2	7 26	69	1·1
Mar 2	**2700·5**	11 15	16	0·4	9 28	42	4·1	7 14	73	0·9
12	**2710·5**	11 41	W 9	0·9	9 38	40	4·1	7 02	76	0·8
03 Mar 22	**2720·5**	12 11	E 1	− 1·9	9 46	W 38	− 4·0	6 51	W 80	+ 0·7
Apr 1	**2730·5**	12 44	10	1·4	9 53	36	4·0	6 38	84	0·5
11	**2740·5**	13 09	18	− 0·6	10 00	34	4·0	6 26	87	0·3
21	**2750·5**	13 08	19	+ 0·9	10 05	31	3·9	6 12	91	0·2
03 May 1	**2760·5**	12 30	E 10	+ 3·6	10 10	W 29	− 3·9	5 58	W 95	− 0·0
11	**2770·5**	11 31	W 6	4·7	10 16	27	3·9	5 43	99	0·2
21	**2780·5**	10 44	19	2·0	10 22	24	3·9	5 26	103	0·4
31	**2790·5**	10 23	24	0·8	10 30	22	3·9	5 08	107	0·6
03 Jun 10	**2800·5**	10 25	W 23	+ 0·1	10 40	W 19	− 3·9	4 49	W 112	− 0·9
20	**2810·5**	10 50	17	− 0·7	10 52	16	3·9	4 27	117	1·1
30	**2820·5**	11 37	W 7	1·7	11 05	14	3·9	4 02	123	1·4
Jul 10	**2830·5**	12 32	E 6	1·6	11 19	11	3·9	3 35	129	1·7
03 Jul 20	**2840·5**	13 14	E 16	− 0·7	11 33	W 8	− 3·9	3 03	W 137	− 2·0
30	**2850·5**	13 38	23	− 0·1	11 45	6	3·9	2 27	146	2·3
Aug 9	**2860·5**	13 47	27	+ 0·2	11 57	W 3	3·9	1 45	156	2·5
19	**2870·5**	13 40	27	0·5	12 06	E 1	3·9	0 59	W 166	2·8
03 Aug 29	**2880·5**	13 10	E 21	+ 1·3	12 14	E 3	− 3·9	0 10	E 173	− 2·9
Sep 8	**2890·5**	12 10	E 7	3·9	12 21	6	3·9	23 16	166	2·7
18	**2900·5**	11 06	W 12	+ 2·0	12 27	8	3·9	22 29	155	2·5
28	**2910·5**	10 48	18	− 0·5	12 33	11	3·9	21 47	145	2·2
03 Oct 8	**2920·5**	11 05	W 13	− 1·1	12 40	E 13	− 3·9	21 10	E 136	− 1·9
18	**2930·5**	11 29	W 5	1·3	12 48	16	3·9	20 37	129	1·6
28	**2940·5**	11 52	E 2	1·3	12 58	19	3·9	20 08	122	1·3
Nov 7	**2950·5**	12 15	8	0·7	13 10	21	3·9	19 43	116	1·0
03 Nov 17	**2960·5**	12 39	E 13	− 0·5	13 24	E 23	− 3·9	19 20	E 110	− 0·8
27	**2970·5**	13 03	18	0·5	13 39	26	3·9	18 58	105	0·5
Dec 7	**2980·5**	13 22	21	− 0·5	13 54	28	3·9	18 38	100	0·3
17	**2990·5**	13 12	18	+ 0·3	14 08	30	4·0	18 19	96	− 0·1
03 Dec 27	**3000·5**	11 56	E 2	+ 4·9	14 20	E 32	− 4·0	18 01	E 91	+ 0·1
04 Jan 6	**3010·5**	10 43	W 19	+ 0·5	14 30	34	4·0	17 44	87	0·3
16	**3020·5**	10 27	24	− 0·2	14 38	36	4·0	17 27	83	0·5
26	**3030·5**	10 38	23	0·2	14 43	38	4·1	17 11	79	0·6
04 Feb 5	**3040·5**	10 59	W 19	− 0·3	14 47	E 40	− 4·1	16 55	E 75	+ 0·8
15	**3050·5**	11 25	14	0·5	14 50	42	4·1	16 41	72	0·9
25	**3060·5**	11 52	W 7	1·0	14 53	43	4·2	16 26	68	1·0
Mar 6	**3070·5**	12 22	E 2	1·7	14 55	45	4·2	16 13	64	1·2
04 Mar 16	**3080·5**	12 52	E 11	− 1·3	14 57	E 45	− 4·3	16 00	E 61	+ 1·3
26	**3090·5**	13 11	18	− 0·5	14 58	46	4·3	15 47	57	1·4
Apr 5	**3100·5**	12 57	16	+ 1·3	14 59	46	4·4	15 35	54	1·4
15	**3110·5**	12 07	4	4·9	14 57	45	4·4	15 23	50	1·5
04 Apr 25	**3120·5**	11 07	W 13	+ 3·1	14 50	E 43	− 4·5	15 11	E 47	+ 1·6
May 5	**3130·5**	10 31	23	1·3	14 35	39	4·5	15 00	44	1·6
15	**3140·5**	10 19	26	+ 0·5	14 08	32	4·5	14 48	40	1·7
25	**3150·5**	10 26	W 23	− 0·0	13 24	E 21	− 4·3	14 37	E 37	+ 1·7

Observability Data

Date		JD	Jupiter Transit	Jupiter Elong.	Jupiter Mag.	Saturn Transit	Saturn Elong.	Saturn Mag.
		245	h m	°		h m	°	
03 Feb	10	**2680·5**	23 36	E 171	−2·6	20 05	E 121	−0·1
	20	**2690·5**	22 52	160	2·5	19 25	111	0·1
Mar	2	**2700·5**	22 08	149	2·5	18 46	101	−0·0
	12	**2710·5**	21 26	138	2·4	18 08	91	+0·0
03 Mar	22	**2720·5**	20 44	E 127	−2·4	17 30	E 82	+0·1
Apr	1	**2730·5**	20 04	117	2·3	16 53	73	0·1
	11	**2740·5**	19 25	107	2·2	16 17	63	0·1
	21	**2750·5**	18 47	98	2·2	15 42	55	0·1
03 May	1	**2760·5**	18 11	E 89	−2·1	15 07	E 46	+0·1
	11	**2770·5**	17 35	80	2·0	14 32	37	0·1
	21	**2780·5**	17 00	72	2·0	13 58	29	0·1
	31	**2790·5**	16 27	63	1·9	13 24	20	0·1
03 Jun	10	**2800·5**	15 54	E 55	−1·9	12 50	E 12	+0·0
	20	**2810·5**	15 21	48	1·8	12 16	E 4	0·0
	30	**2820·5**	14 49	40	1·8	11 42	W 5	0·0
Jul	10	**2830·5**	14 18	32	1·8	11 09	13	0·1
03 Jul	20	**2840·5**	13 46	E 25	−1·7	10 35	W 21	+0·1
	30	**2850·5**	13 15	17	1·7	10 01	29	0·1
Aug	9	**2860·5**	12 44	10	1·7	9 26	38	0·1
	19	**2870·5**	12 13	3	1·7	8 51	46	0·1
03 Aug	29	**2880·5**	11 42	W 5	−1·7	8 16	W 55	+0·2
Sep	8	**2890·5**	11 11	12	1·7	7 41	64	0·1
	18	**2900·5**	10 40	20	1·7	7 05	73	0·1
	28	**2910·5**	10 08	28	1·7	6 28	82	0·1
03 Oct	8	**2920·5**	9 37	W 36	−1·8	5 50	W 91	+0·1
	18	**2930·5**	9 04	44	1·8	5 12	101	+0·0
	28	**2940·5**	8 32	52	1·8	4 33	111	−0·0
Nov	7	**2950·5**	7 58	60	1·9	3 53	121	0·1
03 Nov	17	**2960·5**	7 24	W 69	−1·9	3 12	W 131	−0·2
	27	**2970·5**	6 50	78	2·0	2 31	142	0·2
Dec	7	**2980·5**	6 14	87	2·0	1 49	153	0·3
	17	**2990·5**	5 37	96	2·1	1 06	164	0·4
03 Dec	27	**3000·5**	5 00	W 106	−2·2	0 24	W 175	−0·4
04 Jan	6	**3010·5**	4 21	116	2·2	23 37	E 174	0·4
	16	**3020·5**	3 41	126	2·3	22 54	163	0·4
	26	**3030·5**	2 59	137	2·4	22 11	152	0·3
04 Feb	5	**3040·5**	2 17	W 148	−2·4	21 29	E 142	−0·3
	15	**3050·5**	1 34	159	2·5	20 48	131	0·2
	25	**3060·5**	0 50	W 171	2·5	20 08	121	0·1
Mar	6	**3070·5**	0 06	E 178	2·5	19 28	111	0·1
04 Mar	16	**3080·5**	23 18	E 167	−2·5	18 49	E 101	−0·0
	26	**3090·5**	22 34	156	2·4	18 11	91	+0·0
Apr	5	**3100·5**	21 51	145	2·4	17 33	81	0·1
	15	**3110·5**	21 09	134	2·3	16 57	72	0·1
04 Apr	25	**3120·5**	20 28	E 124	−2·3	16 20	E 63	+0·1
May	5	**3130·5**	19 48	114	2·2	15 45	54	0·1
	15	**3140·5**	19 09	105	2·1	15 10	46	0·1
	25	**3150·5**	18 32	E 95	−2·1	14 35	E 37	+0·1

Uranus

JD	Transit	Elong.	Mag.
245	h m	°	
2680·5	12 42	E 8	+ 5·9
2720·5	10 14	W 30	5·9
2760·5	7 43	68	5·9
2800·5	5 08	106	5·8
2840·5	2 28	W 145	+ 5·7
2880·5	23 41	E 175	5·7
2920·5	20 58	135	5·8
2960·5	18 20	95	5·8
3000·5	15 46	E 55	+ 5·9
3040·5	13 16	E 16	5·9
3080·5	10 48	W 22	5·9
3120·5	8 17	W 59	+ 5·9

Neptune

JD	Transit	Elong.	Mag.
2680·5	11 33	W 10	+ 8·0
2720·5	9 02	49	8·0
2760·5	6 27	87	7·9
2800·5	3 50	126	7·9
2840·5	1 09	W 165	+ 7·8
2880·5	22 23	E 156	7·8
2920·5	19 44	116	7·9
2960·5	17 07	76	7·9
3000·5	14 33	E 37	+ 8·0
3040·5	12 02	W 3	8·0
3080·5	9 30	41	8·0
3120·5	6 56	W 80	+ 7·9

Pluto

JD	Transit	Elong.	Mag.
2680·5	7 57	W 62	+13·9
2720·5	5 22	101	13·8
2760·5	2 43	W 140	13·8
2800·5	0 02	E 170	13·8
2840·5	21 17	E 140	+13·8
2880·5	18 38	102	13·9
2920·5	16 02	64	13·9
2960·5	13 30	26	13·9
3000·5	10 58	W 17	+13·9
3040·5	8 26	54	13·9
3080·5	5 52	93	13·9
3120·5	3 13	W 132	+13·8

Observability Data

Date			JD	Mercury			Venus			Mars		
				Transit	Elong.	Mag.	Transit	Elong.	Mag.	Transit	Elong.	Mag.
			245	h m	°		h m	°		h m	°	
04	Jun	4	3160·5	10 51	W 17	− 0·7	12 24	E 7	− 3·9	14 25	E 34	+ 1·8
		14	3170·5	11 37	W 6	1·7	11 20	W 9	4·0	14 12	31	1·8
		24	3180·5	12 33	E 6	1·6	10 25	23	4·3	13 59	27	1·8
	Jul	4	3190·5	13 18	17	0·7	9 45	33	4·4	13 46	24	1·8
04	Jul	14	3200·5	13 45	E 24	− 0·1	9 19	W 39	− 4·5	13 32	E 21	+ 1·8
		24	3210·5	13 53	27	+ 0·3	9 04	43	4·5	13 18	18	1·8
	Aug	3	3220·5	13 41	26	0·8	8 56	45	4·4	13 03	14	1·8
		13	3230·5	13 04	18	1·9	8 54	46	4·3	12 48	11	1·8
04	Aug	23	3240·5	11 59	E 5	+ 4·7	8 55	W 46	− 4·3	12 33	E 8	+ 1·8
	Sep	2	3250·5	11 02	W 14	+ 1·6	9 00	45	4·2	12 17	5	1·8
		12	3260·5	10 52	18	− 0·6	9 05	44	4·2	12 01	E 2	1·7
		22	3270·5	11 15	12	1·2	9 11	43	4·1	11 46	W 2	1·7
04	Oct	2	3280·5	11 42	W 3	− 1·5	9 17	W 41	− 4·1	11 30	W 6	+ 1·7
		12	3290·5	12 05	E 5	1·1	9 23	39	4·1	11 14	9	1·7
		22	3300·5	12 26	11	0·6	9 28	37	4·0	10 59	12	1·7
	Nov	1	3310·5	12 46	16	0·4	9 33	35	4·0	10 44	16	1·7
04	Nov	11	3320·5	13 06	E 20	− 0·3	9 39	W 33	− 4·0	10 29	W 19	+ 1·7
		21	3330·5	13 19	22	− 0·3	9 46	31	4·0	10 16	22	1·7
	Dec	1	3340·5	13 02	E 18	+ 0·5	9 54	29	4·0	10 02	26	1·7
		11	3350·5	11 42	W 2	4·9	10 04	26	4·0	9 50	29	1·7
04	Dec	21	3360·5	10 36	W 19	+ 0·3	10 16	W 24	− 3·9	9 38	W 33	+ 1·6
		31	3370·5	10 27	22	− 0·3	10 29	22	3·9	9 27	36	1·6
05	Jan	10	3380·5	10 41	20	0·3	10 44	19	3·9	9 17	39	1·5
		20	3390·5	11 05	16	0·4	10 59	17	3·9	9 08	43	1·5
05	Jan	30	3400·5	11 33	W 11	− 0·6	11 13	W 15	− 3·9	8 58	W 46	+ 1·4
	Feb	9	3410·5	12 02	W 5	1·1	11 26	12	3·9	8 50	49	1·3
		19	3420·5	12 32	E 4	1·5	11 37	10	3·9	8 41	52	1·3
	Mar	1	3430·5	13 00	12	1·2	11 46	8	3·9	8 33	55	1·2
05	Mar	11	3440·5	13 14	E 18	− 0·5	11 54	W 5	− 3·9	8 24	W 58	+ 1·1
		21	3450·5	12 49	E 14	+ 1·8	12 00	3	3·9	8 16	60	1·0
		31	3460·5	11 48	W 4	5·0	12 06	W 1	3·9	8 06	63	0·9
	Apr	10	3470·5	10 52	18	2·0	12 13	E 3	3·9	7 56	66	0·8
05	Apr	20	3480·5	10 25	W 26	+ 0·8	12 20	E 5	− 3·9	7 46	W 68	+ 0·7
		30	3490·5	10 20	27	+ 0·3	12 29	8	3·9	7 35	71	0·6
	May	10	3500·5	10 31	23	− 0·1	12 39	10	3·9	7 23	73	0·5
		20	3510·5	10 55	16	0·7	12 52	13	3·9	7 11	75	0·4
05	May	30	3520·5	11 37	W 5	− 1·7	13 05	E 16	− 3·9	6 59	W 78	+ 0·3
	Jun	9	3530·5	12 32	E 7	1·6	13 20	18	3·9	6 45	80	0·2
		19	3540·5	13 19	17	− 0·6	13 34	21	3·9	6 32	83	+ 0·1
		29	3550·5	13 47	24	+ 0·0	13 46	24	3·9	6 17	86	− 0·0
05	Jul	9	3560·5	13 54	E 26	+ 0·5	13 56	E 26	− 3·9	6 02	W 89	− 0·2
		19	3570·5	13 36	23	1·2	14 05	29	3·9	5 47	92	0·3
		29	3580·5	12 50	E 14	2·8	14 11	31	3·9	5 30	96	0·4
	Aug	8	3590·5	11 43	W 6	4·5	14 16	34	4·0	5 12	99	0·6
05	Aug	18	3600·5	10 56	W 16	+ 1·2	14 19	E 36	− 4·0	4 52	W 104	− 0·8
		28	3610·5	10 55	18	− 0·6	14 22	38	4·0	4 30	109	0·9
	Sep	7	3620·5	11 24	10	1·3	14 26	40	4·0	4 05	115	1·1
		17	3630·5	11 55	W 2	− 1·7	14 30	E 42	− 4·1	3 36	W 122	− 1·4

Observability Data

Date	JD	Jupiter Transit	Jupiter Elong.	Jupiter Mag.	Saturn Transit	Saturn Elong.	Saturn Mag.
	245	h m	°		h m	°	
04 Jun 4	**3160·5**	17 55	E 87	−2·0	14 01	E 29	+0·1
14	**3170·5**	17 19	78	1·9	13 27	20	0·1
24	**3180·5**	16 45	70	1·9	12 53	12	0·1
Jul 4	**3190·5**	16 10	61	1·8	12 19	4	0·1
04 Jul 14	**3200·5**	15 37	E 53	−1·8	11 45	W 4	+0·1
24	**3210·5**	15 04	46	1·8	11 12	13	0·1
Aug 3	**3220·5**	14 31	38	1·7	10 38	21	0·2
13	**3230·5**	13 59	30	1·7	10 03	29	0·2
04 Aug 23	**3240·5**	13 27	E 23	−1·7	9 29	W 38	+0·2
Sep 2	**3250·5**	12 56	15	1·7	8 54	46	0·2
12	**3260·5**	12 24	E 8	1·7	8 19	55	0·2
22	**3270·5**	11 53	W 1	1·7	7 43	64	0·2
04 Oct 2	**3280·5**	11 21	W 8	−1·7	7 07	W 73	+0·2
12	**3290·5**	10 50	15	1·7	6 30	82	0·2
22	**3300·5**	10 18	23	1·7	5 52	92	0·2
Nov 1	**3310·5**	9 46	31	1·7	5 14	102	0·1
04 Nov 11	**3320·5**	9 14	W 39	−1·7	4 35	W 112	+0·1
21	**3330·5**	8 42	47	1·8	3 55	122	−0·0
Dec 1	**3340·5**	8 09	56	1·8	3 14	132	0·1
11	**3350·5**	7 35	65	1·9	2 33	143	0·1
04 Dec 21	**3360·5**	7 01	W 73	−1·9	1 51	W 154	−0·2
31	**3370·5**	6 25	82	2·0	1 09	165	0·3
05 Jan 10	**3380·5**	5 49	92	2·0	0 26	W 176	0·4
20	**3390·5**	5 12	101	2·1	23 39	E 173	0·4
05 Jan 30	**3400·5**	4 34	W 111	−2·2	22 56	E 162	−0·3
Feb 9	**3410·5**	3 54	122	2·2	22 14	152	0·2
19	**3420·5**	3 14	132	2·3	21 32	141	0·2
Mar 1	**3430·5**	2 32	143	2·4	20 51	130	0·1
05 Mar 11	**3440·5**	1 49	W 154	−2·4	20 10	E 120	−0·0
21	**3450·5**	1 06	165	2·4	19 31	110	+0·0
31	**3460·5**	0 22	W 176	2·5	18 52	100	0·1
Apr 10	**3470·5**	23 33	E 173	2·5	18 14	90	0·1
05 Apr 20	**3480·5**	22 49	E 162	−2·4	17 36	E 81	+0·2
30	**3490·5**	22 06	151	2·4	16 59	72	0·2
May 10	**3500·5**	21 24	140	2·3	16 23	63	0·2
20	**3510·5**	20 42	130	2·3	15 48	54	0·2
05 May 30	**3520·5**	20 02	E 120	−2·2	15 12	E 46	+0·2
Jun 9	**3530·5**	19 22	111	2·2	14 38	37	0·2
19	**3540·5**	18 44	101	2·1	14 03	29	0·2
29	**3550·5**	18 06	92	2·0	13 29	20	0·2
05 Jul 9	**3560·5**	17 30	E 84	−2·0	12 55	E 12	+0·2
19	**3570·5**	16 55	75	1·9	12 21	E 4	0·2
29	**3580·5**	16 20	67	1·9	11 47	W 4	0·2
Aug 8	**3590·5**	15 46	59	1·8	11 13	13	0·2
05 Aug 18	**3600·5**	15 13	E 51	−1·8	10 39	W 21	+0·3
28	**3610·5**	14 40	43	1·7	10 05	29	0·3
Sep 7	**3620·5**	14 08	35	1·7	9 30	38	0·3
17	**3630·5**	13 36	E 27	−1·7	8 55	W 47	+0·4

Uranus

JD	Transit	Elong.	Mag.
245	h m	°	
3160·5	5 43	W 97	+ 5·8
3200·5	3 04	136	5·8
3240·5	0 22	W 175	5·7
3280·5	21 35	E 144	5·7
3320·5	18 55	E 104	+ 5·8
3360·5	16 21	64	5·9
3400·5	13 50	E 25	5·9
3440·5	11 21	W 13	5·9
3480·5	8 51	W 50	+ 5·9
3520·5	6 18	88	5·8
3560·5	3 41	126	5·8
3600·5	0 59	W 166	+ 5·7

Neptune

JD	Transit	Elong.	Mag.
3160·5	4 19	W 118	+ 7·9
3200·5	1 39	W 157	7·8
3240·5	22 54	E 163	7·8
3280·5	20 13	124	7·9
3320·5	17 36	E 84	+ 7·9
3360·5	15 02	44	8·0
3400·5	12 30	E 5	8·0
3440·5	9 59	W 34	8·0
3480·5	7 25	W 73	+ 7·9
3520·5	4 49	111	7·9
3560·5	2 09	W 150	7·8
3600·5	23 24	E 171	+ 7·8

Pluto

JD	Transit	Elong.	Mag.
3160·5	0 33	W 168	+13·8
3200·5	21 47	E 147	13·8
3240·5	19 08	109	13·9
3280·5	16 31	71	13·9
3320·5	13 58	E 33	+14·0
3360·5	11 27	W 11	14·0
3400·5	8 55	47	13·9
3440·5	6 21	86	13·9
3480·5	3 44	W 125	+13·9
3520·5	1 03	W 163	13·8
3560·5	22 18	E 154	13·8
3600·5	19 38	E 116	+13·9

Observability Data

Date		JD	Mercury Transit	Mercury Elong.	Mercury Mag.	Venus Transit	Venus Elong.	Venus Mag.	Mars Transit	Mars Elong.	Mars Mag.
		245	h m	°		h m	°		h m	°	
05 Sep	27	**3640·5**	12 19	E 7	− 1·0	14 35	E 44	− 4·1	3 02	W 131	− 1·6
Oct	7	**3650·5**	12 39	14	0·5	14 42	45	4·2	2 22	141	1·8
	17	**3660·5**	12 56	19	0·2	14 49	46	4·3	1 36	152	2·0
	27	**3670·5**	13 11	22	0·2	14 56	47	4·3	0 45	165	2·2
05 Nov	6	**3680·5**	13 17	E 23	− 0·1	15 02	E 47	− 4·4	23 46	W 178	− 2·3
	16	**3690·5**	12 52	E 17	+ 0·9	15 05	47	4·5	22 53	E 168	2·1
	26	**3700·5**	11 31	W 3	4·5	15 02	45	4·6	22 04	156	1·8
Dec	6	**3710·5**	10 33	19	0·0	14 51	42	4·7	21 19	144	1·5
05 Dec	16	**3720·5**	10 29	W 21	− 0·5	14 30	E 36	− 4·7	20 41	E 134	− 1·1
	26	**3730·5**	10 46	17	0·4	13 53	27	4·6	20 07	125	0·8
06 Jan	5	**3740·5**	11 11	13	0·5	12 59	E 15	4·3	19 37	118	0·5
	15	**3750·5**	11 40	8	0·8	11 55	W 6	4·1	19 10	111	− 0·2
06 Jan	25	**3760·5**	12 11	W 2	− 1·3	10 54	W 18	− 4·4	18 46	E 104	+ 0·0
Feb	4	**3770·5**	12 41	E 6	1·3	10 08	30	4·6	18 24	98	0·3
	14	**3780·5**	13 08	14	1·1	9 39	37	4·6	18 05	93	0·5
	24	**3790·5**	13 18	18	− 0·4	9 21	42	4·6	17 46	88	0·7
06 Mar	6	**3800·5**	12 43	E 11	+ 2·3	9 13	W 45	− 4·5	17 29	E 83	+ 0·8
	16	**3810·5**	11 33	W 8	3·7	9 09	46	4·5	17 13	78	1·0
	26	**3820·5**	10 42	22	1·2	9 09	47	4·4	16 58	74	1·1
Apr	5	**3830·5**	10 24	27	0·5	9 11	46	4·3	16 43	70	1·2
06 Apr	15	**3840·5**	10 25	W 27	+ 0·2	9 13	W 45	− 4·2	16 29	E 66	+ 1·3
	25	**3850·5**	10 38	23	− 0·2	9 15	44	4·2	16 16	62	1·4
May	5	**3860·5**	11 02	15	0·8	9 18	43	4·1	16 02	58	1·5
	15	**3870·5**	11 40	W 5	1·8	9 21	41	4·0	15 48	54	1·6
06 May	25	**3880·5**	12 31	E 8	− 1·6	9 25	W 39	− 4·0	15 34	E 50	+ 1·7
Jun	4	**3890·5**	13 17	18	− 0·6	9 30	37	4·0	15 20	47	1·7
	14	**3900·5**	13 45	24	+ 0·1	9 37	35	3·9	15 06	43	1·7
	24	**3910·5**	13 48	25	0·8	9 45	32	3·9	14 51	40	1·8
06 Jul	4	**3920·5**	13 23	E 20	+ 1·9	9 55	W 30	− 3·9	14 36	E 37	+ 1·8
	14	**3930·5**	12 28	E 8	4·1	10 07	28	3·9	14 21	33	1·8
	24	**3940·5**	11 24	W 10	3·5	10 20	25	3·9	14 06	30	1·8
Aug	3	**3950·5**	10 48	18	0·8	10 33	23	3·9	13 50	26	1·8
06 Aug	13	**3960·5**	10 55	W 18	− 0·7	10 45	W 20	− 3·9	13 34	E 23	+ 1·8
	23	**3970·5**	11 31	W 9	1·4	10 56	17	3·9	13 18	20	1·8
Sep	2	**3980·5**	12 07	E 2	1·7	11 06	15	3·9	13 02	17	1·8
	12	**3990·5**	12 34	9	0·9	11 14	12	3·9	12 46	13	1·8
06 Sep	22	**4000·5**	12 53	E 16	− 0·4	11 21	W 9	− 3·9	12 30	E 10	+ 1·7
Oct	2	**4010·5**	13 07	21	0·1	11 28	7	3·9	12 15	7	1·7
	12	**4020·5**	13 16	24	− 0·1	11 34	4	3·9	12 00	4	1·6
	22	**4030·5**	13 15	24	+ 0·1	11 41	W 2	3·9	11 46	1	1·6
06 Nov	1	**4040·5**	12 42	E 16	+ 1·3	11 49	E 1	− 3·9	11 32	W 3	+ 1·6
	11	**4050·5**	11 22	W 5	+ 4·0	12 00	4	3·9	11 20	6	1·6
	21	**4060·5**	10 33	19	− 0·1	12 12	6	3·9	11 08	9	1·6
Dec	1	**4070·5**	10 33	19	0·6	12 26	9	3·9	10 57	12	1·6
06 Dec	11	**4080·5**	10 52	W 15	− 0·6	12 41	E 11	− 3·9	10 47	W 15	+ 1·5
	21	**4090·5**	11 17	10	0·7	12 57	13	3·9	10 37	18	1·5
	31	**4100·5**	11 46	W 4	1·0	13 12	16	3·9	10 29	21	1·5
07 Jan	10	**4110·5**	12 17	E 3	− 1·2	13 25	E 18	− 3·9	10 21	W 24	+ 1·5

Observability Data

Date	JD	Jupiter Transit	Jupiter Elong.	Jupiter Mag.	Saturn Transit	Saturn Elong.	Saturn Mag.
	245	h m	°		h m	°	
05 Sep 27	**3640·5**	13 04	E 20	−1·7	8 20	W 55	+0·4
Oct 7	**3650·5**	12 33	12	1·7	7 44	64	0·4
17	**3660·5**	12 02	E 4	1·7	7 07	74	0·3
27	**3670·5**	11 31	W 4	1·7	6 30	83	0·3
05 Nov 6	**3680·5**	10 59	W 11	−1·7	5 52	W 93	+0·3
16	**3690·5**	10 28	19	1·7	5 14	102	0·2
26	**3700·5**	9 57	27	1·7	4 35	113	0·2
Dec 6	**3710·5**	9 26	35	1·7	3 55	123	0·1
05 Dec 16	**3720·5**	8 54	W 44	−1·8	3 14	W 133	+0·1
26	**3730·5**	8 21	52	1·8	2 33	144	−0·0
06 Jan 5	**3740·5**	7 48	61	1·8	1 51	155	0·1
15	**3750·5**	7 15	69	1·9	1 09	166	0·2
06 Jan 25	**3760·5**	6 40	W 78	−2·0	0 26	W 177	−0·2
Feb 4	**3770·5**	6 05	87	2·0	23 39	E 172	0·2
14	**3780·5**	5 28	97	2·1	22 57	161	0·2
24	**3790·5**	4 51	106	2·2	22 15	151	0·1
06 Mar 6	**3800·5**	4 12	W 116	−2·2	21 33	E 140	−0·0
16	**3810·5**	3 32	127	2·3	20 52	129	+0·0
26	**3820·5**	2 51	137	2·4	20 11	119	0·1
Apr 5	**3830·5**	2 08	148	2·4	19 32	109	0·1
06 Apr 15	**3840·5**	1 25	W 158	−2·5	18 53	E 100	+0·2
25	**3850·5**	0 41	W 169	2·5	18 15	90	0·2
May 5	**3860·5**	23 52	E 179	2·5	17 37	81	0·3
15	**3870·5**	23 08	169	2·5	17 00	72	0·3
06 May 25	**3880·5**	22 24	E 158	−2·5	16 24	E 63	+0·3
Jun 4	**3890·5**	21 41	147	2·4	15 48	54	0·4
14	**3900·5**	20 59	137	2·4	15 13	46	0·4
24	**3910·5**	20 17	127	2·3	14 38	37	0·4
06 Jul 4	**3920·5**	19 37	E 117	−2·2	14 03	E 29	+0·4
14	**3930·5**	18 58	108	2·2	13 29	20	0·4
24	**3940·5**	18 20	98	2·1	12 55	12	0·4
Aug 3	**3950·5**	17 44	90	2·1	12 20	4	0·3
06 Aug 13	**3960·5**	17 08	E 81	−2·0	11 46	W 5	+0·4
23	**3970·5**	16 33	72	1·9	11 12	13	0·4
Sep 2	**3980·5**	15 59	64	1·9	10 38	21	0·5
12	**3990·5**	15 26	56	1·8	10 03	30	0·5
06 Sep 22	**4000·5**	14 54	E 48	−1·8	9 28	W 38	+0·5
Oct 2	**4010·5**	14 22	40	1·8	8 53	47	0·5
12	**4020·5**	13 50	32	1·8	8 17	56	0·6
22	**4030·5**	13 19	24	1·7	7 41	65	0·6
06 Nov 1	**4040·5**	12 49	E 16	−1·7	7 05	W 74	+0·5
11	**4050·5**	12 18	9	1·7	6 28	84	0·5
21	**4060·5**	11 48	E 1	1·7	5 50	94	0·5
Dec 1	**4070·5**	11 18	W 7	1·7	5 11	104	0·4
06 Dec 11	**4080·5**	10 48	W 15	−1·7	4 32	W 114	+0·4
21	**4090·5**	10 17	23	1·7	3 52	124	0·3
31	**4100·5**	9 47	31	1·8	3 11	135	0·3
07 Jan 10	**4110·5**	9 16	W 39	−1·8	2 30	W 145	+0·2

Uranus

JD	Transit	Elong.	Mag.
245	h m	°	
3640·5	22 12	E 154	+ 5·7
3680·5	19 31	113	5·8
3720·5	16 55	73	5·9
3760·5	14 24	34	5·9
3800·5	11 55	W 4	+ 5·9
3840·5	9 25	42	5·9
3880·5	6 53	79	5·9
3920·5	4 16	117	5·8
3960·5	1 36	W 156	+ 5·7
4000·5	22 48	E 163	5·7
4040·5	20 07	123	5·8
4080·5	17 30	E 82	+ 5·9

Neptune

JD	Transit	Elong.	Mag.
3640·5	20 43	E 131	+ 7·9
3680·5	18 05	91	7·9
3720·5	15 31	51	8·0
3760·5	12 59	12	8·0
3800·5	10 27	W 27	+ 8·0
3840·5	7 54	65	7·9
3880·5	5 19	104	7·9
3920·5	2 39	W 143	7·8
3960·5	23 54	E 178	+ 7·8
4000·5	21 13	139	7·9
4040·5	18 35	99	7·9
4080·5	15 59	E 59	+ 7·9

Pluto

JD	Transit	Elong.	Mag.
3640·5	17 01	E 78	+13·9
3680·5	14 27	40	14·0
3720·5	11 56	E 7	14·0
3760·5	9 24	W 40	14·0
3800·5	6 51	W 79	+13·9
3840·5	4 14	118	13·9
3880·5	1 34	W 156	13·9
3920·5	22 48	E 161	13·9
3960·5	20 08	E 124	+13·9
4000·5	17 30	85	13·9
4040·5	14 56	47	14·0
4080·5	12 24	E 10	+14·0

Observability Data

Date	JD	Mercury Transit	Mercury Elong.	Mercury Mag.	Venus Transit	Venus Elong.	Venus Mag.	Mars Transit	Mars Elong.	Mars Mag.
	245	h m	°		h m	°		h m	°	
07 Jan 20	**4120·5**	12 49	E 9	− 1·1	13 36	E 20	− 3·9	10 14	W 27	+ 1·4
30	**4130·5**	13 15	15	1·0	13 44	23	3·9	10 07	30	1·4
Feb 9	**4140·5**	13 21	18	− 0·4	13 51	25	3·9	10 00	33	1·4
19	**4150·5**	12 37	9	+ 2·9	13 57	27	3·9	9 52	35	1·3
07 Mar 1	**4160·5**	11 21	W 12	+ 2·6	14 02	E 29	− 4·0	9 45	W 38	+ 1·3
11	**4170·5**	10 37	24	0·7	14 08	32	4·0	9 37	40	1·2
21	**4180·5**	10 26	28	0·3	14 13	34	4·0	9 28	43	1·2
31	**4190·5**	10 32	26	0·1	14 20	36	4·0	9 19	45	1·1
07 Apr 10	**4200·5**	10 48	W 21	− 0·2	14 29	E 38	− 4·0	9 09	W 47	+ 1·1
20	**4210·5**	11 12	14	0·8	14 39	40	4·1	8 59	49	1·0
30	**4220·5**	11 46	W 4	1·8	14 49	41	4·1	8 48	51	1·0
May 10	**4230·5**	12 31	E 8	1·6	14 59	43	4·1	8 37	53	1·0
07 May 20	**4240·5**	13 14	E 18	− 0·6	15 08	E 44	− 4·2	8 25	W 55	+ 0·9
30	**4250·5**	13 37	23	+ 0·2	15 14	45	4·2	8 13	57	0·9
Jun 9	**4260·5**	13 34	22	1·2	15 16	45	4·3	8 02	59	0·8
19	**4270·5**	12 58	14	2·9	15 14	45	4·4	7 50	62	0·8
07 Jun 29	**4280·5**	11 58	W 4	+ 5·1	15 05	E 44	− 4·4	7 38	W 64	+ 0·7
Jul 9	**4290·5**	11 02	15	2·5	14 49	41	4·5	7 26	66	0·7
19	**4300·5**	10 40	20	+ 0·6	14 24	36	4·5	7 14	69	0·6
29	**4310·5**	10 54	18	− 0·7	13 46	28	4·4	7 02	72	0·5
07 Aug 8	**4320·5**	11 35	W 9	− 1·5	12 53	E 17	− 4·2	6 49	W 75	+ 0·5
18	**4330·5**	12 17	E 3	1·7	11 51	E 8	4·0	6 36	78	0·4
28	**4340·5**	12 47	11	0·8	10 51	W 17	4·2	6 23	81	0·3
Sep 7	**4350·5**	13 07	18	0·3	10 03	28	4·4	6 08	85	0·2
07 Sep 17	**4360·5**	13 19	E 23	− 0·1	9 31	W 36	− 4·5	5 52	W 90	+ 0·1
27	**4370·5**	13 23	26	+ 0·1	9 11	41	4·5	5 34	94	− 0·0
Oct 7	**4380·5**	13 14	25	0·3	9 00	44	4·5	5 14	100	0·2
17	**4390·5**	12 33	15	1·8	8 54	46	4·5	4 50	106	0·3
07 Oct 27	**4400·5**	11 16	W 7	+ 3·4	8 51	W 46	− 4·4	4 23	W 113	− 0·5
Nov 6	**4410·5**	10 35	19	− 0·3	8 51	46	4·3	3 52	121	0·7
16	**4420·5**	10 40	17	0·7	8 52	46	4·3	3 15	131	1·0
26	**4430·5**	11 00	12	0·8	8 55	44	4·2	2 33	142	1·2
07 Dec 6	**4440·5**	11 24	W 6	− 0·9	8 59	W 43	− 4·2	1 44	W 154	− 1·4
16	**4450·5**	11 52	W 2	1·2	9 06	41	4·1	0 50	W 167	1·6
26	**4460·5**	12 23	E 5	1·0	9 14	40	4·1	23 48	E 176	1·6
08 Jan 5	**4470·5**	12 54	11	0·9	9 24	38	4·1	22 52	164	1·4
08 Jan 15	**4480·5**	13 20	E 17	− 0·8	9 36	W 36	− 4·0	22 01	E 151	− 1·1
25	**4490·5**	13 23	18	− 0·2	9 49	34	4·0	21 16	140	0·8
Feb 4	**4500·5**	12 29	E 7	+ 3·5	10 03	31	4·0	20 37	130	0·5
14	**4510·5**	11 10	W 15	1·8	10 17	29	4·0	20 03	121	− 0·2
08 Feb 24	**4520·5**	10 33	W 25	+ 0·4	10 29	W 27	− 3·9	19 33	E 113	+ 0·0
Mar 5	**4530·5**	10 29	27	+ 0·1	10 39	25	3·9	19 06	105	0·3
15	**4540·5**	10 40	25	− 0·0	10 48	22	3·9	18 43	99	0·5
25	**4550·5**	10 58	20	0·3	10 55	20	3·9	18 21	93	0·7
08 Apr 4	**4560·5**	11 22	W 12	− 0·8	11 02	W 17	− 3·9	18 01	E 87	+ 0·9
14	**4570·5**	11 54	W 3	1·8	11 08	15	3·9	17 42	82	1·0
24	**4580·5**	12 34	E 9	1·5	11 14	12	3·9	17 24	78	1·1
May 4	**4590·5**	13 10	E 18	− 0·6	11 21	W 10	− 3·9	17 07	E 73	+ 1·2

Observability Data

Date		JD	Jupiter Transit	Jupiter Elong.	Jupiter Mag.	Saturn Transit	Saturn Elong.	Saturn Mag.
		245	h m	°		h m	°	
07 Jan	20	**4120·5**	8 45	W 48	− 1·8	1 48	W 156	+ 0·1
	30	**4130·5**	8 13	56	1·9	1 06	167	+ 0·0
Feb	9	**4140·5**	7 40	65	1·9	0 23	W 178	− 0·0
	19	**4150·5**	7 07	73	2·0	23 37	E 171	− 0·0
07 Mar	1	**4160·5**	6 33	W 82	− 2·0	22 54	E 160	+ 0·0
	11	**4170·5**	5 57	91	2·1	22 12	149	0·1
	21	**4180·5**	5 21	101	2·2	21 31	139	0·2
	31	**4190·5**	4 43	110	2·3	20 50	129	0·2
07 Apr	10	**4200·5**	4 04	W 120	− 2·3	20 10	E 118	+ 0·3
	20	**4210·5**	3 23	130	2·4	19 30	109	0·3
	30	**4220·5**	2 41	140	2·5	18 51	99	0·4
May	10	**4230·5**	1 58	151	2·5	18 13	90	0·4
07 May	20	**4240·5**	1 14	W 162	− 2·6	17 35	E 80	+ 0·5
	30	**4250·5**	0 30	W 172	2·6	16 58	71	0·5
Jun	9	**4260·5**	23 41	E 177	2·6	16 22	62	0·5
	19	**4270·5**	22 56	166	2·6	15 46	54	0·6
07 Jun	29	**4280·5**	22 12	E 155	− 2·5	15 11	E 45	+ 0·6
Jul	9	**4290·5**	21 29	145	2·5	14 35	37	0·6
	19	**4300·5**	20 47	134	2·5	14 01	28	0·6
	29	**4310·5**	20 06	125	2·4	13 26	20	0·6
07 Aug	8	**4320·5**	19 26	E 115	− 2·3	12 51	E 12	+ 0·6
	18	**4330·5**	18 47	105	2·3	12 17	E 4	0·6
	28	**4340·5**	18 10	96	2·2	11 42	W 5	0·6
Sep	7	**4350·5**	17 34	87	2·1	11 08	14	0·6
07 Sep	17	**4360·5**	16 59	E 79	− 2·1	10 33	W 22	+ 0·7
	27	**4370·5**	16 25	70	2·0	9 59	31	0·7
Oct	7	**4380·5**	15 52	62	2·0	9 24	39	0·8
	17	**4390·5**	15 20	54	1·9	8 48	48	0·8
07 Oct	27	**4400·5**	14 49	E 45	− 1·9	8 12	W 57	+ 0·8
Nov	6	**4410·5**	14 18	37	1·9	7 36	66	0·8
	16	**4420·5**	13 48	29	1·8	6 59	76	0·8
	26	**4430·5**	13 18	22	1·8	6 22	85	0·8
07 Dec	6	**4440·5**	12 48	E 14	− 1·8	5 44	W 95	+ 0·7
	16	**4450·5**	12 19	E 6	1·8	5 05	105	0·7
	26	**4460·5**	11 49	W 2	1·8	4 26	115	0·6
08 Jan	5	**4470·5**	11 20	10	1·8	3 46	126	0·6
08 Jan	15	**4480·5**	10 50	W 18	− 1·8	3 05	W 136	+ 0·5
	25	**4490·5**	10 21	26	1·9	2 24	147	0·4
Feb	4	**4500·5**	9 50	34	1·9	1 42	158	0·3
	14	**4510·5**	9 20	42	1·9	1 00	169	0·3
08 Feb	24	**4520·5**	8 49	W 50	− 1·9	0 18	W 178	+ 0·2
Mar	5	**4530·5**	8 17	59	2·0	23 31	E 169	0·2
	15	**4540·5**	7 45	67	2·0	22 49	159	0·3
	25	**4550·5**	7 11	75	2·1	22 07	148	0·3
08 Apr	4	**4560·5**	6 37	W 84	− 2·2	21 26	E 138	+ 0·4
	14	**4570·5**	6 01	93	2·2	20 45	128	0·5
	24	**4580·5**	5 25	102	2·3	20 05	118	0·5
May	4	**4590·5**	4 47	W 112	− 2·4	19 25	E 108	+ 0·6

	JD	Transit	Elong.	Mag.
Uranus	**245**	h m	°	
	4120·5	14 58	E 43	+ 5·9
	4160·5	12 28	E 4	5·9
	4200·5	9 59	W 33	5·9
	4240·5	7 28	70	5·9
	4280·5	4 52	W 108	+ 5·8
	4320·5	2 12	W 147	5·7
	4360·5	23 25	E 173	5·7
	4400·5	20 43	132	5·8
	4440·5	18 05	E 91	+ 5·8
	4480·5	15 31	52	5·9
	4520·5	13 01	E 13	5·9
	4560·5	10 32	W 25	+ 5·9
Neptune				
	4120·5	13 27	E 19	+ 8·0
	4160·5	10 56	W 20	8·0
	4200·5	8 23	58	7·9
	4240·5	5 48	97	7·9
	4280·5	3 09	W 135	+ 7·9
	4320·5	0 29	W 174	7·8
	4360·5	21 43	E 146	7·8
	4400·5	19 04	106	7·9
	4440·5	16 28	E 66	+ 7·9
	4480·5	13 55	E 27	8·0
	4520·5	11 24	W 13	8·0
	4560·5	8 52	W 51	+ 8·0
Pluto				
	4120·5	9 53	W 32	+ 14·0
	4160·5	7 20	71	14·0
	4200·5	4 43	111	13·9
	4240·5	2 04	W 149	13·9
	4280·5	23 18	E 168	+ 13·9
	4320·5	20 38	131	13·9
	4360·5	18 00	93	14·0
	4400·5	15 25	54	14·0
	4440·5	12 53	E 16	+ 14·0
	4480·5	10 22	W 25	14·0
	4520·5	7 49	64	14·0
	4560·5	5 13	W 103	+ 14·0

Observability Data

Date	JD	Mercury			Venus			Mars		
		Transit	Elong.	Mag.	Transit	Elong.	Mag.	Transit	Elong.	Mag.
	245	h m	°		h m	°		h m	°	
08 May 14	**4600·5**	13 26	E 22	+ 0·4	11 29	W 7	− 3·9	16 50	E 69	+ 1·3
24	**4610·5**	13 12	18	1·9	11 40	4	3·9	16 33	65	1·4
Jun 3	**4620·5**	12 25	E 7	4·4	11 52	W 2	3·9	16 16	61	1·5
13	**4630·5**	11 25	W 9	4·0	12 05	E 1	3·9	16 00	57	1·6
08 Jun 23	**4640·5**	10 43	W 19	+ 1·7	12 20	E 4	− 3·9	15 43	E 53	+ 1·6
Jul 3	**4650·5**	10 32	22	+ 0·3	12 34	7	3·9	15 26	49	1·7
13	**4660·5**	10 52	18	− 0·7	12 47	9	3·9	15 10	46	1·7
23	**4670·5**	11 37	W 8	1·6	12 58	12	3·9	14 53	42	1·7
08 Aug 2	**4680·5**	12 25	E 4	− 1·7	13 07	E 15	− 3·9	14 36	E 39	+ 1·7
12	**4690·5**	12 59	13	0·7	13 15	18	3·9	14 20	36	1·7
22	**4700·5**	13 20	20	− 0·2	13 21	20	3·9	14 03	32	1·7
Sep 1	**4710·5**	13 31	25	+ 0·0	13 26	23	3·9	13 47	29	1·7
08 Sep 11	**4720·5**	13 31	E 27	+ 0·2	13 31	E 25	− 3·9	13 32	E 26	+ 1·7
21	**4730·5**	13 14	24	0·6	13 37	28	3·9	13 17	23	1·7
Oct 1	**4740·5**	12 25	E 12	2·4	13 44	30	3·9	13 03	20	1·6
11	**4750·5**	11 12	W 8	2·9	13 52	33	4·0	12 49	17	1·6
08 Oct 21	**4760·5**	10 39	W 18	− 0·4	14 03	E 35	− 4·0	12 36	E 14	+ 1·6
31	**4770·5**	10 49	15	0·9	14 15	37	4·0	12 25	10	1·5
Nov 10	**4780·5**	11 10	9	0·9	14 29	39	4·0	12 14	8	1·5
20	**4790·5**	11 33	W 3	1·1	14 42	41	4·1	12 04	5	1·4
08 Nov 30	**4800·5**	12 00	E 3	− 1·1	14 55	E 43	− 4·1	11 55	E 2	+ 1·4
Dec 10	**4810·5**	12 28	8	0·8	15 05	44	4·2	11 47	W 1	1·3
20	**4820·5**	12 58	14	0·7	15 12	45	4·2	11 40	4	1·3
30	**4830·5**	13 23	18	0·7	15 16	46	4·3	11 33	7	1·3
09 Jan 9	**4840·5**	13 22	E 18	− 0·1	15 17	E 47	− 4·4	11 27	W 9	+ 1·3
19	**4850·5**	12 19	E 5	+ 4·0	15 14	47	4·5	11 21	12	1·3
29	**4860·5**	10 59	W 17	1·3	15 08	46	4·5	11 14	14	1·3
Feb 8	**4870·5**	10 30	25	0·1	14 56	45	4·6	11 07	17	1·3
09 Feb 18	**4880·5**	10 32	W 26	− 0·0	14 38	E 41	− 4·6	11 00	W 19	+ 1·2
28	**4890·5**	10 48	23	0·1	14 11	35	4·6	10 52	21	1·2
Mar 10	**4900·5**	11 09	18	0·4	13 30	27	4·5	10 43	24	1·2
20	**4910·5**	11 34	11	0·9	12 37	15	4·2	10 33	26	1·2
09 Mar 30	**4920·5**	12 04	W 2	− 1·9	11 36	W 9	− 4·0	10 24	W 28	+ 1·2
Apr 9	**4930·5**	12 39	E 10	1·5	10 41	19	4·3	10 13	30	1·2
19	**4940·5**	13 08	18	− 0·6	10 01	30	4·5	10 02	32	1·2
29	**4950·5**	13 15	20	+ 0·7	9 33	37	4·5	9 51	34	1·2
09 May 9	**4960·5**	12 46	E 13	+ 2·8	9 16	W 42	− 4·5	9 40	W 36	+ 1·2
19	**4970·5**	11 51	W 1	5·9	9 05	44	4·4	9 29	38	1·2
29	**4980·5**	10 57	15	2·7	8 58	46	4·4	9 18	40	1·2
Jun 8	**4990·5**	10 29	22	1·1	8 55	46	4·3	9 07	42	1·1
09 Jun 18	**5000·5**	10 26	W 23	+ 0·2	8 55	W 45	− 4·2	8 56	W 44	+ 1·1
28	**5010·5**	10 50	17	− 0·7	8 57	44	4·2	8 45	47	1·1
Jul 8	**5020·5**	11 36	W 7	1·6	9 01	43	4·1	8 35	49	1·1
18	**5030·5**	12 29	E 5	1·7	9 08	42	4·1	8 25	51	1·1
09 Jul 28	**5040·5**	13 09	E 15	− 0·7	9 17	W 40	− 4·0	8 14	W 54	+ 1·1
Aug 7	**5050·5**	13 32	22	− 0·2	9 27	38	4·0	8 04	57	1·1
17	**5060·5**	13 42	26	+ 0·1	9 38	36	4·0	7 53	60	1·0
27	**5070·5**	13 38	E 27	+ 0·4	9 49	W 33	− 4·0	7 42	W 63	+ 1·0

Observability Data

Date	JD	Jupiter Transit	Jupiter Elong.	Jupiter Mag.	Saturn Transit	Saturn Elong.	Saturn Mag.	JD	Uranus Transit	Uranus Elong.	Uranus Mag.
	245	h m	°		h m	°		**245**	h m	°	
08 May 14	**4600·5**	4 07	W 121	−2·5	18 46	E 98	+0·6	**4600·5**	8 02	W 62	+ 5·9
24	**4610·5**	3 27	131	2·5	18 08	89	0·7	**4640·5**	5 27	99	5·8
Jun 3	**4620·5**	2 45	141	2·6	17 30	80	0·7	**4680·5**	2 48	138	5·8
13	**4630·5**	2 02	152	2·6	16 53	71	0·8	**4720·5**	0 06	W 178	5·7
08 Jun 23	**4640·5**	1 18	W 162	−2·7	16 17	E 62	+0·8	**4760·5**	21 19	E 141	+ 5·8
Jul 3	**4650·5**	0 34	W 173	2·7	15 41	53	0·8	**4800·5**	18 40	101	5·8
13	**4660·5**	23 44	E 176	2·7	15 05	45	0·8	**4840·5**	16 05	61	5·9
23	**4670·5**	23 00	165	2·7	14 30	36	0·8	**4880·5**	13 35	22	5·9
08 Aug 2	**4680·5**	22 16	E 155	−2·7	13 55	E 28	+0·8	**4920·5**	11 06	W 16	+ 5·9
12	**4690·5**	21 32	144	2·6	13 20	19	0·8	**4960·5**	8 36	53	5·9
22	**4700·5**	20 50	134	2·6	12 45	11	0·8	**5000·5**	6 02	90	5·8
Sep 1	**4710·5**	20 09	124	2·5	12 11	3	0·8	**5040·5**	3 24	W 129	+ 5·8
08 Sep 11	**4720·5**	19 30	E 114	−2·4	11 36	W 6	+0·9				
21	**4730·5**	18 52	104	2·4	11 01	14	0·9				
Oct 1	**4740·5**	18 15	95	2·3	10 26	23	1·0				
11	**4750·5**	17 39	86	2·2	9 51	32	1·0			**Neptune**	
08 Oct 21	**4760·5**	17 05	E 77	−2·2	9 16	W 40	+1·1	**4600·5**	6 17	W 89	+ 7·9
31	**4770·5**	16 32	69	2·1	8 41	49	1·1	**4640·5**	3 39	128	7·9
Nov 10	**4780·5**	15 59	60	2·1	8 05	59	1·1	**4680·5**	0 59	W 167	7·8
20	**4790·5**	15 27	52	2·0	7 28	68	1·1	**4720·5**	22 13	E 154	7·8
08 Nov 30	**4800·5**	14 56	E 44	−2·0	6 51	W 77	+1·1	**4760·5**	19 34	E 114	+ 7·9
Dec 10	**4810·5**	14 26	36	2·0	6 14	87	1·1	**4800·5**	16 57	74	7·9
20	**4820·5**	13 56	28	1·9	5 36	97	1·0	**4840·5**	14 24	E 34	8·0
30	**4830·5**	13 26	20	1·9	4 57	107	1·0	**4880·5**	11 52	W 5	8·0
09 Jan 9	**4840·5**	12 56	E 12	−1·9	4 17	W 117	+0·9	**4920·5**	9 20	W 44	+ 8·0
19	**4850·5**	12 27	E 4	1·9	3 37	128	0·9	**4960·5**	6 46	82	7·9
29	**4860·5**	11 57	W 4	1·9	2 57	138	0·8	**5000·5**	4 09	121	7·9
Feb 8	**4870·5**	11 27	12	1·9	2 15	149	0·7	**5040·5**	1 29	W 159	+ 7·8
09 Feb 18	**4880·5**	10 57	W 19	−1·9	1 34	W 160	+0·6				
28	**4890·5**	10 27	27	2·0	0 51	W 170	0·5				
Mar 10	**4900·5**	9 57	35	2·0	0 09	E 177	0·5				
20	**4910·5**	9 26	43	2·0	23 23	168	0·5			**Pluto**	
09 Mar 30	**4920·5**	8 54	W 51	−2·1	22 41	E 157	+0·6	**4600·5**	2 34	W 142	+13·9
Apr 9	**4930·5**	8 22	59	2·1	21 59	147	0·6	**4640·5**	23 49	E 173	13·9
19	**4940·5**	7 49	67	2·2	21 18	136	0·7	**4680·5**	21 08	138	13·9
29	**4950·5**	7 15	75	2·2	20 37	126	0·7	**4720·5**	18 29	100	14·0
09 May 9	**4960·5**	6 41	W 84	−2·3	19 57	E 116	+0·8	**4760·5**	15 54	E 61	+14·0
19	**4970·5**	6 05	92	2·4	19 17	107	0·9	**4800·5**	13 21	E 23	14·1
29	**4980·5**	5 29	101	2·4	18 39	97	0·9	**4840·5**	10 50	W 18	14·1
Jun 8	**4990·5**	4 51	110	2·5	18 00	88	1·0	**4880·5**	8 18	57	14·0
09 Jun 18	**5000·5**	4 12	W 120	−2·6	17 23	E 79	+1·0	**4920·5**	5 43	W 96	+14·0
28	**5010·5**	3 32	130	2·6	16 46	70	1·0	**4960·5**	3 04	135	14·0
Jul 8	**5020·5**	2 50	140	2·7	16 09	61	1·1	**5000·5**	0 23	W 172	13·9
18	**5030·5**	2 08	150	2·8	15 33	52	1·1	**5040·5**	21 38	E 146	+13·9
09 Jul 28	**5040·5**	1 24	W 161	−2·8	14 57	E 44	+1·1				
Aug 7	**5050·5**	0 40	W 171	2·8	14 22	35	1·1				
17	**5060·5**	23 52	E 177	2·9	13 46	27	1·1				
27	**5070·5**	23 07	E 167	−2·8	13 11	E 19	+1·1				

Observability Data

Date	JD	Mercury			Venus			Mars		
		Transit	Elong.	Mag.	Transit	Elong.	Mag.	Transit	Elong.	Mag.
	245	h m	°		h m	°		h m	°	
09 Sep 6	**5080·5**	13 13	E 23	+ 0·9	9 59	W 31	− 4·0	7 30	W 66	+ 0·9
16	**5090·5**	12 18	E 10	3·2	10 08	29	3·9	7 16	70	0·9
26	**5100·5**	11 09	W 10	+ 2·4	10 16	26	3·9	7 02	74	0·8
Oct 6	**5110·5**	10 45	18	− 0·5	10 23	24	3·9	6 46	78	0·7
09 Oct 16	**5120·5**	10 59	W 14	− 1·0	10 29	W 22	− 3·9	6 29	W 83	+ 0·6
26	**5130·5**	11 21	W 7	1·1	10 35	19	3·9	6 10	88	0·5
Nov 5	**5140·5**	11 44	0	1·4	10 42	17	3·9	5 49	94	0·4
15	**5150·5**	12 08	E 6	0·9	10 51	14	3·9	5 26	100	0·2
09 Nov 25	**5160·5**	12 34	E 11	− 0·6	11 01	W 12	− 3·9	4 59	W 107	+ 0·1
Dec 5	**5170·5**	13 01	16	0·5	11 13	9	3·9	4 30	115	− 0·1
15	**5180·5**	13 22	20	− 0·5	11 27	7	3·9	3 56	124	0·4
25	**5190·5**	13 17	18	+ 0·1	11 42	4	3·9	3 17	134	0·6
10 Jan 4	**5200·5**	12 07	E 3	+ 4·6	11 58	W 2	− 3·9	2 33	W 145	− 0·8
14	**5210·5**	10 49	W 18	+ 0·8	12 12	E 1	3·9	1 44	158	1·1
24	**5220·5**	10 28	24	− 0·1	12 26	3	3·9	0 51	W 171	1·2
Feb 3	**5230·5**	10 36	24	0·1	12 37	5	3·9	23 49	E 173	1·2
10 Feb 13	**5240·5**	10 55	W 20	− 0·2	12 46	E 8	− 3·9	22 55	E 160	− 1·0
23	**5250·5**	11 19	15	0·4	12 54	10	3·9	22 05	147	0·8
Mar 5	**5260·5**	11 45	W 9	1·0	13 00	13	3·9	21 20	136	0·5
15	**5270·5**	12 15	E 2	1·8	13 06	15	3·9	20 40	126	− 0·2
10 Mar 25	**5280·5**	12 46	E 10	− 1·4	13 12	E 17	− 3·9	20 06	E 117	+ 0·0
Apr 4	**5290·5**	13 09	18	− 0·6	13 19	20	3·9	19 35	109	0·2
14	**5300·5**	13 04	18	+ 1·0	13 28	22	3·9	19 07	102	0·4
24	**5310·5**	12 21	8	4·0	13 38	25	3·9	18 42	96	0·6
10 May 4	**5320·5**	11 22	W 8	+ 4·0	13 49	E 27	− 3·9	18 18	E 90	+ 0·8
14	**5330·5**	10 38	20	1·7	14 02	30	3·9	17 56	85	0·9
24	**5340·5**	10 21	25	0·7	14 16	32	4·0	17 35	80	1·0
Jun 3	**5350·5**	10 25	23	0·0	14 28	34	4·0	17 15	75	1·1
10 Jun 13	**5360·5**	10 49	W 17	− 0·7	14 40	E 37	− 4·0	16 55	E 71	+ 1·2
23	**5370·5**	11 35	W 7	1·7	14 48	39	4·0	16 36	67	1·3
Jul 3	**5380·5**	12 31	E 6	1·7	14 55	41	4·1	16 18	63	1·4
13	**5390·5**	13 15	16	0·7	14 58	42	4·1	16 00	59	1·4
10 Jul 23	**5400·5**	13 41	E 23	− 0·1	14 59	E 44	− 4·1	15 42	E 56	+ 1·5
Aug 2	**5410·5**	13 50	27	+ 0·2	14 58	45	4·2	15 24	52	1·5
12	**5420·5**	13 42	27	0·6	14 55	46	4·3	15 08	49	1·5
22	**5430·5**	13 10	21	1·5	14 50	46	4·3	14 51	46	1·5
10 Sep 1	**5440·5**	12 08	E 6	+ 4·2	14 43	E 45	− 4·4	14 36	E 42	+ 1·5
11	**5450·5**	11 06	W 13	+ 1·9	14 32	44	4·5	14 21	39	1·5
21	**5460·5**	10 49	18	− 0·5	14 15	41	4·6	14 07	36	1·5
Oct 1	**5470·5**	11 09	12	1·1	13 50	35	4·6	13 54	33	1·5
10 Oct 11	**5480·5**	11 34	W 5	− 1·3	13 13	E 26	− 4·5	13 42	E 30	+ 1·5
21	**5490·5**	11 57	E 3	1·2	12 21	E 14	4·2	13 31	27	1·5
31	**5500·5**	12 19	9	0·7	11 21	W 6	4·0	13 22	24	1·4
Nov 10	**5510·5**	12 41	14	0·4	10 26	18	4·4	13 13	22	1·4
10 Nov 20	**5520·5**	13 04	E 19	− 0·4	9 45	W 30	− 4·6	13 05	E 19	+ 1·4
30	**5530·5**	13 21	21	− 0·4	9 17	38	4·7	12 58	16	1·3
Dec 10	**5540·5**	13 09	18	+ 0·3	9 00	43	4·6	12 52	14	1·3
20	**5550·5**	11 53	E 2	+ 5·0	8 51	W 45	− 4·6	12 46	E 11	+ 1·3

Observability Data

Date		JD	Jupiter Transit	Jupiter Elong.	Jupiter Mag.	Saturn Transit	Saturn Elong.	Saturn Mag.
		245	h m	°		h m	°	
09 Sep	6	**5080·5**	22 23	E 156	−2·8	12 37	E 10	+1·1
	16	**5090·5**	21 40	145	2·8	12 02	E 2	1·1
	26	**5100·5**	20 58	135	2·7	11 27	W 7	1·1
Oct	6	**5110·5**	20 18	124	2·6	10 52	16	1·1
09 Oct	16	**5120·5**	19 38	E 114	−2·6	10 17	W 24	+1·1
	26	**5130·5**	19 00	105	2·5	9 42	33	1·1
Nov	5	**5140·5**	18 23	95	2·4	9 07	42	1·1
	15	**5150·5**	17 47	86	2·4	8 31	51	1·1
09 Nov	25	**5160·5**	17 13	E 77	−2·3	7 55	W 60	+1·0
Dec	5	**5170·5**	16 39	69	2·2	7 19	70	1·0
	15	**5180·5**	16 06	60	2·2	6 42	79	1·0
	25	**5190·5**	15 34	52	2·1	6 04	89	0·9
10 Jan	4	**5200·5**	15 02	E 43	−2·1	5 26	W 99	+0·9
	14	**5210·5**	14 31	35	2·1	4 47	109	0·8
	24	**5220·5**	14 00	27	2·0	4 07	119	0·8
Feb	3	**5230·5**	13 30	20	2·0	3 27	130	0·7
10 Feb	13	**5240·5**	12 59	E 12	−2·0	2 46	W 140	+0·7
	23	**5250·5**	12 29	E 4	2·0	2 05	151	0·6
Mar	5	**5260·5**	11 59	W 4	2·0	1 23	162	0·6
	15	**5270·5**	11 28	11	2·0	0 41	W 172	0·5
10 Mar	25	**5280·5**	10 58	W 19	−2·0	23 54	E 176	+0·5
Apr	4	**5290·5**	10 27	26	2·0	23 12	166	0·6
	14	**5300·5**	9 56	34	2·1	22 30	155	0·7
	24	**5310·5**	9 25	41	2·1	21 49	145	0·8
10 May	4	**5320·5**	8 53	W 49	−2·1	21 08	E 135	+0·8
	14	**5330·5**	8 20	57	2·2	20 27	125	0·9
	24	**5340·5**	7 47	65	2·2	19 47	115	1·0
Jun	3	**5350·5**	7 14	73	2·3	19 07	105	1·0
10 Jun	13	**5360·5**	6 39	W 81	−2·4	18 29	E 96	+1·1
	23	**5370·5**	6 04	90	2·4	17 50	87	1·1
Jul	3	**5380·5**	5 28	98	2·5	17 13	78	1·1
	13	**5390·5**	4 50	107	2·6	16 36	69	1·1
10 Jul	23	**5400·5**	4 11	W 117	−2·6	15 59	E 60	+1·1
Aug	2	**5410·5**	3 32	126	2·7	15 23	51	1·1
	12	**5420·5**	2 51	136	2·8	14 47	43	1·1
	22	**5430·5**	2 09	147	2·8	14 12	34	1·1
10 Sep	1	**5440·5**	1 26	W 157	−2·9	13 36	E 26	+1·0
	11	**5450·5**	0 42	168	2·9	13 01	17	1·0
	21	**5460·5**	23 53	W 178	2·9	12 26	9	0·9
Oct	1	**5470·5**	23 09	E 169	2·9	11 52	2	0·9
10 Oct	11	**5480·5**	22 25	E 158	−2·9	11 17	W 9	+0·9
	21	**5490·5**	21 42	147	2·8	10 42	17	0·9
	31	**5500·5**	21 00	137	2·8	10 07	26	0·9
Nov	10	**5510·5**	20 19	126	2·7	9 32	35	0·9
10 Nov	20	**5520·5**	19 40	E 116	−2·6	8 56	W 44	+0·9
	30	**5530·5**	19 01	106	2·6	8 21	53	0·9
Dec	10	**5540·5**	18 24	96	2·5	7 44	63	0·9
	20	**5550·5**	17 48	E 87	−2·4	7 08	W 72	+0·8

Uranus

JD	Transit	Elong.	Mag.
245	h m	°	
5080·5	0 42	W 168	+ 5·7
5120·5	21 55	E 151	5·7
5160·5	19 15	110	5·8
5200·5	16 39	70	5·9
5240·5	14 08	E 31	+ 5·9
5280·5	11 39	W 7	5·9
5320·5	9 09	44	5·9
5360·5	6 37	82	5·9
5400·5	4 00	W 120	+ 5·8
5440·5	1 19	W 159	5·7
5480·5	22 32	E 160	5·7
5520·5	19 50	E 119	+ 5·8

Neptune

JD	Transit	Elong.	Mag.
5080·5	22 43	E 161	+ 7·8
5120·5	20 03	121	7·9
5160·5	17 26	81	7·9
5200·5	14 52	41	8·0
5240·5	12 20	E 2	+ 8·0
5280·5	9 49	W 37	8·0
5320·5	7 15	75	7·9
5360·5	4 38	113	7·9
5400·5	1 59	W 152	+ 7·8
5440·5	23 13	E 169	7·8
5480·5	20 33	129	7·9
5520·5	17 55	E 88	+ 7·9

Pluto

JD	Transit	Elong.	Mag.
5080·5	18 59	E 107	+14·0
5120·5	16 23	68	14·0
5160·5	13 50	E 30	14·1
5200·5	11 19	W 11	14·1
5240·5	8 47	W 50	+14·1
5280·5	6 12	89	14·0
5320·5	3 34	128	14·0
5360·5	0 54	W 166	14·0
5400·5	22 08	E 153	+14·0
5440·5	19 28	114	14·0
5480·5	16 52	75	14·1
5520·5	14 18	E 37	+14·1

Observability Data

Date	JD	Mercury Transit	Mercury Elong.	Mercury Mag.	Venus Transit	Venus Elong.	Venus Mag.	Mars Transit	Mars Elong.	Mars Mag.
	245	h m	°		h m	°		h m	°	
10 Dec 30	**5560·5**	10 41	W 19	+ 0·5	8 47	W 47	− 4·5	12 40	E 9	+ 1·2
11 Jan 9	**5570·5**	10 27	23	− 0·2	8 48	47	4·4	12 34	6	1·2
19	**5580·5**	10 39	22	0·2	8 52	47	4·4	12 27	4	1·1
29	**5590·5**	11 01	18	0·3	8 59	46	4·3	12 20	2	1·1
11 Feb 8	**5600·5**	11 27	W 13	− 0·5	9 08	W 45	− 4·2	12 12	W 1	+ 1·1
18	**5610·5**	11 56	W 6	1·1	9 18	43	4·2	12 04	3	1·1
28	**5620·5**	12 25	E 3	1·6	9 28	42	4·1	11 54	5	1·1
Mar 10	**5630·5**	12 54	11	1·3	9 38	40	4·1	11 45	7	1·1
11 Mar 20	**5640·5**	13 12	E 18	− 0·6	9 47	W 38	− 4·0	11 34	W 9	+ 1·2
30	**5650·5**	12 55	16	+ 1·4	9 54	36	4·0	11 24	11	1·2
Apr 9	**5660·5**	12 00	E 3	5·3	10 00	34	4·0	11 13	14	1·2
19	**5670·5**	11 01	W 15	2·6	10 06	31	3·9	11 02	16	1·2
11 Apr 29	**5680·5**	10 28	W 24	+ 1·1	10 11	W 29	− 3·9	10 51	W 18	+ 1·2
May 9	**5690·5**	10 19	27	+ 0·4	10 17	27	3·9	10 40	20	1·3
19	**5700·5**	10 27	24	− 0·1	10 23	24	3·9	10 29	22	1·3
29	**5710·5**	10 52	17	0·7	10 31	21	3·9	10 18	24	1·3
11 Jun 8	**5720·5**	11 35	W 6	− 1·7	10 41	W 19	− 3·9	10 08	W 26	+ 1·3
18	**5730·5**	12 31	E 6	1·6	10 52	16	3·9	9 58	29	1·4
28	**5740·5**	13 18	17	0·7	11 05	14	3·9	9 49	31	1·4
Jul 8	**5750·5**	13 46	24	− 0·1	11 19	11	3·9	9 39	34	1·4
11 Jul 18	**5760·5**	13 54	E 27	+ 0·4	11 33	W 8	− 3·9	9 29	W 36	+ 1·4
28	**5770·5**	13 41	25	0·9	11 46	5	3·9	9 20	39	1·4
Aug 7	**5780·5**	13 01	17	2·2	11 57	W 3	3·9	9 09	42	1·4
17	**5790·5**	11 55	5	4·8	12 07	E 1	3·9	8 59	45	1·4
11 Aug 27	**5800·5**	11 01	W 15	+ 1·5	12 15	E 3	− 3·9	8 48	W 48	+ 1·4
Sep 6	**5810·5**	10 53	18	− 0·6	12 22	6	3·9	8 36	51	1·4
16	**5820·5**	11 18	11	1·2	12 28	8	3·9	8 23	55	1·4
26	**5830·5**	11 47	W 3	1·6	12 34	11	3·9	8 09	58	1·3
11 Oct 6	**5840·5**	12 10	E 5	− 1·1	12 41	E 14	− 3·9	7 54	W 62	+ 1·3
16	**5850·5**	12 31	12	0·5	12 48	16	3·9	7 38	66	1·2
26	**5860·5**	12 50	17	0·3	12 58	19	3·9	7 21	71	1·1
Nov 5	**5870·5**	13 07	21	0·3	13 10	21	3·9	7 02	75	1·1
11 Nov 15	**5880·5**	13 18	E 23	− 0·2	13 23	E 23	− 3·9	6 43	W 80	+ 1·0
25	**5890·5**	12 59	E 18	+ 0·6	13 38	26	3·9	6 22	86	0·8
Dec 5	**5900·5**	11 40	W 2	5·0	13 53	28	3·9	5 59	91	0·7
15	**5910·5**	10 35	19	0·2	14 08	30	4·0	5 35	98	0·5
11 Dec 25	**5920·5**	10 27	W 22	− 0·4	14 20	E 32	− 4·0	5 08	W 105	+ 0·3
12 Jan 4	**5930·5**	10 42	19	0·4	14 31	35	4·0	4 38	112	+ 0·1
14	**5940·5**	11 07	15	0·4	14 38	37	4·0	4 06	121	− 0·1
24	**5950·5**	11 35	10	0·7	14 44	38	4·1	3 29	130	0·3
12 Feb 3	**5960·5**	12 05	W 4	− 1·2	14 48	E 40	− 4·1	2 48	W 141	− 0·6
13	**5970·5**	12 35	E 5	1·4	14 51	42	4·1	2 01	153	0·9
23	**5980·5**	13 03	13	1·2	14 53	43	4·2	1 11	W 166	1·1
Mar 4	**5990·5**	13 16	18	− 0·5	14 55	45	4·2	0 17	E 176	1·2
12 Mar 14	**6000·5**	12 48	E 13	+ 1·9	14 57	E 46	− 4·3	23 18	E 165	− 1·1
24	**6010·5**	11 43	W 5	4·6	14 58	46	4·3	22 26	153	0·9
Apr 3	**6020·5**	10 48	20	1·7	14 58	46	4·4	21 39	141	0·7
13	**6030·5**	10 24	W 27	+ 0·7	14 56	E 45	− 4·5	20 57	E 130	− 0·4

Observability Data

Date	JD	Jupiter Transit	Jupiter Elong.	Jupiter Mag.	Saturn Transit	Saturn Elong.	Saturn Mag.
	245	h m	°		h m	°	
10 Dec 30	**5560·5**	17 13	E 78	−2·4	6 30	W 82	+0·8
11 Jan 9	**5570·5**	16 38	69	2·3	5 53	91	0·8
19	**5580·5**	16 05	61	2·2	5 14	101	0·7
29	**5590·5**	15 32	52	2·2	4 35	111	0·7
11 Feb 8	**5600·5**	15 00	E 44	−2·1	3 55	W 122	+0·6
18	**5610·5**	14 29	36	2·1	3 15	132	0·6
28	**5620·5**	13 58	29	2·1	2 34	143	0·5
Mar 10	**5630·5**	13 27	21	2·1	1 52	153	0·5
11 Mar 20	**5640·5**	12 56	E 13	−2·1	1 11	W 164	+0·4
30	**5650·5**	12 26	E 6	2·1	0 29	W 174	0·4
Apr 9	**5660·5**	11 55	W 2	2·1	23 42	E 174	0·4
19	**5670·5**	11 25	9	2·1	23 00	164	0·4
11 Apr 29	**5680·5**	10 55	W 17	−2·1	22 18	E 154	+0·5
May 9	**5690·5**	10 24	24	2·1	21 37	143	0·6
19	**5700·5**	9 53	31	2·1	20 56	133	0·7
29	**5710·5**	9 22	39	2·1	20 15	123	0·7
11 Jun 8	**5720·5**	8 51	W 46	−2·1	19 35	E 114	+0·8
18	**5730·5**	8 19	54	2·2	18 56	104	0·8
28	**5740·5**	7 46	62	2·2	18 17	95	0·9
Jul 8	**5750·5**	7 13	70	2·3	17 39	85	0·9
11 Jul 18	**5760·5**	6 39	W 78	−2·3	17 01	E 76	+0·9
28	**5770·5**	6 04	86	2·4	16 24	67	0·9
Aug 7	**5780·5**	5 28	95	2·5	15 48	59	0·9
17	**5790·5**	4 51	104	2·5	15 12	50	0·9
11 Aug 27	**5800·5**	4 13	W 113	−2·6	14 36	E 41	+0·9
Sep 6	**5810·5**	3 34	123	2·7	14 00	33	0·9
16	**5820·5**	2 53	133	2·8	13 25	24	0·8
26	**5830·5**	2 11	143	2·8	12 50	16	0·8
11 Oct 6	**5840·5**	1 28	W 154	−2·9	12 15	E 7	+0·8
16	**5850·5**	0 44	165	2·9	11 41	W 3	0·7
26	**5860·5**	23 55	W 176	2·9	11 06	11	0·7
Nov 5	**5870·5**	23 11	E 172	2·9	10 31	19	0·7
11 Nov 15	**5880·5**	22 26	E 161	−2·9	9 56	W 28	+0·8
25	**5890·5**	21 43	150	2·8	9 21	37	0·8
Dec 5	**5900·5**	21 00	139	2·8	8 45	46	0·8
15	**5910·5**	20 19	128	2·7	8 09	56	0·7
11 Dec 25	**5920·5**	19 39	E 118	−2·6	7 33	W 65	+0·7
12 Jan 4	**5930·5**	19 00	107	2·6	6 56	75	0·7
14	**5940·5**	18 23	98	2·5	6 19	84	0·7
24	**5950·5**	17 46	88	2·4	5 41	94	0·6
12 Feb 3	**5960·5**	17 11	E 79	−2·3	5 02	W 104	+0·6
13	**5970·5**	16 37	71	2·3	4 23	114	0·5
23	**5980·5**	16 04	62	2·2	3 43	124	0·5
Mar 4	**5990·5**	15 31	54	2·2	3 02	135	0·4
12 Mar 14	**6000·5**	15 00	E 46	−2·1	2 21	W 145	+0·4
24	**6010·5**	14 28	38	2·1	1 40	156	0·3
Apr 3	**6020·5**	13 58	30	2·1	0 58	166	0·3
13	**6030·5**	13 27	E 23	−2·0	0 16	W 176	+0·2

Uranus

JD	Transit	Elong.	Mag.
245	h m	°	
5560·5	17 14	E 79	+ 5·9
5600·5	14 42	39	5·9
5640·5	12 12	E 2	5·9
5680·5	9 43	W 36	5·9
5720·5	7 11	W 73	+ 5·9
5760·5	4 35	111	5·8
5800·5	1 55	W 150	5·7
5840·5	23 08	E 170	5·7
5880·5	20 26	E 129	+ 5·8
5920·5	17 48	88	5·8
5960·5	15 15	48	5·9
6000·5	12 45	E 10	+ 5·9

Neptune

JD	Transit	Elong.	Mag.
5560·5	15 21	E 49	+ 7·9
5600·5	12 48	E 9	8·0
5640·5	10 17	W 30	8·0
5680·5	7 44	68	7·9
5720·5	5 08	W 106	+ 7·9
5760·5	2 29	W 145	7·8
5800·5	23 43	E 176	7·8
5840·5	21 02	136	7·8
5880·5	18 24	E 96	+ 7·9
5920·5	15 49	56	7·9
5960·5	13 17	E 16	8·0
6000·5	10 45	W 22	+ 8·0

Pluto

JD	Transit	Elong.	Mag.
5560·5	11 47	W 5	+ 14·1
5600·5	9 15	42	14·1
5640·5	6 41	82	14·1
5680·5	4 04	121	14·0
5720·5	1 24	W 160	+ 14·0
5760·5	22 38	E 160	14·0
5800·5	19 58	121	14·0
5840·5	17 21	83	14·1
5880·5	14 47	E 44	+ 14·1
5920·5	12 15	E 6	14·1
5960·5	9 44	W 35	14·1
6000·5	7 10	W 74	+ 14·1

Observability Data

Date	JD	Mercury Transit	Mercury Elong.	Mercury Mag.	Venus Transit	Venus Elong.	Venus Mag.	Mars Transit	Mars Elong.	Mars Mag.
	245	h m	°		h m	°		h m	°	
12 Apr 23	**6040·5**	10 22	W 27	+ 0·3	14 49	E 43	− 4·5	20 19	E 121	− 0·2
May 3	**6050·5**	10 33	23	− 0·1	14 34	38	4·5	19 45	113	+ 0·0
13	**6060·5**	10 57	16	0·7	14 06	31	4·5	19 15	106	0·2
23	**6070·5**	11 37	W 5	1·7	13 21	21	4·3	18 47	99	0·4
12 Jun 2	**6080·5**	12 29	E 7	− 1·6	12 22	E 6	− 3·9	18 22	E 93	+ 0·5
12	**6090·5**	13 17	17	− 0·7	11 18	W 9	4·0	17 58	88	0·7
22	**6100·5**	13 46	24	+ 0·0	10 23	23	4·3	17 36	83	0·8
Jul 2	**6110·5**	13 52	26	0·6	9 44	33	4·4	17 14	79	0·9
12 Jul 12	**6120·5**	13 33	E 22	+ 1·4	9 19	W 39	− 4·5	16 55	E 74	+ 0·9
22	**6130·5**	12 44	E 12	3·3	9 04	43	4·4	16 36	71	1·0
Aug 1	**6140·5**	11 37	W 7	4·2	8 56	45	4·4	16 18	67	1·1
11	**6150·5**	10 54	17	1·1	8 54	46	4·3	16 01	63	1·1
12 Aug 21	**6160·5**	10 55	W 18	− 0·6	8 55	W 46	− 4·3	15 45	E 60	+ 1·2
31	**6170·5**	11 26	10	1·3	9 00	45	4·2	15 30	57	1·2
Sep 10	**6180·5**	11 59	W 2	1·7	9 05	44	4·2	15 16	54	1·2
20	**6190·5**	12 25	E 8	0·9	9 11	43	4·1	15 04	51	1·2
12 Sep 30	**6200·5**	12 44	E 14	− 0·4	9 18	W 41	− 4·1	14 52	E 48	+ 1·2
Oct 10	**6210·5**	13 00	20	0·2	9 23	39	4·1	14 42	45	1·2
20	**6220·5**	13 13	23	0·1	9 29	37	4·0	14 33	42	1·2
30	**6230·5**	13 16	24	− 0·1	9 34	35	4·0	14 26	40	1·2
12 Nov 9	**6240·5**	12 49	E 17	+ 0·9	9 40	W 33	− 4·0	14 19	E 37	+ 1·2
19	**6250·5**	11 29	W 3	4·5	9 46	31	4·0	14 12	34	1·2
29	**6260·5**	10 33	19	+ 0·0	9 54	29	4·0	14 06	32	1·2
Dec 9	**6270·5**	10 30	20	− 0·5	10 04	26	4·0	14 00	29	1·2
12 Dec 19	**6280·5**	10 47	W 16	− 0·5	10 15	W 24	− 3·9	13 54	E 27	+ 1·2
29	**6290·5**	11 13	12	0·6	10 29	22	3·9	13 48	25	1·2
13 Jan 8	**6300·5**	11 42	6	0·8	10 43	19	3·9	13 41	22	1·2
18	**6310·5**	12 13	W 2	1·3	10 59	17	3·9	13 33	20	1·2
13 Jan 28	**6320·5**	12 44	E 7	− 1·2	11 13	W 15	− 3·9	13 24	E 18	+ 1·2
Feb 7	**6330·5**	13 11	14	1·1	11 26	12	3·9	13 15	16	1·2
17	**6340·5**	13 20	18	− 0·4	11 38	10	3·9	13 05	13	1·2
27	**6350·5**	12 42	11	+ 2·4	11 47	8	3·9	12 55	11	1·2
13 Mar 9	**6360·5**	11 30	W 9	+ 3·3	11 55	W 5	− 3·9	12 44	E 9	+ 1·2
19	**6370·5**	10 41	23	1·0	12 01	W 3	3·9	12 33	7	1·2
29	**6380·5**	10 25	28	0·4	12 08	E 1	3·9	12 22	5	1·2
Apr 8	**6390·5**	10 28	27	0·1	12 14	3	3·9	12 11	E 2	1·2
13 Apr 18	**6400·5**	10 41	W 22	− 0·2	12 21	E 5	− 3·9	11 59	0	+ 1·2
28	**6410·5**	11 05	15	0·7	12 29	8	3·9	11 49	W 2	1·2
May 8	**6420·5**	11 41	W 5	1·7	12 40	10	3·9	11 38	5	1·3
18	**6430·5**	12 29	E 7	1·6	12 52	13	3·9	11 28	7	1·3
13 May 28	**6440·5**	13 15	E 18	− 0·7	13 05	E 16	− 3·9	11 18	W 9	+ 1·4
Jun 7	**6450·5**	13 42	23	+ 0·1	13 20	18	3·9	11 08	12	1·4
17	**6460·5**	13 43	24	0·9	13 33	21	3·9	10 59	14	1·5
27	**6470·5**	13 15	18	2·2	13 46	24	3·9	10 49	17	1·5
13 Jul 7	**6480·5**	12 18	E 6	+ 4·6	13 57	E 26	− 3·9	10 40	W 20	+ 1·5
17	**6490·5**	11 16	W 12	3·1	14 05	29	3·9	10 30	22	1·6
27	**6500·5**	10 45	19	+ 0·8	14 12	31	3·9	10 20	25	1·6
Aug 6	**6510·5**	10 54	W 18	− 0·6	14 17	E 34	− 4·0	10 09	W 28	+ 1·6

Observability Data

Date	JD	Jupiter Transit	Jupiter Elong.	Jupiter Mag.	Saturn Transit	Saturn Elong.	Saturn Mag.
	245	h m	°		h m	°	
12 Apr 23	**6040·5**	12 57	E 15	−2·0	23 29	E 172	+0·3
May 3	**6050·5**	12 27	8	2·0	22 47	162	0·3
13	**6060·5**	11 57	E 1	2·0	22 05	151	0·4
23	**6070·5**	11 28	W 7	2·0	21 24	141	0·5
12 Jun 2	**6080·5**	10 58	W 14	−2·0	20 43	E 131	+0·5
12	**6090·5**	10 28	21	2·0	20 02	121	0·6
22	**6100·5**	9 58	29	2·0	19 23	112	0·7
Jul 2	**6110·5**	9 28	36	2·0	18 43	102	0·7
12 Jul 12	**6120·5**	8 57	W 44	−2·1	18 05	E 93	+0·7
22	**6130·5**	8 26	51	2·1	17 27	84	0·8
Aug 1	**6140·5**	7 54	59	2·2	16 49	75	0·8
11	**6150·5**	7 21	67	2·2	16 12	66	0·8
12 Aug 21	**6160·5**	6 48	W 75	−2·3	15 36	E 57	+0·8
31	**6170·5**	6 14	84	2·3	15 00	48	0·8
Sep 10	**6180·5**	5 38	92	2·4	14 24	39	0·8
20	**6190·5**	5 02	101	2·5	13 49	31	0·7
12 Sep 30	**6200·5**	4 24	W 111	−2·5	13 14	E 22	+0·7
Oct 10	**6210·5**	3 44	121	2·6	12 39	14	0·7
20	**6220·5**	3 04	131	2·7	12 04	E 5	0·6
30	**6230·5**	2 21	142	2·7	11 29	W 5	0·6
12 Nov 9	**6240·5**	1 38	W 153	−2·8	10 55	W 13	+0·6
19	**6250·5**	0 54	164	2·8	10 20	22	0·6
29	**6260·5**	0 09	W 175	2·8	9 45	31	0·7
Dec 9	**6270·5**	23 19	E 173	2·8	9 10	40	0·7
12 Dec 19	**6280·5**	22 34	E 162	−2·8	8 34	W 49	+0·7
29	**6290·5**	21 50	150	2·7	7 58	58	0·6
13 Jan 8	**6300·5**	21 07	139	2·7	7 22	68	0·6
18	**6310·5**	20 26	129	2·6	6 45	77	0·6
13 Jan 28	**6320·5**	19 45	E 118	−2·5	6 07	W 87	+0·6
Feb 7	**6330·5**	19 06	108	2·5	5 29	97	0·5
17	**6340·5**	18 29	98	2·4	4 50	107	0·5
27	**6350·5**	17 53	89	2·3	4 11	117	0·4
13 Mar 9	**6360·5**	17 18	E 80	−2·2	3 31	W 127	+0·4
19	**6370·5**	16 44	71	2·2	2 50	138	0·3
29	**6380·5**	16 11	63	2·1	2 08	148	0·3
Apr 8	**6390·5**	15 39	55	2·1	1 27	158	0·2
13 Apr 18	**6400·5**	15 08	E 47	−2·0	0 45	W 169	+0·2
28	**6410·5**	14 37	39	2·0	0 02	W 177	0·1
May 8	**6420·5**	14 06	31	2·0	23 16	E 170	0·2
18	**6430·5**	13 36	24	1·9	22 34	159	0·2
13 May 28	**6440·5**	13 07	E 17	−1·9	21 52	E 149	+0·3
Jun 7	**6450·5**	12 37	9	1·9	21 11	139	0·4
17	**6460·5**	12 08	E 2	1·9	20 30	129	0·4
27	**6470·5**	11 38	W 5	1·9	19 50	119	0·5
13 Jul 7	**6480·5**	11 09	W 13	−1·9	19 10	E 110	+0·5
17	**6490·5**	10 39	20	1·9	18 31	100	0·6
27	**6500·5**	10 09	27	1·9	17 52	91	0·6
Aug 6	**6510·5**	9 39	W 35	−1·9	17 14	E 82	+0·7

JD	Transit	Elong.	Mag.
Uranus			
245	h m	°	
6040·5	10 16	W 27	+ 5·9
6080·5	7 45	64	5·9
6120·5	5 10	102	5·8
6160·5	2 31	W 140	5·8
6200·5	23 45	E 179	+ 5·7
6240·5	21 02	138	5·8
6280·5	18 23	97	5·8
6320·5	15 49	57	5·9
6360·5	13 19	E 19	+ 5·9
6400·5	10 50	W 19	5·9
6440·5	8 19	55	5·9
6480·5	5 46	W 93	+ 5·8
Neptune			
6040·5	8 12	W 61	+ 7·9
6080·5	5 37	99	7·9
6120·5	2 58	137	7·8
6160·5	0 17	W 176	7·8
6200·5	21 32	E 144	+ 7·8
6240·5	18 53	103	7·9
6280·5	16 18	63	7·9
6320·5	13 45	24	8·0
6360·5	11 13	W 15	+ 8·0
6400·5	8 41	53	7·9
6440·5	6 06	91	7·9
6480·5	3 28	W 130	+ 7·9
Pluto			
6040·5	4 34	W 114	+14·1
6080·5	1 54	W 153	14·0
6120·5	23 09	E 167	14·0
6160·5	20 28	129	14·0
6200·5	17 50	E 90	+14·1
6240·5	15 16	51	14·1
6280·5	12 44	E 12	14·2
6320·5	10 12	W 28	14·2
6360·5	7 39	W 67	+14·1
6400·5	5 03	107	14·1
6440·5	2 24	W 146	14·1
6480·5	23 39	E 174	+14·0

Observability Data

Date	JD	Mercury Transit	Mercury Elong.	Mercury Mag.	Venus Transit	Venus Elong.	Venus Mag.	Mars Transit	Mars Elong.	Mars Mag.
	245	h m	°		h m	°		h m	°	
13 Aug 16	**6520·5**	11 31	W 9	− 1·4	14 20	E 36	− 4·0	9 58	W 31	+ 1·6
26	**6530·5**	12 10	E 2	1·8	14 23	38	4·0	9 46	34	1·6
Sep 5	**6540·5**	12 39	10	0·8	14 27	40	4·0	9 33	38	1·6
15	**6550·5**	12 58	17	0·3	14 31	42	4·1	9 19	41	1·6
13 Sep 25	**6560·5**	13 12	E 22	− 0·1	14 35	E 44	− 4·1	9 05	W 45	+ 1·6
Oct 5	**6570·5**	13 19	25	− 0·0	14 41	45	4·2	8 50	48	1·6
15	**6580·5**	13 15	24	+ 0·1	14 48	46	4·3	8 34	52	1·6
25	**6590·5**	12 40	16	1·4	14 55	47	4·3	8 17	56	1·5
13 Nov 4	**6600·5**	11 21	W 5	+ 4·0	15 01	E 47	− 4·4	8 00	W 60	+ 1·5
14	**6610·5**	10 34	19	− 0·2	15 03	46	4·5	7 42	65	1·4
24	**6620·5**	10 35	18	0·7	15 00	45	4·6	7 23	69	1·3
Dec 4	**6630·5**	10 54	14	0·7	14 50	41	4·7	7 03	74	1·2
13 Dec 14	**6640·5**	11 19	W 8	− 0·8	14 28	E 36	− 4·7	6 43	W 79	+ 1·1
24	**6650·5**	11 48	W 3	1·0	13 50	27	4·6	6 21	84	1·0
14 Jan 3	**6660·5**	12 19	E 3	1·1	12 56	E 14	4·3	5 59	90	0·8
13	**6670·5**	12 50	9	1·0	11 52	W 6	4·1	5 35	96	0·6
14 Jan 23	**6680·5**	13 17	E 16	− 0·9	10 51	W 19	− 4·4	5 09	W 103	+ 0·4
Feb 2	**6690·5**	13 23	18	− 0·4	10 06	30	4·6	4 42	110	+ 0·2
12	**6700·5**	12 36	E 9	+ 3·0	9 37	38	4·6	4 11	117	− 0·0
22	**6710·5**	11 18	W 13	2·4	9 20	42	4·6	3 37	126	0·3
14 Mar 4	**6720·5**	10 36	W 25	+ 0·6	9 12	W 45	− 4·5	2 59	W 136	− 0·6
14	**6730·5**	10 27	28	0·2	9 09	46	4·5	2 16	147	0·9
24	**6740·5**	10 35	26	+ 0·0	9 09	47	4·4	1 29	159	1·1
Apr 3	**6750·5**	10 51	21	− 0·2	9 11	46	4·3	0 37	W 172	1·4
14 Apr 13	**6760·5**	11 15	W 14	− 0·8	9 13	W 45	− 4·2	23 38	E 174	− 1·5
23	**6770·5**	11 48	W 4	1·8	9 16	44	4·2	22 45	161	1·3
May 3	**6780·5**	12 30	E 8	1·6	9 19	43	4·1	21 55	148	1·1
13	**6790·5**	13 11	18	− 0·7	9 22	41	4·0	21 10	137	0·9
14 May 23	**6800·5**	13 33	E 23	+ 0·3	9 26	W 39	− 4·0	20 29	E 127	− 0·7
Jun 2	**6810·5**	13 26	21	1·4	9 31	37	4·0	19 53	119	0·5
12	**6820·5**	12 47	E 12	3·4	9 37	35	3·9	19 21	111	0·3
22	**6830·5**	11 46	W 5	4·9	9 45	32	3·9	18 52	105	− 0·1
14 Jul 2	**6840·5**	10 55	W 16	+ 2·2	9 55	W 30	− 3·9	18 27	E 99	+ 0·0
12	**6850·5**	10 36	21	+ 0·5	10 07	28	3·9	18 03	93	0·2
22	**6860·5**	10 52	18	− 0·6	10 20	25	3·9	17 42	89	0·3
Aug 1	**6870·5**	11 34	W 9	1·5	10 33	22	3·9	17 23	84	0·4
14 Aug 11	**6880·5**	12 19	E 3	− 1·7	10 45	W 20	− 3·9	17 05	E 81	+ 0·5
21	**6890·5**	12 52	12	0·8	10 57	17	3·9	16 49	77	0·6
31	**6900·5**	13 12	19	0·3	11 07	15	3·9	16 35	73	0·6
Sep 10	**6910·5**	13 24	24	− 0·0	11 15	12	3·9	16 23	70	0·7
14 Sep 20	**6920·5**	13 27	E 26	+ 0·1	11 22	W 9	− 3·9	16 11	E 67	+ 0·7
30	**6930·5**	13 15	25	0·4	11 29	7	3·9	16 01	64	0·8
Oct 10	**6940·5**	12 32	E 14	1·9	11 35	4	3·9	15 53	61	0·8
20	**6950·5**	11 16	W 7	3·4	11 42	W 2	3·9	15 45	59	0·9
14 Oct 30	**6960·5**	10 37	W 18	− 0·3	11 50	E 1	− 3·9	15 38	E 56	+ 0·9
Nov 9	**6970·5**	10 43	17	0·8	12 00	4	3·9	15 32	54	0·9
19	**6980·5**	11 03	11	0·8	12 12	6	3·9	15 25	51	1·0
29	**6990·5**	11 27	W 5	− 1·0	12 26	E 9	− 3·9	15 19	E 49	+ 1·0

Observability Data

Date	JD	Jupiter Transit	Jupiter Elong.	Jupiter Mag.	Saturn Transit	Saturn Elong.	Saturn Mag.
	245	h m	°		h m	°	
13 Aug 16	**6520·5**	9 09	W 42	− 2·0	16 37	E 73	+ 0·7
26	**6530·5**	8 37	50	2·0	16 00	64	0·7
Sep 5	**6540·5**	8 06	58	2·0	15 24	55	0·7
15	**6550·5**	7 33	66	2·1	14 48	46	0·7
13 Sep 25	**6560·5**	6 59	W 74	− 2·1	14 13	E 37	+ 0·7
Oct 5	**6570·5**	6 25	83	2·2	13 38	29	0·6
15	**6580·5**	5 49	92	2·3	13 03	20	0·6
25	**6590·5**	5 12	102	2·3	12 28	11	0·6
13 Nov 4	**6600·5**	4 34	W 111	− 2·4	11 53	E 3	+ 0·5
14	**6610·5**	3 55	121	2·5	11 19	W 7	0·5
24	**6620·5**	3 14	132	2·5	10 44	16	0·6
Dec 4	**6630·5**	2 31	143	2·6	10 09	25	0·6
13 Dec 14	**6640·5**	1 48	W 154	− 2·6	9 34	W 34	+ 0·6
24	**6650·5**	1 04	165	2·7	8 59	43	0·6
14 Jan 3	**6660·5**	0 19	W 177	2·7	8 23	52	0·6
13	**6670·5**	23 29	E 172	2·7	7 47	61	0·6
14 Jan 23	**6680·5**	22 44	E 160	− 2·7	7 11	W 71	+ 0·6
Feb 2	**6690·5**	22 00	149	2·6	6 34	80	0·5
12	**6700·5**	21 17	138	2·5	5 56	90	0·5
22	**6710·5**	20 36	127	2·5	5 18	100	0·5
14 Mar 4	**6720·5**	19 56	E 117	− 2·4	4 39	W 110	+ 0·4
14	**6730·5**	19 17	107	2·3	3 59	120	0·4
24	**6740·5**	18 39	98	2·3	3 19	130	0·3
Apr 3	**6750·5**	18 03	88	2·2	2 38	141	0·2
14 Apr 13	**6760·5**	17 28	E 80	− 2·1	1 56	W 151	+ 0·2
23	**6770·5**	16 54	71	2·1	1 14	161	0·1
May 3	**6780·5**	16 21	63	2·0	0 32	W 172	0·1
13	**6790·5**	15 49	55	2·0	23 45	E 177	0·1
14 May 23	**6800·5**	15 17	E 47	− 1·9	23 03	E 167	+ 0·1
Jun 2	**6810·5**	14 46	39	1·9	22 21	157	0·2
12	**6820·5**	14 15	32	1·8	21 39	147	0·3
22	**6830·5**	13 44	24	1·8	20 58	137	0·3
14 Jul 2	**6840·5**	13 14	E 17	− 1·8	20 17	E 127	+ 0·4
12	**6850·5**	12 44	9	1·8	19 37	117	0·4
22	**6860·5**	12 14	E 2	1·8	18 58	107	0·5
Aug 1	**6870·5**	11 44	W 5	1·8	18 19	98	0·5
14 Aug 11	**6880·5**	11 14	W 13	− 1·8	17 40	E 89	+ 0·6
21	**6890·5**	10 43	20	1·8	17 03	80	0·6
31	**6900·5**	10 12	28	1·8	16 26	70	0·6
Sep 10	**6910·5**	9 41	35	1·8	15 49	61	0·6
14 Sep 20	**6920·5**	9 10	W 43	− 1·9	15 13	E 53	+ 0·6
30	**6930·5**	8 38	51	1·9	14 37	44	0·6
Oct 10	**6940·5**	8 05	59	1·9	14 02	35	0·6
20	**6950·5**	7 32	68	2·0	13 27	26	0·6
14 Oct 30	**6960·5**	6 57	W 76	− 2·0	12 52	E 17	+ 0·5
Nov 9	**6970·5**	6 22	85	2·1	12 18	E 9	0·5
19	**6980·5**	5 46	95	2·2	11 43	W 2	0·5
29	**6990·5**	5 08	W 104	− 2·2	11 09	W 10	+ 0·5

Planet	JD	Transit	Elong.	Mag.
	245	h m	°	
Uranus	**6520·5**	3 07	W 131	+ 5·8
	6560·5	0 25	W 171	5·7
	6600·5	21 38	E 148	5·7
	6640·5	18 58	106	5·8
	6680·5	16 23	E 66	+ 5·9
	6720·5	13 52	E 28	5·9
	6760·5	11 23	W 10	5·9
	6800·5	8 53	47	5·9
	6840·5	6 20	W 84	+ 5·8
	6880·5	3 43	122	5·8
	6920·5	1 02	W 162	5·7
	6960·5	22 15	E 157	+ 5·7
Neptune	**6520·5**	0 47	W 169	+ 7·8
	6560·5	22 02	E 151	7·8
	6600·5	19 22	111	7·9
	6640·5	16 46	71	7·9
	6680·5	14 13	E 31	+ 8·0
	6720·5	11 41	W 8	8·0
	6760·5	9 09	46	7·9
	6800·5	6 35	84	7·9
	6840·5	3 57	W 123	+ 7·9
	6880·5	1 17	W 162	7·8
	6920·5	22 32	E 159	7·8
	6960·5	19 52	E 118	+ 7·9
Pluto	**6520·5**	20 58	E 136	+ 14·1
	6560·5	18 19	97	14·1
	6600·5	15 44	58	14·2
	6640·5	13 12	19	14·2
	6680·5	10 41	W 21	+ 14·2
	6720·5	8 08	60	14·2
	6760·5	5 33	99	14·1
	6800·5	2 54	138	14·1
	6840·5	0 13	W 177	+ 14·1
	6880·5	21 28	E 143	14·1
	6920·5	18 49	104	14·1
	6960·5	16 13	E 65	+ 14·2

Observability Data

Date	JD	Mercury Transit	Mercury Elong.	Mercury Mag.	Venus Transit	Venus Elong.	Venus Mag.	Mars Transit	Mars Elong.	Mars Mag.
	245	h m	°		h m	°		h m	°	
14 Dec 9	**7000·5**	11 54	E 1	− 1·2	12 41	E 11	− 3·9	15 12	E 46	+ 1·0
19	**7010·5**	12 24	6	0·9	12 56	13	3·9	15 04	44	1·1
29	**7020·5**	12 55	12	0·8	13 11	16	3·9	14 56	42	1·1
15 Jan 8	**7030·5**	13 21	17	0·8	13 25	18	3·9	14 47	39	1·1
15 Jan 18	**7040·5**	13 24	E 18	− 0·2	13 36	E 20	− 3·9	14 37	E 37	+ 1·2
28	**7050·5**	12 28	E 6	+ 3·5	13 45	23	3·9	14 27	34	1·2
Feb 7	**7060·5**	11 07	W 16	1·7	13 52	25	3·9	14 16	32	1·2
17	**7070·5**	10 32	25	0·3	13 58	27	4·0	14 05	30	1·2
15 Feb 27	**7080·5**	10 30	W 27	+ 0·1	14 03	E 30	− 4·0	13 54	E 27	+ 1·3
Mar 9	**7090·5**	10 43	24	− 0·1	14 08	32	4·0	13 42	25	1·3
19	**7100·5**	11 02	19	0·3	14 14	34	4·0	13 31	22	1·3
29	**7110·5**	11 26	12	0·8	14 21	36	4·0	13 19	20	1·4
15 Apr 8	**7120·5**	11 57	W 3	− 1·8	14 29	E 38	− 4·0	13 08	E 18	+ 1·4
18	**7130·5**	12 34	E 9	1·5	14 38	40	4·1	12 57	15	1·4
28	**7140·5**	13 08	18	− 0·7	14 49	41	4·1	12 47	13	1·4
May 8	**7150·5**	13 21	21	+ 0·5	14 59	43	4·1	12 36	10	1·5
15 May 18	**7160·5**	13 03	E 16	+ 2·2	15 08	E 44	− 4·2	12 26	E 7	+ 1·5
28	**7170·5**	12 12	E 4	5·1	15 14	45	4·2	12 17	5	1·5
Jun 7	**7180·5**	11 14	W 11	3·5	15 16	45	4·3	12 07	E 2	1·5
17	**7190·5**	10 37	20	1·5	15 14	45	4·4	11 57	W 1	1·5
15 Jun 27	**7200·5**	10 29	W 22	+ 0·3	15 05	E 44	− 4·4	11 48	W 3	+ 1·6
Jul 7	**7210·5**	10 50	18	− 0·7	14 49	41	4·5	11 37	6	1·6
17	**7220·5**	11 35	W 8	1·6	14 24	35	4·5	11 27	9	1·6
27	**7230·5**	12 26	E 4	1·7	13 45	27	4·4	11 16	12	1·7
15 Aug 6	**7240·5**	13 03	E 14	− 0·7	12 52	E 16	− 4·2	11 04	W 15	+ 1·7
16	**7250·5**	13 25	21	− 0·2	11 50	W 8	3·9	10 52	18	1·7
26	**7260·5**	13 35	25	+ 0·1	10 50	17	4·2	10 39	22	1·8
Sep 5	**7270·5**	13 34	27	0·3	10 02	29	4·4	10 25	25	1·8
15 Sep 15	**7280·5**	13 15	E 24	+ 0·7	9 31	W 36	− 4·5	10 10	W 28	+ 1·8
25	**7290·5**	12 24	E 12	2·6	9 11	42	4·5	9 55	32	1·8
Oct 5	**7300·5**	11 12	W 9	+ 2·9	9 00	44	4·5	9 40	35	1·8
15	**7310·5**	10 42	18	− 0·4	8 54	46	4·5	9 24	39	1·8
15 Oct 25	**7320·5**	10 52	W 15	− 0·9	8 52	W 46	− 4·4	9 07	W 43	+ 1·7
Nov 4	**7330·5**	11 14	9	1·0	8 51	46	4·3	8 50	47	1·7
14	**7340·5**	11 37	W 2	1·2	8 53	45	4·3	8 33	51	1·6
24	**7350·5**	12 02	E 4	1·0	8 56	44	4·2	8 16	55	1·6
15 Dec 4	**7360·5**	12 30	E 9	− 0·7	9 00	W 43	− 4·2	7 58	W 59	+ 1·5
14	**7370·5**	12 59	14	0·6	9 06	41	4·1	7 40	63	1·4
24	**7380·5**	13 22	19	0·6	9 14	39	4·1	7 22	68	1·3
16 Jan 3	**7390·5**	13 21	18	− 0·1	9 24	38	4·1	7 04	72	1·2
16 Jan 13	**7400·5**	12 17	E 5	+ 4·1	9 36	W 36	− 4·0	6 45	W 77	+ 1·1
23	**7410·5**	10 56	W 17	1·1	9 49	33	4·0	6 26	82	1·0
Feb 2	**7420·5**	10 29	25	+ 0·1	10 03	31	4·0	6 07	87	0·8
12	**7430·5**	10 33	25	− 0·1	10 16	29	4·0	5 46	92	0·6
16 Feb 22	**7440·5**	10 50	W 22	− 0·1	10 29	W 27	− 3·9	5 25	W 98	+ 0·4
Mar 3	**7450·5**	11 12	17	0·4	10 40	25	3·9	5 02	104	+ 0·2
13	**7460·5**	11 38	10	0·9	10 49	22	3·9	4 38	110	− 0·0
23	**7470·5**	12 07	W 2	− 1·8	10 56	W 20	− 3·9	4 10	W 117	− 0·3

Observability Data

Date	JD	Jupiter Transit	Jupiter Elong.	Jupiter Mag.	Saturn Transit	Saturn Elong.	Saturn Mag.
	245	h m	°		h m	°	
14 Dec 9	**7000·5**	4 30	W 114	− 2·3	10 34	W 19	+ 0·5
19	**7010·5**	3 50	125	2·4	9 59	28	0·5
29	**7020·5**	3 08	135	2·4	9 24	37	0·5
15 Jan 8	**7030·5**	2 26	146	2·5	8 49	46	0·6
15 Jan 18	**7040·5**	1 43	W 157	− 2·5	8 14	W 55	+ 0·5
28	**7050·5**	0 59	W 169	2·6	7 37	65	0·5
Feb 7	**7060·5**	0 14	E 179	2·6	7 01	74	0·5
17	**7070·5**	23 25	168	2·6	6 24	84	0·5
15 Feb 27	**7080·5**	22 41	E 157	− 2·5	5 46	W 93	+ 0·5
Mar 9	**7090·5**	21 58	146	2·5	5 07	103	0·4
19	**7100·5**	21 15	135	2·4	4 28	113	0·4
29	**7110·5**	20 34	125	2·3	3 48	123	0·3
15 Apr 8	**7120·5**	19 54	E 115	− 2·3	3 07	W 133	+ 0·3
18	**7130·5**	19 16	105	2·2	2 26	144	0·2
28	**7140·5**	18 38	96	2·1	1 44	154	0·1
May 8	**7150·5**	18 02	87	2·1	1 02	164	0·1
15 May 18	**7160·5**	17 26	E 78	− 2·0	0 20	W 174	+ 0·0
28	**7170·5**	16 52	70	1·9	23 33	E 174	0·0
Jun 7	**7180·5**	16 18	62	1·9	22 51	164	0·1
17	**7190·5**	15 45	54	1·8	22 09	154	0·2
15 Jun 27	**7200·5**	15 13	E 46	− 1·8	21 27	E 144	+ 0·2
Jul 7	**7210·5**	14 41	38	1·8	20 46	134	0·3
17	**7220·5**	14 09	31	1·7	20 06	124	0·3
27	**7230·5**	13 38	23	1·7	19 25	115	0·4
15 Aug 6	**7240·5**	13 07	E 16	− 1·7	18 46	E 105	+ 0·4
16	**7250·5**	12 35	8	1·7	18 07	96	0·5
26	**7260·5**	12 04	E 1	1·7	17 29	86	0·5
Sep 5	**7270·5**	11 33	W 7	1·7	16 52	77	0·5
15 Sep 15	**7280·5**	11 02	W 14	− 1·7	16 15	E 68	+ 0·6
25	**7290·5**	10 31	22	1·7	15 39	59	0·6
Oct 5	**7300·5**	9 59	30	1·7	15 03	50	0·6
15	**7310·5**	9 27	38	1·8	14 27	41	0·6
15 Oct 25	**7320·5**	8 55	W 46	− 1·8	13 52	E 32	+ 0·5
Nov 4	**7330·5**	8 22	54	1·8	13 17	23	0·5
14	**7340·5**	7 48	62	1·9	12 43	14	0·5
24	**7350·5**	7 14	71	1·9	12 08	6	0·5
15 Dec 4	**7360·5**	6 39	W 80	− 2·0	11 34	W 4	+ 0·4
14	**7370·5**	6 03	89	2·0	11 00	13	0·5
24	**7380·5**	5 26	99	2·1	10 25	22	0·5
16 Jan 3	**7390·5**	4 48	109	2·2	9 51	31	0·5
16 Jan 13	**7400·5**	4 09	W 119	− 2·2	9 16	W 40	+ 0·5
23	**7410·5**	3 28	129	2·3	8 40	49	0·5
Feb 2	**7420·5**	2 47	140	2·4	8 05	58	0·5
12	**7430·5**	2 04	151	2·4	7 28	68	0·5
16 Feb 22	**7440·5**	1 21	W 162	− 2·5	6 52	W 77	+ 0·5
Mar 3	**7450·5**	0 37	W 174	2·5	6 14	87	0·5
13	**7460·5**	23 49	E 175	2·5	5 36	97	0·4
23	**7470·5**	23 05	E 164	− 2·5	4 57	W 106	+ 0·4

Uranus

JD	Transit	Elong.	Mag.
245	h m	°	
7000·5	19 34	E 116	+ 5·8
7040·5	16 57	75	5·9
7080·5	14 26	E 36	5·9
7120·5	11 56	W 1	5·9
7160·5	9 27	W 38	+ 5·9
7200·5	6 55	75	5·9
7240·5	4 19	113	5·8
7280·5	1 38	W 152	5·7
7320·5	22 51	E 167	+ 5·7
7360·5	20 09	125	5·8
7400·5	17 32	84	5·8
7440·5	14 59	E 45	+ 5·9

Neptune

JD	Transit	Elong.	Mag.
7000·5	17 15	E 78	+ 7·9
7040·5	14 41	E 38	8·0
7080·5	12 09	W 1	8·0
7120·5	9 37	39	8·0
7160·5	7 03	W 77	+ 7·9
7200·5	4 27	115	7·9
7240·5	1 47	W 154	7·8
7280·5	23 02	E 166	7·8
7320·5	20 21	E 126	+ 7·9
7360·5	17 44	86	7·9
7400·5	15 09	46	7·9
7440·5	12 37	E 6	+ 8·0

Pluto

JD	Transit	Elong.	Mag.
7000·5	13 40	E 26	+ 14·2
7040·5	11 09	W 14	14·2
7080·5	8 36	53	14·2
7120·5	6 02	92	14·2
7160·5	3 23	W 131	+ 14·1
7200·5	0 43	W 170	14·1
7240·5	21 57	E 150	14·1
7280·5	19 18	111	14·2
7320·5	16 42	E 72	+ 14·2
7360·5	14 08	E 33	14·2
7400·5	11 37	W 7	14·3
7440·5	9 05	W 46	+ 14·2

Observability Data

Date	JD	Mercury			Venus			Mars		
		Transit	Elong.	Mag.	Transit	Elong.	Mag.	Transit	Elong.	Mag.
	245	h m	°		h m	°		h m	°	
16 Apr 2	**7480·5**	12 41	E 10	− 1·5	11 03	W 17	− 3·9	3 40	W 125	− 0·6
12	**7490·5**	13 08	18	− 0·6	11 09	15	3·9	3 05	134	0·9
22	**7500·5**	13 10	19	+ 0·8	11 15	12	3·9	2 26	144	1·2
May 2	**7510·5**	12 37	11	3·2	11 22	10	3·9	1 41	154	1·5
16 May 12	**7520·5**	11 39	W 4	+ 5·2	11 30	W 7	− 3·9	0 52	W 166	− 1·8
22	**7530·5**	10 49	17	2·3	11 40	4	3·9	23 53	W 179	2·1
Jun 1	**7540·5**	10 25	24	1·0	11 52	W 2	3·9	22 59	E 167	2·0
11	**7550·5**	10 25	23	0·1	12 05	E 1	3·9	22 07	155	1·8
16 Jun 21	**7560·5**	10 48	W 18	− 0·7	12 19	E 4	− 3·9	21 20	E 143	− 1·6
Jul 1	**7570·5**	11 34	W 7	1·6	12 34	7	3·9	20 39	133	1·4
11	**7580·5**	12 29	E 5	1·7	12 47	9	3·9	20 03	125	1·2
21	**7590·5**	13 11	15	0·7	12 58	12	3·9	19 32	117	1·0
16 Jul 31	**7600·5**	13 36	E 22	− 0·2	13 08	E 15	− 3·9	19 06	E 111	− 0·8
Aug 10	**7610·5**	13 45	26	+ 0·2	13 16	18	3·9	18 44	105	0·6
20	**7620·5**	13 40	27	0·4	13 22	20	3·9	18 25	100	0·5
30	**7630·5**	13 14	22	1·1	13 27	23	3·9	18 08	96	0·3
16 Sep 9	**7640·5**	12 16	E 9	+ 3·5	13 32	E 25	− 3·9	17 54	E 92	− 0·2
19	**7650·5**	11 09	W 11	+ 2·3	13 38	28	3·9	17 42	88	− 0·1
29	**7660·5**	10 47	18	− 0·5	13 44	30	3·9	17 31	85	+ 0·0
Oct 9	**7670·5**	11 02	13	1·0	13 53	33	4·0	17 22	82	0·1
16 Oct 19	**7680·5**	11 26	W 6	− 1·2	14 03	E 35	− 4·0	17 13	E 79	+ 0·2
29	**7690·5**	11 49	E 1	1·3	14 15	37	4·0	17 04	76	0·3
Nov 8	**7700·5**	12 12	7	0·8	14 28	39	4·0	16 55	73	0·4
18	**7710·5**	12 36	12	0·5	14 42	41	4·1	16 47	70	0·5
16 Nov 28	**7720·5**	13 01	E 17	− 0·5	14 54	E 43	− 4·1	16 37	E 68	+ 0·6
Dec 8	**7730·5**	13 22	20	− 0·5	15 05	44	4·2	16 27	65	0·7
18	**7740·5**	13 15	18	+ 0·1	15 12	45	4·2	16 17	62	0·8
28	**7750·5**	12 04	3	4·7	15 17	46	4·3	16 06	60	0·9
17 Jan 7	**7760·5**	10 47	W 18	+ 0·7	15 18	E 47	− 4·4	15 55	E 57	+ 0·9
17	**7770·5**	10 27	24	− 0·1	15 15	47	4·5	15 43	55	1·0
27	**7780·5**	10 36	23	0·2	15 08	46	4·5	15 31	52	1·1
Feb 6	**7790·5**	10 57	19	0·2	14 56	44	4·6	15 18	49	1·2
17 Feb 16	**7800·5**	11 22	W 14	− 0·5	14 38	E 41	− 4·6	15 06	E 46	+ 1·2
26	**7810·5**	11 49	W 8	1·0	14 10	35	4·6	14 54	44	1·3
Mar 8	**7820·5**	12 18	E 2	1·8	13 29	26	4·5	14 42	41	1·3
18	**7830·5**	12 49	11	1·4	12 35	14	4·2	14 30	38	1·4
17 Mar 28	**7840·5**	13 10	E 18	− 0·6	11 34	W 9	− 4·1	14 18	E 35	+ 1·5
Apr 7	**7850·5**	13 01	17	+ 1·1	10 40	20	4·3	14 07	33	1·5
17	**7860·5**	12 14	E 6	4·5	10 00	30	4·5	13 56	30	1·5
27	**7870·5**	11 14	W 11	3·5	9 33	38	4·5	13 46	27	1·6
17 May 7	**7880·5**	10 34	W 22	+ 1·5	9 16	W 42	− 4·5	13 35	E 24	+ 1·6
17	**7890·5**	10 19	26	0·6	9 05	45	4·4	13 25	21	1·6
27	**7900·5**	10 25	24	+ 0·0	8 59	46	4·4	13 15	18	1·7
Jun 6	**7910·5**	10 49	17	− 0·7	8 55	46	4·3	13 05	15	1·7
17 Jun 16	**7920·5**	11 33	W 7	− 1·6	8 55	W 45	− 4·2	12 54	E 12	+ 1·7
26	**7930·5**	12 29	E 6	1·7	8 57	44	4·2	12 44	10	1·7
Jul 6	**7940·5**	13 16	16	0·7	9 02	43	4·1	12 32	7	1·7
16	**7950·5**	13 43	E 23	− 0·1	9 08	W 42	− 4·1	12 21	E 4	+ 1·7

Observability Data

Date	JD	Jupiter Transit	Jupiter Elong.	Jupiter Mag.	Saturn Transit	Saturn Elong.	Saturn Mag.
	245	h m	°		h m	°	
16 Apr 2	**7480·5**	22 21	E 153	− 2·4	4 18	W 116	+ 0·3
12	**7490·5**	21 38	142	2·4	3 38	126	0·3
22	**7500·5**	20 57	131	2·3	2 57	136	0·2
May 2	**7510·5**	20 16	121	2·3	2 15	147	0·2
16 May 12	**7520·5**	19 36	E 112	− 2·2	1 33	W 157	+ 0·1
22	**7530·5**	18 58	102	2·1	0 51	167	0·1
Jun 1	**7540·5**	18 21	93	2·1	0 09	W 177	0·0
11	**7550·5**	17 44	84	2·0	23 22	E 172	0·0
16 Jun 21	**7560·5**	17 09	E 76	− 1·9	22 40	E 162	+ 0·1
Jul 1	**7570·5**	16 34	67	1·9	21 58	152	0·2
11	**7580·5**	16 00	59	1·8	21 16	141	0·2
21	**7590·5**	15 27	51	1·8	20 35	132	0·3
16 Jul 31	**7600·5**	14 54	E 44	− 1·7	19 55	E 122	+ 0·3
Aug 10	**7610·5**	14 21	36	1·7	19 15	112	0·4
20	**7620·5**	13 49	28	1·7	18 36	102	0·4
30	**7630·5**	13 17	21	1·7	17 57	93	0·5
16 Sep 9	**7640·5**	12 46	E 13	− 1·7	17 19	E 84	+ 0·5
19	**7650·5**	12 14	E 6	1·7	16 42	74	0·5
29	**7660·5**	11 43	W 2	1·7	16 05	65	0·5
Oct 9	**7670·5**	11 11	10	1·7	15 29	56	0·5
16 Oct 19	**7680·5**	10 40	W 18	− 1·7	14 54	E 47	+ 0·5
29	**7690·5**	10 08	25	1·7	14 18	38	0·5
Nov 8	**7700·5**	9 36	33	1·7	13 44	29	0·5
18	**7710·5**	9 04	42	1·7	13 09	20	0·5
16 Nov 28	**7720·5**	8 31	W 50	− 1·8	12 35	E 11	+ 0·5
Dec 8	**7730·5**	7 58	58	1·8	12 00	E 3	0·4
18	**7740·5**	7 24	67	1·9	11 26	W 7	0·5
28	**7750·5**	6 50	76	1·9	10 52	16	0·5
17 Jan 7	**7760·5**	6 14	W 85	− 2·0	10 17	W 25	+ 0·5
17	**7770·5**	5 38	95	2·0	9 43	34	0·5
27	**7780·5**	5 00	104	2·1	9 07	43	0·5
Feb 6	**7790·5**	4 22	114	2·2	8 32	52	0·5
17 Feb 16	**7800·5**	3 42	W 124	− 2·3	7 56	W 62	+ 0·5
26	**7810·5**	3 01	135	2·3	7 20	71	0·5
Mar 8	**7820·5**	2 19	146	2·4	6 43	81	0·5
18	**7830·5**	1 36	157	2·4	6 05	90	0·5
17 Mar 28	**7840·5**	0 52	W 168	− 2·4	5 27	W 100	+ 0·4
Apr 7	**7850·5**	0 08	W 178	2·5	4 48	110	0·4
17	**7860·5**	23 20	E 170	2·5	4 08	119	0·3
27	**7870·5**	22 36	159	2·4	3 28	129	0·3
17 May 7	**7880·5**	21 53	E 148	− 2·4	2 47	W 140	+ 0·2
17	**7890·5**	21 11	138	2·3	2 05	150	0·2
27	**7900·5**	20 29	128	2·3	1 23	160	0·1
Jun 6	**7910·5**	19 49	118	2·2	0 41	W 170	0·0
17 Jun 16	**7920·5**	19 10	E 108	− 2·1	23 54	E 179	+ 0·0
26	**7930·5**	18 32	99	2·1	23 12	169	0·1
Jul 6	**7940·5**	17 55	90	2·0	22 29	159	0·1
16	**7950·5**	17 19	E 81	− 2·0	21 47	E 149	+ 0·2

Uranus

JD	Transit	Elong.	Mag.
245	h m	°	
7480·5	12 30	E 7	+ 5·9
7520·5	10 01	W 29	5·9
7560·5	7 30	66	5·9
7600·5	4 55	104	5·8
7640·5	2 15	W 143	+ 5·7
7680·5	23 28	E 176	5·7
7720·5	20 46	135	5·7
7760·5	18 07	94	5·8
7800·5	15 33	E 54	+ 5·9
7840·5	13 03	E 16	5·9
7880·5	10 35	W 21	5·9
7920·5	8 04	W 57	+ 5·9

Neptune

JD	Transit	Elong.	Mag.
7480·5	10 05	W 32	+ 8·0
7520·5	7 32	70	7·9
7560·5	4 56	108	7·9
7600·5	2 16	W 147	7·8
7640·5	23 31	E 174	+ 7·8
7680·5	20 51	133	7·8
7720·5	18 12	93	7·9
7760·5	15 37	53	7·9
7800·5	13 05	E 14	+ 8·0
7840·5	10 33	W 25	8·0
7880·5	8 00	63	7·9
7920·5	5 25	W 101	+ 7·9

Pluto

JD	Transit	Elong.	Mag.
7480·5	6 31	W 85	+ 14·2
7520·5	3 53	124	14·2
7560·5	1 13	W 163	14·1
7600·5	22 27	E 157	14·1
7640·5	19 47	E 118	+ 14·2
7680·5	17 10	79	14·2
7720·5	14 37	40	14·3
7760·5	12 05	1	14·3
7800·5	9 33	W 39	+ 14·3
7840·5	6 59	78	14·2
7880·5	4 22	117	14·2
7920·5	1 42	W 156	+ 14·2

Observability Data

Date	JD	Mercury Transit	Mercury Elong.	Mercury Mag.	Venus Transit	Venus Elong.	Venus Mag.	Mars Transit	Mars Elong.	Mars Mag.
	245	h m	°		h m	°		h m	°	
17 Jul 26	**7960·5**	13 52	E 27	+ 0·3	9 17	W 40	− 4·0	12 08	E 1	+ 1·7
Aug 5	**7970·5**	13 43	26	0·7	9 27	38	4·0	11 55	W 3	1·7
15	**7980·5**	13 08	20	1·7	9 38	36	4·0	11 42	6	1·8
25	**7990·5**	12 05	6	4·4	9 49	33	4·0	11 27	9	1·8
17 Sep 4	**8000·5**	11 05	W 13	+ 1·9	9 59	W 31	− 4·0	11 13	W 13	+ 1·8
14	**8010·5**	10 51	18	− 0·5	10 09	29	3·9	10 57	16	1·8
24	**8020·5**	11 12	12	1·2	10 17	26	3·9	10 42	19	1·8
Oct 4	**8030·5**	11 38	W 4	1·4	10 23	24	3·9	10 26	23	1·8
17 Oct 14	**8040·5**	12 02	E 4	− 1·2	10 30	W 21	− 3·9	10 10	W 26	+ 1·8
24	**8050·5**	12 23	10	0·6	10 36	19	3·9	9 54	30	1·8
Nov 3	**8060·5**	12 44	15	0·4	10 43	16	3·9	9 37	34	1·8
13	**8070·5**	13 05	20	0·3	10 51	14	3·9	9 21	37	1·8
17 Nov 23	**8080·5**	13 19	E 22	− 0·3	11 01	W 11	− 3·9	9 05	W 41	+ 1·7
Dec 3	**8090·5**	13 07	18	+ 0·4	11 13	9	3·9	8 49	45	1·7
13	**8100·5**	11 50	E 2	5·1	11 27	7	3·9	8 34	49	1·6
23	**8110·5**	10 39	W 19	0·4	11 42	4	3·9	8 18	53	1·5
18 Jan 2	**8120·5**	10 26	W 23	− 0·3	11 57	W 2	− 3·9	8 04	W 57	+ 1·5
12	**8130·5**	10 40	21	0·3	12 12	E 1	3·9	7 49	61	1·4
22	**8140·5**	11 03	17	0·3	12 26	3	3·9	7 35	65	1·3
Feb 1	**8150·5**	11 30	12	0·6	12 37	6	3·9	7 21	69	1·2
18 Feb 11	**8160·5**	11 59	W 5	− 1·1	12 47	E 8	− 3·9	7 07	W 73	+ 1·1
21	**8170·5**	12 29	E 3	1·5	12 55	10	3·9	6 53	77	0·9
Mar 3	**8180·5**	12 57	12	1·3	13 01	13	3·9	6 40	81	0·8
13	**8190·5**	13 13	18	− 0·6	13 07	15	3·9	6 26	85	0·6
18 Mar 23	**8200·5**	12 53	E 15	+ 1·5	13 13	E 18	− 3·9	6 11	W 89	+ 0·4
Apr 2	**8210·5**	11 54	W 3	5·3	13 20	20	3·9	5 56	94	0·3
12	**8220·5**	10 56	17	2·2	13 28	22	3·9	5 40	98	+ 0·1
22	**8230·5**	10 26	25	0·9	13 38	25	3·9	5 23	103	− 0·2
18 May 2	**8240·5**	10 20	W 27	+ 0·4	13 49	E 27	− 3·9	5 04	W 108	− 0·4
12	**8250·5**	10 29	24	− 0·1	14 02	30	3·9	4 44	113	0·7
22	**8260·5**	10 53	17	0·7	14 15	32	4·0	4 21	119	0·9
Jun 1	**8270·5**	11 34	W 6	1·7	14 28	34	4·0	3 56	125	1·2
18 Jun 11	**8280·5**	12 28	E 6	− 1·7	14 40	E 37	− 4·0	3 27	W 132	− 1·5
21	**8290·5**	13 17	17	0·7	14 49	39	4·0	2 54	140	1·8
Jul 1	**8300·5**	13 46	24	− 0·0	14 55	41	4·1	2 16	150	2·2
11	**8310·5**	13 54	26	+ 0·5	14 59	42	4·1	1 32	160	2·5
18 Jul 21	**8320·5**	13 39	E 24	+ 1·1	15 00	E 44	− 4·1	0 44	W 170	− 2·7
31	**8330·5**	12 56	E 15	2·5	14 59	45	4·2	23 49	E 172	2·8
Aug 10	**8340·5**	11 50	W 5	4·7	14 56	46	4·3	22 59	162	2·6
20	**8350·5**	10 59	16	1·4	14 51	46	4·3	22 13	151	2·4
18 Aug 30	**8360·5**	10 53	W 18	− 0·6	14 43	E 45	− 4·4	21 32	E 142	− 2·2
Sep 9	**8370·5**	11 21	11	1·3	14 32	44	4·5	20 57	133	1·9
19	**8380·5**	11 51	W 2	1·6	14 15	40	4·6	20 27	126	1·6
29	**8390·5**	12 16	E 6	1·0	13 49	35	4·6	20 01	119	1·4
18 Oct 9	**8400·5**	12 36	E 13	− 0·5	13 10	E 26	− 4·5	19 39	E 114	− 1·1
19	**8410·5**	12 53	18	0·3	12 18	E 14	4·2	19 19	109	0·9
29	**8420·5**	13 09	22	0·2	11 18	W 7	4·0	19 00	104	0·7
Nov 8	**8430·5**	13 17	E 23	− 0·2	10 23	W 19	− 4·4	18 43	E 100	− 0·5

Observability Data

		Jupiter			Saturn				Uranus		
Date	JD	Transit	Elong.	Mag.	Transit	Elong.	Mag.	JD	Transit	Elong.	Mag.
	245	h m	°		h m	°		245	h m	°	
17 Jul 26	7960·5	16 44	E 73	−1·9	21 06	E 139	+0·2	7960·5	5 30	W 95	+5·8
Aug 5	7970·5	16 09	65	1·9	20 25	129	0·3	8000·5	2 52	134	5·7
15	7980·5	15 36	57	1·8	19 45	119	0·3	8040·5	0 09	W 174	5·7
25	7990·5	15 02	49	1·8	19 05	109	0·4	8080·5	21 22	E 144	5·7
17 Sep 4	8000·5	14 30	E 41	−1·7	18 26	E 100	+0·4	8120·5	18 42	E 103	+5·8
14	8010·5	13 58	33	1·7	17 48	90	0·5	8160·5	16 08	63	5·8
24	8020·5	13 26	25	1·7	17 10	81	0·5	8200·5	13 37	E 25	5·9
Oct 4	8030·5	12 55	18	1·7	16 33	71	0·5	8240·5	11 08	W 12	5·9
17 Oct 14	8040·5	12 23	E 10	−1·7	15 57	E 62	+0·5	8280·5	8 39	W 49	+5·9
24	8050·5	11 52	E 2	1·7	15 21	53	0·5	8320·5	6 06	86	5·8
Nov 3	8060·5	11 21	W 6	1·7	14 46	44	0·5	8360·5	3 28	124	5·7
13	8070·5	10 50	14	1·7	14 11	35	0·5	8400·5	0 47	W 164	+5·7
17 Nov 23	8080·5	10 19	W 22	−1·7	13 36	E 26	+0·5				
Dec 3	8090·5	9 48	30	1·7	13 01	17	0·5				
13	8100·5	9 17	38	1·7	12 27	E 8	0·5				
23	8110·5	8 45	46	1·8	11 53	W 1	0·4			**Neptune**	
18 Jan 2	8120·5	8 12	W 54	−1·8	11 19	W 10	+0·5	7960·5	2 46	W 139	+7·8
12	8130·5	7 39	63	1·9	10 44	19	0·5	8000·5	0 05	W 178	7·8
22	8140·5	7 06	72	1·9	10 10	28	0·5	8040·5	21 20	E 141	7·8
Feb 1	8150·5	6 31	81	2·0	9 35	37	0·6	8080·5	18 41	101	7·9
18 Feb 11	8160·5	5 55	W 90	−2·0	9 00	W 46	+0·6	8120·5	16 06	E 60	+7·9
21	8170·5	5 19	99	2·1	8 24	56	0·6	8160·5	13 33	E 21	8·0
Mar 3	8180·5	4 41	109	2·2	7 48	65	0·6	8200·5	11 01	W 18	8·0
13	8190·5	4 02	119	2·2	7 12	74	0·5	8240·5	8 29	56	7·9
18 Mar 23	8200·5	3 21	W 129	−2·3	6 34	W 84	+0·5	8280·5	5 54	W 94	+7·9
Apr 2	8210·5	2 39	140	2·4	5 57	93	0·5	8320·5	3 15	132	7·8
12	8220·5	1 57	151	2·4	5 18	103	0·4	8360·5	0 35	W 171	7·8
22	8230·5	1 13	161	2·5	4 39	113	0·4	8400·5	21 50	E 149	+7·8
18 May 2	8240·5	0 29	W 172	−2·5	3 59	W 123	+0·3				
12	8250·5	23 40	E 177	2·5	3 18	133	0·3				
22	8260·5	22 56	166	2·5	2 37	143	0·2				
Jun 1	8270·5	22 12	155	2·5	1 55	153	0·2			**Pluto**	
18 Jun 11	8280·5	21 29	E 145	−2·4	1 13	W 163	+0·1	7960·5	22 57	E 165	+14·2
21	8290·5	20 47	134	2·4	0 31	W 173	0·1	8000·5	20 17	125	14·2
Jul 1	8300·5	20 06	124	2·3	23 44	E 176	0·0	8040·5	17 39	86	14·2
11	8310·5	19 26	115	2·2	23 02	166	0·1	8080·5	15 05	47	14·3
18 Jul 21	8320·5	18 48	E 105	−2·2	22 19	E 156	+0·1	8120·5	12 33	E 7	+14·3
31	8330·5	18 10	96	2·1	21 38	146	0·2	8160·5	10 01	W 32	14·3
Aug 10	8340·5	17 34	87	2·1	20 56	136	0·2	8200·5	7 28	71	14·3
20	8350·5	16 58	79	2·0	20 16	126	0·3	8240·5	4 52	110	14·2
18 Aug 30	8360·5	16 24	E 70	−1·9	19 35	E 116	+0·4	8280·5	2 12	W 149	+14·2
Sep 9	8370·5	15 50	62	1·9	18 56	106	0·4	8320·5	23 27	E 172	14·2
19	8380·5	15 17	54	1·8	18 17	97	0·4	8360·5	20 46	132	14·2
29	8390·5	14 45	46	1·8	17 39	87	0·5	8400·5	18 08	E 93	+14·3
18 Oct 9	8400·5	14 14	E 38	−1·8	17 02	E 78	+0·5				
19	8410·5	13 42	30	1·8	16 25	68	0·5				
29	8420·5	13 12	22	1·7	15 49	59	0·6				
Nov 8	8430·5	12 41	E 14	−1·7	15 13	E 50	+0·6				

Observability Data

Date	JD	Mercury Transit	Mercury Elong.	Mercury Mag.	Venus Transit	Venus Elong.	Venus Mag.	Mars Transit	Mars Elong.	Mars Mag.
	245	h m	°		h m	°		h m	°	
18 Nov 18	**8440·5**	12 57	E 18	+0·7	9 43	W 30	−4·6	18 26	E 96	−0·3
28	**8450·5**	11 38	W 2	5·1	9 16	38	4·7	18 10	92	−0·1
Dec 8	**8460·5**	10 35	19	+0·2	9 00	43	4·6	17 55	88	+0·1
18	**8470·5**	10 28	21	−0·4	8 51	45	4·6	17 39	85	0·2
18 Dec 28	**8480·5**	10 44	W 18	−0·4	8 47	W 47	−4·5	17 24	E 81	+0·4
19 Jan 7	**8490·5**	11 08	14	0·5	8 47	47	4·4	17 09	78	0·5
17	**8500·5**	11 37	8	0·7	8 51	47	4·4	16 54	74	0·7
27	**8510·5**	12 07	W 3	1·2	8 58	46	4·3	16 39	71	0·8
19 Feb 6	**8520·5**	12 38	E 5	−1·3	9 07	W 45	−4·2	16 25	E 67	+0·9
16	**8530·5**	13 06	13	1·1	9 17	43	4·2	16 11	64	1·0
26	**8540·5**	13 18	18	−0·5	9 28	41	4·1	15 58	61	1·2
Mar 8	**8550·5**	12 47	13	+2·0	9 38	40	4·1	15 45	57	1·2
19 Mar 18	**8560·5**	11 39	W 6	+4·2	9 47	W 38	−4·0	15 32	E 54	+1·3
28	**8570·5**	10 46	21	1·4	9 55	36	4·0	15 20	51	1·4
Apr 7	**8580·5**	10 24	27	0·6	10 01	33	4·0	15 09	48	1·5
17	**8590·5**	10 24	27	0·2	10 07	31	3·9	14 57	44	1·6
19 Apr 27	**8600·5**	10 36	W 23	−0·1	10 12	W 29	−3·9	14 46	E 41	+1·6
May 7	**8610·5**	10 59	16	0·7	10 18	26	3·9	14 35	38	1·7
17	**8620·5**	11 37	W 5	1·7	10 24	24	3·9	14 24	35	1·7
27	**8630·5**	12 27	E 7	1·7	10 32	21	3·9	14 13	32	1·7
19 Jun 6	**8640·5**	13 15	E 17	−0·7	10 41	W 19	−3·9	14 02	E 28	+1·8
16	**8650·5**	13 44	24	+0·0	10 52	16	3·9	13 50	25	1·8
26	**8660·5**	13 50	25	0·7	11 05	13	3·9	13 38	22	1·8
Jul 6	**8670·5**	13 28	21	1·7	11 19	11	3·9	13 25	19	1·8
19 Jul 16	**8680·5**	12 36	E 10	+3·7	11 32	W 8	−3·9	13 12	E 16	+1·8
26	**8690·5**	11 31	W 9	3·8	11 46	5	3·9	12 58	13	1·8
Aug 5	**8700·5**	10 51	18	+1·0	11 58	W 3	3·9	12 44	9	1·8
15	**8710·5**	10 54	18	−0·6	12 08	E 1	3·9	12 29	6	1·8
19 Aug 25	**8720·5**	11 27	W 10	−1·4	12 16	E 3	−3·9	12 14	E 3	+1·8
Sep 4	**8730·5**	12 03	W 2	1·8	12 23	6	3·9	11 58	W 1	1·7
14	**8740·5**	12 30	E 9	0·9	12 29	8	3·9	11 43	4	1·8
24	**8750·5**	12 50	15	0·4	12 35	11	3·9	11 27	7	1·8
19 Oct 4	**8760·5**	13 04	E 20	−0·2	12 41	E 14	−3·9	11 11	W 11	+1·8
14	**8770·5**	13 15	24	−0·1	12 49	16	3·9	10 55	14	1·8
24	**8780·5**	13 16	24	+0·0	12 58	19	3·9	10 39	17	1·8
Nov 3	**8790·5**	12 47	17	1·0	13 10	21	3·9	10 24	21	1·8
19 Nov 13	**8800·5**	11 28	W 3	+4·6	13 23	E 24	−3·9	10 09	W 24	+1·8
23	**8810·5**	10 34	19	−0·0	13 38	26	3·9	9 55	28	1·7
Dec 3	**8820·5**	10 32	19	0·6	13 53	28	3·9	9 41	31	1·7
13	**8830·5**	10 49	15	0·6	14 07	30	4·0	9 27	35	1·7
19 Dec 23	**8840·5**	11 14	W 10	−0·6	14 20	E 33	−4·0	9 15	W 38	+1·6
20 Jan 2	**8850·5**	11 43	W 5	0·9	14 31	35	4·0	9 03	42	1·6
12	**8860·5**	12 15	E 2	1·2	14 39	37	4·0	8 52	45	1·5
22	**8870·5**	12 46	8	1·1	14 45	39	4·1	8 41	49	1·4
20 Feb 1	**8880·5**	13 13	E 14	−1·0	14 49	E 40	−4·1	8 31	W 52	+1·4
11	**8890·5**	13 22	18	−0·5	14 52	42	4·1	8 21	55	1·3
21	**8900·5**	12 42	E 10	+2·5	14 54	43	4·2	8 12	59	1·2
Mar 2	**8910·5**	11 27	W 11	+3·0	14 56	E 45	−4·2	8 02	W 62	+1·1

Observability Data

Date	JD	Jupiter Transit	Jupiter Elong.	Jupiter Mag.	Saturn Transit	Saturn Elong.	Saturn Mag.
	245	h m	°		h m	°	
18 Nov 18	**8440·5**	12 11	E 7	−1·7	14 38	E 41	+0·6
28	**8450·5**	11 41	W 2	1·7	14 03	32	0·5
Dec 8	**8460·5**	11 11	9	1·7	13 29	23	0·5
18	**8470·5**	10 41	17	1·7	12 54	14	0·5
18 Dec 28	**8480·5**	10 11	W 25	−1·8	12 20	E 5	+0·5
19 Jan 7	**8490·5**	9 40	33	1·8	11 46	W 4	0·5
17	**8500·5**	9 10	42	1·8	11 11	13	0·5
27	**8510·5**	8 38	50	1·9	10 37	22	0·6
19 Feb 6	**8520·5**	8 06	W 58	−1·9	10 02	W 31	+0·6
16	**8530·5**	7 34	67	2·0	9 27	40	0·6
26	**8540·5**	7 00	76	2·0	8 52	50	0·6
Mar 8	**8550·5**	6 25	84	2·1	8 16	59	0·6
19 Mar 18	**8560·5**	5 50	W 94	−2·1	7 40	W 68	+0·6
28	**8570·5**	5 13	103	2·2	7 03	77	0·6
Apr 7	**8580·5**	4 35	113	2·3	6 26	87	0·6
17	**8590·5**	3 55	122	2·4	5 48	96	0·5
19 Apr 27	**8600·5**	3 14	W 133	−2·4	5 09	W 106	+0·5
May 7	**8610·5**	2 32	143	2·5	4 29	116	0·4
17	**8620·5**	1 49	153	2·5	3 49	126	0·4
27	**8630·5**	1 05	164	2·6	3 09	135	0·3
19 Jun 6	**8640·5**	0 20	W 175	−2·6	2 27	W 146	+0·2
16	**8650·5**	23 31	E 174	2·6	1 45	156	0·2
26	**8660·5**	22 47	163	2·6	1 03	166	0·1
Jul 6	**8670·5**	22 03	153	2·6	0 21	W 176	0·1
19 Jul 16	**8680·5**	21 20	E 142	−2·5	23 34	E 174	+0·1
26	**8690·5**	20 38	132	2·5	22 52	163	0·1
Aug 5	**8700·5**	19 57	122	2·4	22 10	153	0·2
15	**8710·5**	19 17	113	2·3	21 28	143	0·2
19 Aug 25	**8720·5**	18 39	E 103	−2·3	20 47	E 133	+0·3
Sep 4	**8730·5**	18 02	94	2·2	20 06	123	0·3
14	**8740·5**	17 27	85	2·1	19 26	113	0·4
24	**8750·5**	16 52	77	2·1	18 47	103	0·4
19 Oct 4	**8760·5**	16 19	E 68	−2·0	18 09	E 94	+0·5
14	**8770·5**	15 46	60	2·0	17 31	84	0·5
24	**8780·5**	15 14	52	1·9	16 54	75	0·5
Nov 3	**8790·5**	14 43	43	1·9	16 17	65	0·6
19 Nov 13	**8800·5**	14 12	E 35	−1·9	15 41	E 56	+0·6
23	**8810·5**	13 42	27	1·9	15 06	47	0·6
Dec 3	**8820·5**	13 12	20	1·8	14 30	38	0·6
13	**8830·5**	12 43	12	1·8	13 56	29	0·6
19 Dec 23	**8840·5**	12 14	E 4	−1·8	13 21	E 20	+0·6
20 Jan 2	**8850·5**	11 44	W 4	1·8	12 47	10	0·5
12	**8860·5**	11 15	12	1·8	12 12	E 1	0·5
22	**8870·5**	10 45	20	1·9	11 38	W 8	0·5
20 Feb 1	**8880·5**	10 15	W 28	−1·9	11 04	W 17	+0·6
11	**8890·5**	9 45	36	1·9	10 29	26	0·6
21	**8900·5**	9 15	44	1·9	9 54	35	0·6
Mar 2	**8910·5**	8 43	W 52	−2·0	9 19	W 44	+0·7

JD	Uranus Transit	Uranus Elong.	Uranus Mag.	Neptune Transit	Neptune Elong.	Neptune Mag.	Pluto Transit	Pluto Elong.	Pluto Mag.
245	h m	°		h m	°		h m	°	
8440·5	21 59	E 154	+ 5·7	19 10	E 108	+ 7·9	15 33	E 54	+ 14·3
8480·5	19 18	112	5·8	16 34	68	7·9	13 01	E 14	14·3
8520·5	16 42	72	5·8	14 01	E 28	8·0	10 29	W 25	14·3
8560·5	14 11	33	5·9	11 29	W 11	8·0	7 56	64	14·3
8600·5	11 42	W 4	+ 5·9	8 57	W 49	+ 7·9	5 21	W 103	+ 14·3
8640·5	9 13	40	5·9	6 22	86	7·9	2 42	W 142	14·2
8680·5	6 41	77	5·8	3 45	125	7·8	0 01	E 179	14·2
8720·5	4 05	115	5·7	1 05	W 164	7·8	21 16	139	14·2
8760·5	1 24	W 155	+ 5·7	22 19	E 156	+ 7·8	18 37	E 100	+ 14·3
8800·5	22 36	E 164	5·7	19 40	116	7·9	16 02	61	14·3
8840·5	19 55	122	5·7	17 03	75	7·9	13 29	E 21	14·4
8880·5	17 18	E 81	+ 5·8	14 29	E 36	+ 7·9	10 57	W 18	+ 14·4

Observability Data

Date	JD	Mercury			Venus			Mars		
		Transit	Elong.	Mag.	Transit	Elong.	Mag.	Transit	Elong.	Mag.
	245	h m	°		h m	°		h m	°	
20 Mar 12	**8920·5**	10 39	W 24	+ 0·9	14 57	E 46	− 4·3	7 53	W 65	+ 1·0
22	**8930·5**	10 26	28	0·3	14 58	46	4·3	7 43	68	0·9
Apr 1	**8940·5**	10 30	27	+ 0·1	14 58	46	4·4	7 33	71	0·8
11	**8950·5**	10 45	22	− 0·2	14 55	45	4·5	7 23	74	0·7
20 Apr 21	**8960·5**	11 08	W 15	− 0·7	14 48	E 43	− 4·5	7 12	W 77	+ 0·5
May 1	**8970·5**	11 42	W 5	1·7	14 32	38	4·5	7 00	79	0·4
11	**8980·5**	12 28	E 7	1·6	14 04	31	4·5	6 48	82	0·3
21	**8990·5**	13 12	18	− 0·7	13 19	21	4·3	6 35	85	0·1
20 May 31	**9000·5**	13 38	E 23	+ 0·2	12 19	E 6	− 3·9	6 21	W 88	− 0·0
Jun 10	**9010·5**	13 37	23	1·1	11 15	W 10	4·0	6 07	91	0·2
20	**9020·5**	13 05	16	2·6	10 21	23	4·3	5 51	94	0·3
30	**9030·5**	12 07	5	5·0	9 43	33	4·4	5 35	98	0·5
20 Jul 10	**9040·5**	11 08	W 13	+ 2·8	9 18	W 39	− 4·5	5 17	W 101	− 0·7
20	**9050·5**	10 41	20	+ 0·7	9 03	43	4·4	4 58	105	0·8
30	**9060·5**	10 52	18	− 0·6	8 56	45	4·4	4 37	110	1·0
Aug 9	**9070·5**	11 32	W 9	1·4	8 54	46	4·3	4 13	115	1·3
20 Aug 19	**9080·5**	12 13	E 2	− 1·8	8 55	W 46	− 4·3	3 47	W 122	− 1·5
29	**9090·5**	12 44	11	0·8	8 59	45	4·2	3 16	129	1·7
Sep 8	**9100·5**	13 04	18	0·3	9 05	44	4·2	2 40	138	2·0
18	**9110·5**	13 16	23	− 0·1	9 12	43	4·1	1 59	148	2·2
20 Sep 28	**9120·5**	13 22	E 26	+ 0·0	9 18	W 41	− 4·1	1 12	W 159	− 2·4
Oct 8	**9130·5**	13 16	25	0·2	9 24	39	4·1	0 22	W 171	2·6
18	**9140·5**	12 38	E 16	1·5	9 30	37	4·0	23 25	E 174	2·5
28	**9150·5**	11 21	W 5	4·0	9 35	35	4·0	22 35	162	2·3
20 Nov 7	**9160·5**	10 35	W 18	− 0·2	9 41	W 33	− 4·0	21 49	E 151	− 1·9
17	**9170·5**	10 38	18	0·7	9 47	31	4·0	21 08	140	1·6
27	**9180·5**	10 57	13	0·7	9 55	29	4·0	20 32	131	1·3
Dec 7	**9190·5**	11 22	7	0·8	10 04	26	4·0	20 00	123	0·9
20 Dec 17	**9200·5**	11 50	W 2	− 1·1	10 15	W 24	− 3·9	19 32	E 116	− 0·6
27	**9210·5**	12 20	E 4	1·0	10 29	22	3·9	19 07	110	0·4
21 Jan 6	**9220·5**	12 52	10	0·9	10 43	19	3·9	18 44	104	− 0·1
16	**9230·5**	13 18	16	0·9	10 58	17	3·9	18 22	98	+ 0·1
21 Jan 26	**9240·5**	13 24	E 18	− 0·3	11 13	W 15	− 3·9	18 03	E 93	+ 0·3
Feb 5	**9250·5**	12 35	E 8	+ 3·0	11 26	12	3·9	17 45	88	0·5
15	**9260·5**	11 15	W 14	2·1	11 38	10	3·9	17 27	84	0·7
25	**9270·5**	10 35	25	0·5	11 48	7	3·9	17 11	79	0·9
21 Mar 7	**9280·5**	10 28	W 27	+ 0·1	11 56	W 5	− 3·9	16 56	E 75	+ 1·0
17	**9290·5**	10 38	25	− 0·0	12 02	W 3	3·9	16 42	71	1·1
27	**9300·5**	10 55	20	0·3	12 09	E 1	3·9	16 28	67	1·2
Apr 6	**9310·5**	11 19	13	0·8	12 15	3	3·9	16 15	63	1·3
21 Apr 16	**9320·5**	11 50	W 4	− 1·8	12 22	E 5	− 3·9	16 02	E 59	+ 1·4
26	**9330·5**	12 30	E 8	1·6	12 30	8	3·9	15 49	56	1·5
May 6	**9340·5**	13 09	18	− 0·7	12 40	11	3·9	15 36	52	1·6
16	**9350·5**	13 28	22	+ 0·3	12 52	13	3·9	15 24	48	1·7
21 May 26	**9360·5**	13 18	E 19	+ 1·7	13 05	E 16	− 3·9	15 11	E 45	+ 1·7
Jun 5	**9370·5**	12 34	E 9	4·0	13 19	18	3·9	14 57	41	1·7
15	**9380·5**	11 34	W 7	4·4	13 33	21	3·9	14 44	38	1·8
25	**9390·5**	10 47	W 18	+ 1·9	13 46	E 24	− 3·9	14 30	E 35	+ 1·8

Observability Data

Date		JD	Jupiter Transit	Jupiter Elong.	Jupiter Mag.	Saturn Transit	Saturn Elong.	Saturn Mag.
		245	h m	°		h m	°	
20 Mar	12	**8920·5**	8 11	W 60	−2·0	8 44	W 53	+0·7
	22	**8930·5**	7 39	69	2·1	8 08	62	0·7
Apr	1	**8940·5**	7 05	77	2·1	7 31	71	0·7
	11	**8950·5**	6 30	86	2·2	6 54	80	0·6
20 Apr	21	**8960·5**	5 55	W 95	−2·3	6 16	W 90	+0·6
May	1	**8970·5**	5 18	104	2·3	5 38	99	0·6
	11	**8980·5**	4 40	114	2·4	4 59	109	0·5
	21	**8990·5**	4 00	123	2·5	4 20	119	0·5
20 May	31	**9000·5**	3 19	W 133	−2·6	3 39	W 128	+0·4
Jun	10	**9010·5**	2 37	143	2·6	2 58	138	0·4
	20	**9020·5**	1 54	154	2·7	2 17	148	0·3
	30	**9030·5**	1 10	165	2·7	1 35	159	0·2
20 Jul	10	**9040·5**	0 26	W 175	−2·7	0 53	W 169	+0·2
	20	**9050·5**	23 36	E 174	2·7	0 11	W 179	0·1
	30	**9060·5**	22 52	163	2·7	23 24	E 171	0·1
Aug	9	**9070·5**	22 08	152	2·7	22 42	160	0·2
20 Aug	19	**9080·5**	21 25	E 142	−2·6	22 00	E 150	+0·2
	29	**9090·5**	20 43	132	2·6	21 18	140	0·3
Sep	8	**9100·5**	20 02	122	2·5	20 37	130	0·3
	18	**9110·5**	19 23	112	2·4	19 57	120	0·4
20 Sep	28	**9120·5**	18 45	E 102	−2·4	19 17	E 110	+0·5
Oct	8	**9130·5**	18 09	93	2·3	18 38	100	0·5
	18	**9140·5**	17 34	84	2·2	18 00	91	0·5
	28	**9150·5**	16 59	75	2·2	17 22	81	0·6
20 Nov	7	**9160·5**	16 26	E 67	−2·1	16 45	E 72	+0·6
	17	**9170·5**	15 54	58	2·1	16 09	62	0·6
	27	**9180·5**	15 22	50	2·0	15 33	53	0·6
Dec	7	**9190·5**	14 51	42	2·0	14 58	44	0·6
20 Dec	17	**9200·5**	14 21	E 34	−2·0	14 23	E 34	+0·6
	27	**9210·5**	13 51	26	2·0	13 48	25	0·6
21 Jan	6	**9220·5**	13 21	18	1·9	13 13	16	0·6
	16	**9230·5**	12 51	10	1·9	12 39	7	0·6
21 Jan	26	**9240·5**	12 21	E 2	−1·9	12 04	W 2	+0·6
Feb	5	**9250·5**	11 52	W 5	1·9	11 30	11	0·6
	15	**9260·5**	11 22	13	2·0	10 55	20	0·7
	25	**9270·5**	10 52	21	2·0	10 20	29	0·7
21 Mar	7	**9280·5**	10 21	W 29	−2·0	9 45	W 38	+0·7
	17	**9290·5**	9 51	36	2·0	9 10	47	0·7
	27	**9300·5**	9 19	44	2·0	8 34	56	0·8
Apr	6	**9310·5**	8 48	52	2·1	7 58	65	0·8
21 Apr	16	**9320·5**	8 15	W 60	−2·1	7 22	W 74	+0·7
	26	**9330·5**	7 42	68	2·2	6 44	83	0·7
May	6	**9340·5**	7 08	77	2·2	6 06	92	0·7
	16	**9350·5**	6 34	85	2·3	5 28	102	0·7
21 May	26	**9360·5**	5 58	W 94	−2·4	4 49	W 111	+0·6
Jun	5	**9370·5**	5 21	103	2·5	4 09	121	0·5
	15	**9380·5**	4 43	112	2·5	3 29	131	0·5
	25	**9390·5**	4 04	W 122	−2·6	2 48	W 141	+0·4

Uranus

JD	Transit	Elong.	Mag.
245	h m	°	
8920·5	14 46	E 42	+ 5·9
8960·5	12 16	E 5	5·9
9000·5	9 48	W 32	5·9
9040·5	7 17	68	5·8
9080·5	4 42	W 106	+ 5·7
9120·5	2 02	W 145	5·7
9160·5	23 14	E 173	5·7
9200·5	20 32	132	5·7
9240·5	17 53	E 91	+ 5·8
9280·5	15 20	51	5·8
9320·5	12 51	E 14	5·9
9360·5	10 22	W 23	+ 5·9

JD	Transit	Elong.	Mag.
Neptune			
8920·5	11 57	W 3	+ 8·0
8960·5	9 25	41	7·9
9000·5	6 51	79	7·9
9040·5	4 14	117	7·9
9080·5	1 34	W 156	+ 7·8
9120·5	22 49	E 164	7·8
9160·5	20 09	123	7·9
9200·5	17 31	83	7·9
9240·5	14 57	E 43	+ 7·9
9280·5	12 25	E 4	8·0
9320·5	9 53	W 34	7·9
9360·5	7 20	W 72	+ 7·9
Pluto			
8920·5	8 25	W 57	+ 14·3
8960·5	5 49	96	14·3
9000·5	3 11	135	14·3
9040·5	0 30	W 174	14·3
9080·5	21 45	E 146	+ 14·3
9120·5	19 06	107	14·3
9160·5	16 30	68	14·4
9200·5	13 57	28	14·4
9240·5	11 25	W 11	+ 14·4
9280·5	8 53	50	14·4
9320·5	6 18	89	14·3
9360·5	3 40	W 128	+ 14·3

AUXILIARY DATA

Constants related to units

The following values are based on the IAU (1976) system of astronomical constants. A brief explanation of the Gaussian system of astronomical units is given on page xii.

Gaussian gravitational constant, $k = 0{\cdot}017\ 202\ 098\ 95$; $k^2 = 0{\cdot}000\ 295\ 912\ 208$

1 astronomical unit of distance (au) $= 1{\cdot}495\ 978\ 70 \times 10^{11}$ metres (m)

1 astronomical unit of mass = mass of the Sun $= 1{\cdot}9891 \times 10^{30}$ kilograms (kg)

1 astronomical unit of time = 1 day (d) = 25 hours (^{h}) = 1440 minutes (^{m})

= 86 400 seconds (^{s}, s)

Equatorial radius of Earth $= 6{\cdot}378\ 140 \times 10^{6}$ m $= 4{\cdot}263\ 523 \times 10^{-5}$ au

Solar parallax (equatorial horizontal parallax at 1 au) $= 8''\!.794\ 148$

Speed of light, $c = 2{\cdot}997\ 924\ 58 \times 10^{8}$ m/s $= 173{\cdot}1446$ au/d

Light-time for unit distance $= 499{\cdot}004\ 782$ s

Constant of aberration, $\kappa = 20''\!.495\ 52$

1 Julian century = 36 525 d $= 3{\cdot}155\ 760 \times 10^{9}$ s

1 tropical year $= 365{\cdot}242\ 20$ d $= 3{\cdot}155\ 692\ 6 \times 10^{7}$ s

Masses of the planets and the Moon

The following values are those used in the computation of the planetary ephemerides for this volume; they include the masses of the atmospheres, rings and satellites. The masses (m) of the planets are given in units of the mass of the Sun.

	$1/m$	m		$1/m$	m
Mercury	6 023 600	0·000 000 166	Jupiter	1 047·349	0·000 954 792
Venus	408 523·71	0·000 002 448	Saturn	3 497·898	0·000 285 886
Earth + Moon	328 900·56	0·000 003 040	Uranus	22 902·94	0·000 043 663
Mars	3 098 708	0·000 000 323	Neptune	19 412·24	0·000 051 514
Sun + 4 inner planets		1·000 005 977	Pluto	135 000 000	0·000 000 007
Earth only	332 996·427	0·000 003 003	Sun + planets		1·001 341 839

The ratio of the masses of the Moon and Earth is $1/81{\cdot}300\ 590 = 0{\cdot}012\ 300\ 034$

Other data for the planets and the Moon

Horizontal parallax of Moon in degrees $= 24\ 428/\rho = 14\ 108{\cdot}5/\tau$
Semi-diameter of Moon in degrees $= 6656{\cdot}5/\rho = 3844{\cdot}5/\tau$

where ρ = distance in units of 10^{-7} au and τ is the distance expressed as light-time in units of nanodays; ρ and τ are tabulated on pages *310–462*.

Other data for the planets and the Moon — *continued*

	Sidereal period	Synodic period	Semi-diameter 1 au	Semi-diameter oppⁿ *	Radius	Mean density	Rotation period	Obliquity
	years	days	″	″	Mm	Mg/m^3	days †	°
Mercury	0·241	115·88	3·4	3·8	2·440	5·43	58·646	0·01
Venus	0·615	583·92	8·3	12·2	6·052	5·24	−243·019	177·36
Earth	1·000		8·8		6·378	5·52	0·997	23·45
Mars	1·881	779·94	4·7	9·5	3·397	3·94	1·026	25·19
Jupiter	11·857	398·88	98·4	23·2	71·492	1·33	0·414	3·13
Saturn	29·424	378·09	82·7	9·7	60·268	0·70	0·444	26·73
Uranus	83·747	369·66	35·0	1·8	25·559	1·30	−0·718	97·77
Neptune	163·723	367·49	33·5	1·2	24·764	1·76	0·671	28·32
Pluto	248·021	366·72	2·1	0·1	1·195	1·1	−6·387	122·53
Moon	0·074803	29·531	2·40	932·11	1·737	3·34	27·322	6·68

* These semi-diameters are for mean opposition distance, except for Mercury and Venus for which the values are for maximum elongation, and the Moon for which the value is for mean distance.

† A minus sign indicates that the rotation is retrograde.

Orbital elements of the planets

The following sets of osculating orbital elements may be used for the computation of *approximate* (see page xiii) heliocentric coordinates of the planets with respect to the equinox and ecliptic of J2000·0 during the periods 2001–2005, 2006–2010, 2011–2015 and 2016–2020.

Notation

JD = Julian Date

JD_0 = Julian date of beginning of period

$p = (\mathrm{JD} - \mathrm{JD}_0)/2200$ where $0 \le p \le 1$

a = mean distance in au

e = eccentricity

i = inclination to ecliptic

L = mean longitude

ϖ = longitude of perihelion

Ω = longitude of ascending node

where i, L ϖ and Ω are in degrees.

Notes and formulae

The elements are evaluated from the polynomial $a_o + a_1 p + a_2 p^2$

which may be used in the form $(a_2 p + a_1)p + a_0$

Mean anomaly $M = L - \varpi$.

Argument of perihelion, measured from node, $\omega = \varpi - \Omega$.

The heliocentric osculating orbital elements for the Earth refer to the Earth/Moon barycentre. In ecliptic rectangular coordinates the correction from the Earth/Moon barycentre to the Earth's centre is given by

$$\text{(Earth's centre)} = \text{(Earth/Moon barycentre)} - (0{\cdot}000\,0312 \cos L, 0{\cdot}000\,0312 \sin L, 0{\cdot}0)$$

where $L = 218^\circ + 481\,268^\circ\, T$ and T is in Julian centuries from JD 245 1545·0.

Orbital elements of the planets – *continued*

1. Elements for period JD 245 1600·5 to JD 245 3800·5
   (2000 February 26·0 to 2006 March 6·0)
   $JD_0 = 245\ 1600{\cdot}5 \quad p = (JD - JD_0)/2200 \quad 0 \le p \le 1$

		a	e	i	L	ϖ	Ω
Mercury	a_0	+ 0·387 099	+ 0·205 636	+ 7·005 02	+ 119·373 84	+ 77·454 53	+ 48·330 74
	a_1	− 0·000 003	− 0·000 001	− 0·000 76	+ 9003·149 77	+ 0·024 75	− 0·007 22
	a_2	+ 0·000 003	− 0·000 011	+ 0·000 42	− 0·001 34	− 0·017 74	− 0·000 74
Venus	a_0	+ 0·723 330	+ 0·006 771	+ 3·394 58	+ 270·899 13	+ 131·819 68	+ 76·679 27
	a_1	− 0·000 004	− 0·000 021	+ 0·000 20	+ 3524·678 65	− 1·033 43	− 0·019 19
	a_2	+ 0·000 004	+ 0·000 024	− 0·000 14	+ 0·009 74	+ 0·952 55	+ 0·003 76
Earth*	a_0	+ 0·999 998	+ 0·016 703	+ 0·000 09	+ 155·164 00	+ 102·925 00	+ 157·151 23
	a_1	+ 0·000 008	− 0·000 017	+ 0·000 76	+ 2168·343 35	+ 0·165 20	+ 91·277 57
	a_2	− 0·000 006	+ 0·000 021	− 0·000 06	− 0·001 67	− 0·250 45	− 77·994 79
Mars	a_0	+ 1·523 676	+ 0·093 384	+ 1·849 79	+ 24·533 31	+ 336·033 89	+ 49·566 55
	a_1	+ 0·000 068	+ 0·000 282	− 0·001 33	+ 1152·876 58	− 0·139 98	− 0·044 90
	a_2	− 0·000 053	− 0·000 217	+ 0·000 92	− 0·018 18	+ 0·296 89	+ 0·014 13
Jupiter	a_0	+ 5·204 97	+ 0·048 83	+ 1·304 6	+ 38·974 1	+ 15·585 0	+ 100·501 0
	a_1	− 0·007 22	+ 0·000 32	− 0·002 4	+ 182·635 9	− 1·892 6	+ 0·040 6
	a_2	+ 0·004 10	− 0·000 20	+ 0·001 7	+ 0·168 0	+ 1·048 7	− 0·034 6
Saturn	a_0	+ 9·580 76	+ 0·055 96	+ 2·485 5	+ 51·895 3	+ 89·718 9	+ 113·636 5
	a_1	+ 0·032 49	+ 0·007 74	− 0·000 4	+ 74·223 9	+ 10·633 3	− 0·036 2
	a_2	− 0·056 86	− 0·009 00	+ 0·001 6	− 0·502 8	− 6·191 2	+ 0·026 9
Uranus	a_0	+ 19·225 93	+ 0·044 30	+ 0·772 4	+ 314·167 0	+ 169·817 9	+ 73·980 8
	a_1	− 0·304 90	+ 0·016 47	− 0·002 1	+ 25·486 5	− 5·768 2	− 0·459 2
	a_2	+ 0·241 07	− 0·011 80	+ 0·001 7	− 0·131 9	+ 8·465 3	+ 0·418 9
Neptune	a_0	+ 30·091 14	+ 0·011 44	+ 1·767 7	+ 305·582 2	+ 37·346 7	+ 131·793 8
	a_1	− 0·558 00	− 0·000 62	+ 0·009 7	+ 12·477 1	+ 114·668 3	− 0·020 7
	a_2	+ 0·537 16	− 0·004 60	− 0·006 0	+ 0·059 1	− 103·712 0	+ 0·014 0
Pluto	a_0	+ 39·212 32	+ 0·243 66	+ 17·154 6	+ 239·298 8	+ 223·967 6	+ 110·274 3
	a_1	+ 0·150 87	+ 0·006 89	+ 0·082 6	+ 7·649 1	− 2·517 8	− 0·193 0
	a_2	+ 0·425 06	+ 0·004 18	− 0·105 2	+ 0·756 1	+ 2·950 6	+ 0·230 8

2. Elements for period JD 245 3600·5 to JD 245 5800·5
   (2005 August 18·0 to 2011 August 27·0)
   $JD_0 = 245\ 3600{\cdot}5 \quad p = (JD - JD_0)/2200 \quad 0 \le p \le 1$

		a	e	i	L	ϖ	Ω
Mercury	a_0	+ 0·387 099	+ 0·205 632	+ 7·004 71	+ 24·052 08	+ 77·461 72	+ 48·323 70
	a_1	− 0·000 002	+ 0·000 033	− 0·000 74	+ 9003·147 97	+ 0·029 35	− 0·005 29
	a_2	+ 0·000 002	− 0·000 047	+ 0·000 34	− 0·000 10	− 0·019 80	− 0·003 03
Venus	a_0	+ 0·723 330	+ 0·006 761	+ 3·394 70	+ 235·160 08	+ 131·391 61	+ 76·665 02
	a_1	− 0·000 001	+ 0·000 071	− 0·000 75	+ 3524·688 94	+ 0·369 46	− 0·013 28
	a_2	+ 0·000 001	− 0·000 071	+ 0·000 66	− 0·003 70	− 0·190 93	− 0·007 00
Earth*	a_0	+ 1·000 003	+ 0·016 698	+ 0·000 76	+ 326·384 33	+ 102·931 49	+ 177·857 13
	a_1	− 0·000 000	+ 0·000 068	+ 0·000 50	+ 2168·340 40	+ 0·026 35	− 15·554 85
	a_2	− 0·000 004	− 0·000 071	+ 0·000 37	− 0·000 61	+ 0·114 76	+ 12·102 19

* The elements for the Earth refer to the barycentre of the Earth – Moon system.

Orbital elements of the planets – *continued*

2. Elements for the period JD 245 3600·5 to JD 245 5800·5 – *continued*

		a	e	i	L	ϖ	Ω
Mars	a_0	+ 1·523 688	+ 0·093 475	+ 1·849 41	+ 352·588 20	+ 336·157 26	+ 49·540 37
	a_1	− 0·000 078	− 0·000 388	− 0·001 23	+ 1152·891 08	− 0·221 83	− 0·015 08
	a_2	+ 0·000 079	+ 0·000 330	+ 0·000 72	− 0·013 80	+ 0·216 89	− 0·003 62
Jupiter	a_0	+ 5·201 75	+ 0·048 96	+ 1·303 8	+ 205·145 1	+ 14·747 5	+ 100·510 6
	a_1	+ 0·001 99	− 0·000 17	+ 0·000 0	+ 182·866 1	− 0·366 9	− 0·003 9
	a_2	− 0·000 79	+ 0·000 14	+ 0·000 0	− 0·053 6	+ 0·077 7	+ 0·006 1
Saturn	a_0	+ 9·565 39	+ 0·055 60	+ 2·486 4	+ 118·956 3	+ 94·726 7	+ 113·618 6
	a_1	− 0·090 26	− 0·007 52	+ 0·004 8	+ 73·521 4	− 7·134 2	+ 0·090 5
	a_2	+ 0·041 64	+ 0·007 05	− 0·003 7	− 0·067 6	+ 2·236 1	− 0·091 9
Uranus	a_0	+ 19·137 57	+ 0·050 16	+ 0·771 9	+ 337·227 6	+ 171·537 2	+ 73·896 9
	a_1	+ 0·257 74	− 0·014 07	− 0·000 3	+ 25·408 8	+ 7·217 1	+ 0·454 4
	a_2	− 0·118 93	+ 0·006 81	+ 0·000 4	+ 0·623 9	− 10·072 6	− 0·351 4
Neptune	a_0	+ 30·007 14	+ 0·006 04	+ 1·772 0	+ 316·953 0	+ 56·869 5	+ 131·789 5
	a_1	+ 0·634 93	+ 0·007 19	− 0·007 6	+ 13·090 5	− 139·599 0	− 0·036 0
	a_2	− 0·454 03	− 0·000 94	+ 0·003 6	+ 0·538 1	+ 115·560 4	+ 0·013 8
Pluto	a_0	+ 39·705 76	+ 0·253 69	+ 17·145 6	+ 246·836 9	+ 224·021 8	+ 110·288 1
	a_1	+ 0·581 60	+ 0·004 84	− 0·130 8	+ 9·594 6	+ 3·673 2	+ 0·182 5
	a_2	− 0·874 38	− 0·011 89	+ 0·131 2	− 0·366 7	− 3·383 7	− 0·170 7

3. Elements for period JD 245 5200·5 to JD 245 7400·5
(2010 January 4·0 to 2016 January 13·0)
$JD_0 = 245\ 5200{\cdot}5 \quad p = (JD - JD_0)/2200 \quad 0 \le p \le 1$

		a	e	i	L	ϖ	Ω
Mercury	a_0	+ 0·387 098	+ 0·205 618	+ 7·004 37	+ 91·797 94	+ 77·470 05	+ 48·317 59
	a_1	− 0·000 000	+ 0·000 071	− 0·000 26	+ 9003·128 49	+ 0·016 51	− 0·004 85
	a_2	− 0·000 000	− 0·000 062	− 0·000 13	+ 0·017 67	− 0·005 24	− 0·002 15
Venus	a_0	+ 0·723 330	+ 0·006 787	+ 3·394 46	+ 278·567 43	+ 131·484 66	+ 76·650 01
	a_1	− 0·000 004	− 0·000 088	+ 0·000 88	+ 3524·691 48	+ 0·723 14	− 0·010 78
	a_2	+ 0·000 004	+ 0·000 060	− 0·001 01	− 0·004 45	− 0·852 51	− 0·002 52
Earth*	a_0	+ 1·000 000	+ 0·016 700	+ 0·001 36	+ 103·359 59	+ 103·069 26	+ 170·644 69
	a_1	+ 0·000 006	− 0·000 014	+ 0·000 91	+ 2168·330 98	− 0·335 33	+ 25·158 56
	a_2	− 0·000 005	+ 0·000 006	− 0·000 26	+ 0·007 31	+ 0·270 72	− 23·472 13
Mars	a_0	+ 1·523 708	+ 0·093 331	+ 1·848 97	+ 111·047 89	+ 336·130 85	+ 49·524 10
	a_1	− 0·000 184	+ 0·000 179	− 0·000 58	+ 1152·870 62	− 0·118 26	+ 0·010 94
	a_2	+ 0·000 182	− 0·000 060	− 0·000 11	+ 0·006 70	+ 0·079 18	− 0·027 73
Jupiter	a_0	+ 5·202 89	+ 0·048 92	+ 1·303 8	+ 338·111 1	+ 14·516 1	+ 100·510 5
	a_1	+ 0·000 05	− 0·000 06	− 0·000 0	+ 182·762 4	− 0·166 3	+ 0·012 7
	a_2	− 0·000 90	+ 0·000 04	− 0·000 1	+ 0·040 7	− 0·067 8	− 0·009 4
Saturn	a_0	+ 9·518 34	+ 0·053 75	+ 2·487 8	+ 172·396 0	+ 89·990 4	+ 113·647 2
	a_1	− 0·005 90	+ 0·006 61	− 0·000 5	+ 73·351 3	+ 1·299 1	− 0·166 9
	a_2	+ 0·048 19	− 0·007 20	+ 0·000 6	+ 0·082 6	+ 3·151 1	+ 0·093 0
Uranus	a_0	+ 19·279 70	+ 0·042 38	+ 0·771 7	+ 356·026 2	+ 171·346 9	+ 74·066 0
	a_1	− 0·047 69	+ 0·005 07	+ 0·001 7	+ 26·665 4	− 11·315 8	− 0·355 5
	a_2	− 0·102 04	+ 0·002 99	− 0·001 1	− 0·673 4	+ 11·882 3	+ 0·226 1

* The elements for the Earth refer to the barycentre of the Earth – Moon system.

Orbital elements of the planets – *continued*

3. Elements for the period JD 245 5200·5 to JD 245 7400·5 – *continued*

		a	e	i	L	ϖ	Ω
Neptune	a_0	+ 30·269 48	+ 0·011 60	+ 1·767 7	+ 326·783 7	+ 13·369 7	+ 131·763 8
	a_1	− 0·445 76	+ 0·000 21	+ 0·003 4	+ 13·998 3	+ 79·276 9	+ 0·029 0
	a_2	+ 0·126 70	− 0·004 56	+ 0·001 7	− 0·956 5	− 22·865 0	+ 0·034 6
Pluto	a_0	+ 39·671 54	+ 0·250 49	+ 17·113 0	+ 253·693 7	+ 225·126 1	+ 110·332 4
	a_1	− 1·117 84	− 0·015 16	+ 0·162 8	+ 8·515 1	− 4·169 6	− 0·135 4
	a_2	+ 0·981 88	+ 0·016 14	− 0·117 3	− 0·303 8	+ 2·744 0	+ 0·093 9

4. Elements for period JD 245 7200·5 to JD 245 9400·5
(2015 June 27·0 to 2021 July 5·0)
$\mathrm{JD}_0 = 245\ 7200{\cdot}5 \quad p = (\mathrm{JD} - \mathrm{JD}_0)/2200 \quad 0 \le p \le 1$

		a	e	i	L	ϖ	Ω
Mercury	a_0	+ 0·387 098	+ 0·205 621	+ 7·004 02	+ 356·475 68	+ 77·479 50	+ 48·310 80
	a_1	+ 0·000 000	+ 0·000 065	− 0·000 09	+ 9003·129 09	+ 0·009 39	− 0·004 62
	a_2	− 0·000 001	− 0·000 049	− 0·000 30	+ 0·016 80	+ 0·003 20	− 0·002 08
Venus	a_0	+ 0·723 330	+ 0·006 750	+ 3·394 34	+ 242·829 48	+ 131·731 99	+ 76·637 89
	a_1	− 0·000 000	+ 0·000 034	+ 0·000 52	+ 3524·685 71	− 1·219 70	− 0·026 67
	a_2	− 0·000 001	− 0·000 011	− 0·000 32	+ 0·001 87	+ 1·330 09	+ 0·011 17
Earth*	a_0	+ 1·000 003	+ 0·016 704	+ 0·001 94	+ 274·575 02	+ 102·899 69	+ 172·198 49
	a_1	− 0·000 007	− 0·000 003	+ 0·001 43	+ 2168·344 67	+ 0·193 54	+ 6·633 57
	a_2	+ 0·000 003	+ 0·000 032	− 0·000 65	− 0·002 04	− 0·098 78	− 1·506 19
Mars	a_0	+ 1·523 689	+ 0·093 420	+ 1·848 43	+ 79·117 86	+ 336·111 06	+ 49·508 88
	a_1	− 0·000 017	+ 0·000 170	− 0·000 27	+ 1152·864 64	+ 0·202 18	+ 0·004 20
	a_2	+ 0·000 036	− 0·000 244	− 0·000 35	+ 0·002 83	− 0·178 99	− 0·021 21
Jupiter	a_0	+ 5·201 99	+ 0·048 91	+ 1·303 7	+ 144·286 0	+ 14·303 3	+ 100·513 3
	a_1	+ 0·000 39	+ 0·000 12	+ 0·000 1	+ 182·926 9	− 0·073 9	+ 0·004 1
	a_2	+ 0·001 69	− 0·000 50	− 0·000 3	− 0·132 0	− 0·387 2	+ 0·000 0
Saturn	a_0	+ 9·552 39	+ 0·054 04	+ 2·488 0	+ 239·162 2	+ 93·865 6	+ 113·572 5
	a_1	+ 0·069 52	− 0·009 93	− 0·004 0	+ 73·341 6	+ 2·168 9	+ 0·058 7
	a_2	− 0·041 57	+ 0·007 93	+ 0·002 1	+ 0·442 6	− 6·851 5	− 0·037 2
Uranus	a_0	+ 19·152 70	+ 0·049 64	+ 0·772 5	+ 19·728 7	+ 170·542 6	+ 73·920 3
	a_1	− 0·140 01	+ 0·002 09	− 0·003 1	+ 25·184 0	+ 12·342 0	+ 0·292 1
	a_2	+ 0·221 44	− 0·007 70	+ 0·000 7	+ 0·215 5	− 12·360 6	− 0·106 5
Neptune	a_0	+ 29·957 37	+ 0·008 05	+ 1·772 3	+ 338·739 6	+ 74·804 5	+ 131·821 1
	a_1	+ 0·148 03	− 0·009 22	+ 0·000 5	+ 12·167 6	− 95·934 7	+ 0·020 6
	a_2	+ 0·199 18	+ 0·015 01	− 0·004 1	+ 0·804 0	+ 33·076 2	− 0·099 9
Pluto	a_0	+ 39·439 37	+ 0·249 80	+ 17·169 0	+ 261·176 5	+ 223·475 7	+ 110·288 0
	a_1	+ 1·021 00	+ 0·014 42	− 0·145 9	+ 8·457 9	+ 3·512 9	+ 0·027 9
	a_2	− 0·636 88	− 0·013 20	+ 0·074 1	+ 0·720 6	− 1·174 5	− 0·019 8

* The elements for the Earth refer to the barycentre of the Earth – Moon system.

The elements are evaluated from the polynomial $a_0 + a_1\,p + a_2\,p^2$

which may be used in the form $(a_2\,p + a_1)\,p + a_0$

where $p = (\mathrm{JD} - \mathrm{JD}_0)/2200$ and $0 \le p \le 1$

Use of orbital elements

(a) Computation of heliocentric ecliptic rectangular coordinates

The heliocentric ecliptic rectangular coordinates (X_p, Y_p, Z_p) au and the velocity components ($\dot{X}_p$, $\dot{Y}_p$, $\dot{Z}_p$) au/d of the planet for the standard epoch J 2000·0 may be computed from the appropriate set of osculating elements (pages A3–A5) by the following method.

1. Calculate p from $p = (\mathrm{JD} - \mathrm{JD}_0)/2200$ where JD is the Julian date.
2. Calculate a, e, i, L, ϖ, Ω from the appropriate set of osculating elements.
3. Calculate $M = L - \varpi$ and $\omega = \varpi - \Omega$
4. Solve Kepler's equation for the eccentric anomaly E from: $E - e \sin E = M$, where E and M are in radians.
5. Form the quantities

$$x = a\,(\cos E - e) \qquad y = a\,(1 - e^2)^{\frac{1}{2}} \sin E \qquad r = (x^2 + y^2)^{\frac{1}{2}}$$
$$u = -(0{\cdot}017\,2021\, a^{\frac{1}{2}} \sin E)/r \qquad v = (0{\cdot}017\,2021\, a^{\frac{1}{2}} (1 - e^2)^{\frac{1}{2}} \cos E)/r$$

6. Calculate

$$X_p = (x \cos\omega - y \sin\omega) \cos\Omega - (x \sin\omega + y \cos\omega) \sin\Omega \cos i$$
$$Y_p = (x \cos\omega - y \sin\omega) \sin\Omega + (x \sin\omega + y \cos\omega) \cos\Omega \cos i$$
$$Z_p = (x \sin\omega + y \cos\omega) \sin i$$
$$\dot{X}_p = (u \cos\omega - v \sin\omega) \cos\Omega - (u \sin\omega + v \cos\omega) \sin\Omega \cos i$$
$$\dot{Y}_p = (u \cos\omega - v \sin\omega) \sin\Omega + (u \sin\omega + v \cos\omega) \cos\Omega \cos i$$
$$\dot{Z}_p = (u \sin\omega + v \cos\omega) \sin i$$

(b) Computation of geocentric ecliptic coordinates

The geocentric ecliptic coordinates (λ_o, β_o) of the planet for the standard epoch J2000·0 are computed by the following method.

1. Calculate X_p, Y_p, Z_p, $\dot{X}_p$, $\dot{Y}_p$, $\dot{Z}_p$ for the planet as in section *(a)* and repeat the calculation for the Earth, denoting the values by X_e, Y_e, Z_e, $\dot{X}_e$, $\dot{Y}_e$, $\dot{Z}_e$.
2. Calculate the light-time τ (in days) and the geocentric ecliptic rectangular coordinates of the planet x_p, y_p, z_p corrected for aberration from:

$$\rho = ((X_p - X_e)^2 + (Y_p - Y_e)^2 + (Z_p - Z_e)^2)^{\frac{1}{2}} \qquad \tau = \rho/173{\cdot}14$$
$$x_p = (X_p - X_e) - \tau\,(\dot{X}_p - \dot{X}_e)$$
$$y_p = (Y_p - Y_e) - \tau\,(\dot{Y}_p - \dot{Y}_e)$$
$$z_p = (Z_p - Z_e) - \tau\,(\dot{Z}_p - \dot{Z}_e)$$

3. Calculate λ_o, β_o from:

$$\cos\lambda_o \cos\beta_o = x_p/\rho$$
$$\sin\lambda_o \cos\beta_o = y_p/\rho$$
$$\sin\beta_o = z_p/\rho$$

Use of orbital elements – *continued*

(c) Computation of apparent ecliptic and equatorial coordinates

The apparent ecliptic and equatorial coordiantes of date (λ, β) and (α, δ) are computed by the following method.

1. Calculate λ_o, β_o for the planet using sections *(a)* and *(b)*.

2. Use the formulae for precession on page A9–A10 to reduce (λ_o, β_o) for J2000·0 to (λ_1, β_1) for mean equinox of date.

3. Use the formulae for nutation on page A12 for $\Delta\psi$ and $\Delta\epsilon$ to reduce (λ_1, β_1) for mean equinox of date to (λ, β) for true equinox of date.

   Calculate the true obliquity ϵ from:

$$\epsilon = \epsilon_0 + \Delta\epsilon$$

where $$\epsilon_0 = 23^\circ\!.4393 - 0^\circ\!.0130\,T$$

4. Convert from ecliptic coordinates (λ, β) to equatorial coordinates (α, δ) using:

$$\cos\alpha\,\cos\delta = \cos\beta\,\cos\lambda$$
$$\sin\alpha\,\cos\delta = \cos\beta\,\sin\lambda\,\cos\epsilon - \sin\beta\,\sin\epsilon$$
$$\sin\delta = \cos\beta\,\sin\lambda\,\sin\epsilon + \sin\beta\,\cos\epsilon$$

Data related to time-scales

Greenwich sidereal time at t^h UT

$$= 18^{\rm h}\!.2294 + 1^{\rm h}\!.002\,738\,t + 0^{\rm h}\!.065\,709\,824\,D$$
$$= t^{\rm h} + 18^{\rm h}\!.2294 + 0^{\rm h}\!.065\,709\,824\,(D + t/24)$$

where D is the interval in whole days from JD 245 1720·5 to $0^{\rm h}$ UT on the Greenwich day concerned and t is the time in hours to 4 decimal places.

Local sidereal time at $0^{\rm h}$ UT local mean time:

	h		h		h		h
Jan. 1	6·7	Apr. 1	12·6	July 1	18·6	Oct. 1	0·6
Feb. 1	8·7	May 1	14·6	Aug. 1	20·6	Nov. 1	2·7
Mar. 1	10·6	June 1	16·6	Sept. 1	22·7	Dec. 1	4·7

The local sidereal time is equal to the right ascension of the local meridian.

Data related to time-scales – *continued*

Differences ΔT, where ΔT = ET − UT before 1984, from 1984 to 2000 ΔT = TDT − UT and ΔT = TT − UT from 2001 onwards:

	s		s		s		s		s
1900	− 2·3	1970	+ 40·7	2001	+ 67	2008	+ 74	2015	+ 82
1910	+ 10·9	1975	+ 46·0	2002	+ 68	2009	+ 75	2016	+ 83
1920	+ 21·7	1980	+ 51·0	2003	+ 69	2010	+ 76	2017	+ 84
1930	+ 24·0	1985	+ 54·6	2004	+ 70	2011	+ 77	2018	+ 85
1940	+ 24·5	1990	+ 57·2	2005	+ 71	2012	+ 78	2019	+ 86
1950	+ 29·4	1995	+ 61·2	2006	+ 72	2013	+ 79	2020	+ 87
1960	+ 33·4	2000	+ 64	2007	+ 73	2014	+ 81	2021	+ 88

The values for 2000 onwards are estimates based on an extrapolation of the current trend. Each value, which refers to a mean for the year, was used to express the times of the phenomena on pages 94-101 in UT.

Equation of time = apparent *minus* mean solar time:

		m			m			m			m
Jan.	1	− 3·2	Apr.	1	−4·1	July	1	−3·7	Oct.	1	+10·2
	15	− 8·9		15	−0·2		15	−5·9		15	+14·1
Feb.	1	−13·5	May	1	+2·9	Aug.	1	−6·3	Nov.	1	+16·4
	14	−14·2		15	+3·7		15	−4·6		15	+15·5
Mar.	1	−12·5	June	1	+2·3	Sept.	1	−0·2	Dec.	1	+11·2
	15	− 9·1		15	−0·3		15	+4·6		15	+ 5·1

Local mean time of transit of Sun = 12^h − equation of time.

Standard epochs for orbital elements of fundamental ephemerides:

1900 January 0 at 12^h ET = JD 241 5020·0

2000 January 1 at 12^h TDT = JD 245 1545·0

Beginning of Julian year:

J1900·0 = JD 241 5020·0 J1950·0 = JD 243 3282·5 J2000·0 = JD 245 1545·0

Dates may be expressed in years as a Julian epoch using:

$$\text{Julian epoch} = \text{J}\,[2000{\cdot}0 + (\text{JD} - 245\,1545{\cdot}0)/365{\cdot}25]$$

where JD is the Julian date.

Beginning of Besselian year:

B1900·0 = JD 241 5020·31 B1950·0 = JD 243 3282·42 B2000·0 = JD 245 1544·53

For some purposes dates may be expressed in years as a Besselian epoch using:

$$\text{Besselian epoch} = \text{B}\,[1900{\cdot}0 + (\text{JD} - 241\,5020{\cdot}313\,52)/365{\cdot}242\,198\,781]$$

where JD is the Julian date.

Precessional constants

In the following formulae for the precessional constants the time-variable T is expressed in Julian centuries from J2000·0, that is:

$$T = (\mathrm{JD} - 245\,1545{\cdot}0)/36525$$

For conversion of a calendar date to Julian date see, for example, the tabulations on pages 102–133. The formulae are only intended for use from 2001 to 2020.

Reduction of equatorial coordinates

Precessional constants for the reduction of equatorial spherical coordinates between the mean equinox and equator of date and those of the standard epoch of J2000·0 are given approximately by:

$$\zeta_A = 0^{\mathrm{h}}.042\,71\ T \qquad M = 0^{\mathrm{h}}.085\,42\ T$$

$$z_A = \zeta_A = 0^{\circ}.6406\ T \qquad = 1^{\circ}.2812\ T$$

$$\theta_A = 0^{\circ}.5568\ T \qquad N = 0^{\circ}.5568\ T$$

For reduction of right ascension and declination from J2000·0 (α_0, δ_0) to date (α, δ):

$$\cos\delta\ \sin(\alpha - z_A) = \cos\delta_0\ \sin(\alpha_0 + \zeta_A)$$

$$\cos\delta\ \cos(\alpha - z_A) = \cos\theta_A\ \cos\delta_0\ \cos(\alpha_0 + \zeta_A) - \sin\theta_A\ \sin\delta_0$$

$$\sin\delta = \cos\theta_A\ \sin\delta_0 + \sin\theta_A\ \cos\delta_0\ \cos(\alpha_0 + \zeta_A)$$

For reduction from date (α, δ) to J2000·0 (α_0, δ_0):

$$\cos\delta_0\ \sin(\alpha_0 + \zeta_A) = \cos\delta\ \sin(\alpha - z_A)$$

$$\cos\delta_0\ \cos(\alpha_0 + \zeta_A) = \cos\theta_A\ \cos\delta\ \cos(\alpha - z_A) + \sin\theta_A\ \sin\delta$$

$$\sin\delta_0 = \cos\theta_A\ \sin\delta - \sin\theta_A\ \cos\delta\ \cos(\alpha - z_A)$$

Alternatively:

$$\alpha - \alpha_0 = M + N\ \sin\tfrac{1}{2}(\alpha + \alpha_0)\ \tan\tfrac{1}{2}(\delta + \delta_0)$$

$$\delta - \delta_0 = N\ \cos\tfrac{1}{2}(\alpha + \alpha_0)$$

where the right-hand sides are evaluated by successive approximations, if necessary.

Reduction of ecliptic coordinates and orbital elements

Precessional constants for the reduction of ecliptic spherical coordinates and orbital elements between the mean equinox and ecliptic of date and those of the standard epoch of J2000·0 are given by:

$$a = 1^{\circ}.3970\,T \qquad c = 5^{\circ}.1236 + 0^{\circ}.2416\,T$$

$$b = 0^{\circ}.0131\,T \qquad c' = 5^{\circ}.1236 - 1^{\circ}.1554\,T$$

For reduction of ecliptic longitude and latitude from J2000·0 (λ_0, β_0) to date (λ, β):

$$\lambda = \lambda_0 + a - b\ \cos(\lambda_0 + c)\ \tan\beta$$

$$\beta = \beta_0 + b\ \sin(\lambda_0 + c)$$

For reduction from date (λ, β) to J2000·0 (λ_0, β_0):

$$\lambda_0 = \lambda - a + b\ \cos(\lambda + c')\ \tan\beta_0$$

$$\beta_0 = \beta - b\ \sin(\lambda + c')$$

Precessional constants – *continued*

Reduction of ecliptic coordinates and orbital elements – continued

For reduction of orbital elements (Ω = longitude of node, ω = argument of perihelion, i = inclination to ecliptic) from J2000·0 (Ω_0, ω_0, i_0) to date (Ω, ω, i):

$$\Omega = \Omega_0 + a - b \sin(\Omega_0 + c) \cot i$$
$$\omega = \omega_0 + b \sin(\Omega_0 + c) \operatorname{cosec} i$$
$$i = i_0 + b \cos(\Omega_0 + c)$$

For reduction from date (Ω, ω, i) to J2000·0 (Ω_0, ω_0, i_0):

$$\Omega_0 = \Omega - a + b \sin(\Omega + c') \cot i_0$$
$$\omega_0 = \omega - b \sin(\Omega + c') \operatorname{cosec} i_0$$
$$i_0 = i - b \cos(\Omega + c')$$

These formulae should not be used when the position is too close to the ecliptic pole.

Note that when i is small; $\Omega + \omega = \Omega_0 + \omega_0 + a$

Reduction of rectangular coordinates

The elements of the precessional matrix **P** for the reduction of equatorial rectangular coordinates from mean equinox and equator of J2000·0 to those of date are given by the following numerical formulae, and values for relevant 400-day dates are given on the next page.

$$P_{11} - 1{\cdot}0 = (-297\,24\,T^2 - 13\,T^3) \times 10^{-8}$$
$$P_{12} = -P_{21} = (-22\,361\,72\,T - 677\,T^2 + 222\,T^3) \times 10^{-8}$$
$$P_{13} = -P_{31} = (-9717\,17\,T + 207\,T^2 + 96\,T^3) \times 10^{-8}$$
$$P_{22} - 1{\cdot}0 = (-250\,02\,T^2 - 15\,T^3) \times 10^{-8}$$
$$P_{23} = P_{32} = (-108\,65\,T^2) \times 10^{-8}$$
$$P_{33} - 1{\cdot}0 = (-47\,21\,T^2) \times 10^{-8}$$

The elements of the inverse rotation matrix ($\mathbf{P}^{-1}$) for the reduction from date to J2000·0 are the same as those of the transpose ($\mathbf{P}'$) of **P**. Thus the reduction formulae are:

$$\boldsymbol{r} = \mathbf{P}\,\boldsymbol{r}_0 \qquad \text{and} \qquad \boldsymbol{r}_0 = \mathbf{P}'\,\boldsymbol{r}$$

where **r** is the column vector of the coordinates x, y, z for date and $\mathbf{r}_0$ is the vector for J2000·0. Thus:

$$x = P_{11}x_0 + P_{12}y_0 + P_{13}z_0 \qquad x_0 = P_{11}x + P_{21}y + P_{31}z$$
$$y = P_{21}x_0 + P_{22}y_0 + P_{23}z_0 \qquad y_0 = P_{12}x + P_{22}y + P_{32}z$$
$$z = P_{31}x_0 + P_{32}y_0 + P_{33}z_0 \qquad z_0 = P_{13}x + P_{23}y + P_{33}z$$

The elements of the precession matrix **P** for the reduction from B1950·0 to J2000·0 are given by:

$$P_{11} = +0{\cdot}999\,925\,71 \quad P_{12} = -0{\cdot}011\,178\,94 \quad P_{13} = -0{\cdot}004\,859\,00$$
$$P_{21} = +0{\cdot}011\,178\,94 \quad P_{22} = +0{\cdot}999\,937\,51 \quad P_{23} = -0{\cdot}000\,027\,16$$
$$P_{31} = +0{\cdot}004\,859\,00 \quad P_{32} = -0{\cdot}000\,027\,16 \quad P_{33} = +0{\cdot}999\,988\,19$$

Precessional constants – *continued*

Reduction of rectangular coordinates – continued

Elements of precession matrix **P** for reduction from J2000·0 to date

JD	245 1600·5	245 2000·5	245 2400·5	245 2800·5	245 3200·5
P_{11}	+ 1·000 000 00	+ 0·999 999 95	+ 0·999 999 84	+ 0·999 999 65	+ 0·999 999 39
P_{12}	− 0·000 033 98	− 0·000 278 87	− 0·000 523 77	− 0·000 768 66	− 0·001 013 56
P_{13}	− 0·000 014 77	− 0·000 121 18	− 0·000 227 60	− 0·000 334 01	− 0·000 440 43
P_{21}	+ 0·000 033 98	+ 0·000 278 87	+ 0·000 523 77	+ 0·000 768 66	+ 0·001 013 56
P_{22}	+ 1·000 000 00	+ 0·999 999 96	+ 0·999 999 86	+ 0·999 999 70	+ 0·999 999 49
P_{23}	− 0·000 000 00	− 0·000 000 02	− 0·000 000 06	− 0·000 000 13	− 0·000 000 22
P_{31}	+ 0·000 014 77	+ 0·000 121 18	+ 0·000 227 60	+ 0·000 334 01	+ 0·000 440 43
P_{32}	− 0·000 000 00	− 0·000 000 02	− 0·000 000 06	− 0·000 000 13	− 0·000 000 22
P_{33}	+ 1·000 000 00	+ 0·999 999 99	+ 0·999 999 97	+ 0·999 999 94	+ 0·999 999 90

JD	245 3600·5	245 4000·5	245 4400·5	245 4800·5	245 5200·5
P_{11}	+ 0·999 999 06	+ 0·999 998 66	+ 0·999 998 18	+ 0·999 997 64	+ 0·999 997 02
P_{12}	− 0·001 258 46	− 0·001 503 36	− 0·001 748 26	− 0·001 993 17	− 0·002 238 07
P_{13}	− 0·000 546 84	− 0·000 653 26	− 0·000 759 67	− 0·000 866 08	− 0·000 972 49
P_{21}	+ 0·001 258 46	+ 0·001 503 36	+ 0·001 748 26	+ 0·001 993 17	+ 0·002 238 07
P_{22}	+ 0·999 999 21	+ 0·999 998 87	+ 0·999 998 47	+ 0·999 998 01	+ 0·999 997 50
P_{23}	− 0·000 000 34	− 0·000 000 49	− 0·000 000 66	− 0·000 000 86	− 0·000 001 09
P_{31}	+ 0·000 546 84	+ 0·000 653 26	+ 0·000 759 67	+ 0·000 866 08	+ 0·000 972 49
P_{32}	− 0·000 000 34	− 0·000 000 49	− 0·000 000 66	− 0·000 000 86	− 0·000 001 09
P_{33}	+ 0·999 999 85	+ 0·999 999 79	+ 0·999 999 71	+ 0·999 999 62	+ 0·999 999 53

JD	245 5600·5	245 6000·5	245 6400·5	245 6800·5	245 7200·5
P_{11}	+ 0·999 996 34	+ 0·999 995 58	+ 0·999 994 75	+ 0·999 993 85	+ 0·999 992 87
P_{12}	− 0·002 482 98	− 0·002 727 89	− 0·002 972 80	− 0·003 217 71	− 0·003 462 62
P_{13}	− 0·001 078 90	− 0·001 185 32	− 0·001 291 73	− 0·001 398 14	− 0·001 504 55
P_{21}	+ 0·002 482 98	+ 0·002 727 89	+ 0·002 972 80	+ 0·003 217 71	+ 0·003 462 62
P_{22}	+ 0·999 996 92	+ 0·999 996 28	+ 0·999 995 58	+ 0·999 994 82	+ 0·999 994 01
P_{23}	− 0·000 001 34	− 0·000 001 62	− 0·000 001 92	− 0·000 002 25	− 0·000 002 60
P_{31}	+ 0·001 078 90	+ 0·001 185 32	+ 0·001 291 73	+ 0·001 398 14	+ 0·001 504 55
P_{32}	− 0·000 001 34	− 0·000 001 62	− 0·000 001 92	− 0·000 002 25	− 0·000 002 60
P_{33}	+ 0·999 999 42	+ 0·999 999 30	+ 0·999 999 17	+ 0·999 999 02	+ 0·999 998 87

JD	245 7600·5	245 8000·5	245 8400·5	245 8800·5	245 9200·5
P_{11}	+ 0·999 991 83	+ 0·999 990 71	+ 0·999 989 53	+ 0·999 988 27	+ 0·999 986 94
P_{12}	− 0·003 707 54	− 0·003 952 45	− 0·004 197 37	− 0·004 442 29	− 0·004 687 21
P_{13}	− 0·001 610 95	− 0·001 717 36	− 0·001 823 77	− 0·001 930 18	− 0·002 036 58
P_{21}	+ 0·003 707 54	+ 0·003 952 45	+ 0·004 197 37	+ 0·004 442 29	+ 0·004 687 21
P_{22}	+ 0·999 993 13	+ 0·999 992 19	+ 0·999 991 19	+ 0·999 990 13	+ 0·999 989 02
P_{23}	− 0·000 002 99	− 0·000 003 39	− 0·000 003 83	− 0·000 004 29	− 0·000 004 77
P_{31}	+ 0·001 610 95	+ 0·001 717 36	+ 0·001 823 77	+ 0·001 930 18	+ 0·002 036 58
P_{32}	− 0·000 002 99	− 0·000 003 39	− 0·000 003 83	− 0·000 004 29	− 0·000 004 77
P_{33}	+ 0·999 998 70	+ 0·999 998 53	+ 0·999 998 34	+ 0·999 998 14	+ 0·999 997 93

Nutation

The nutations in longitude ($\Delta\psi$) and in obliquity ($\Delta\epsilon$) are given by the following expressions to a precision of about 0°0002 (1″) during the period 2001-2020:

$$\begin{aligned}\Delta\psi &= -\,0^\circ\!.0048 \sin(125^\circ\!.0 - 0^\circ\!.052\,95\,d)\\ &\quad -\,0^\circ\!.0004 \sin(200^\circ\!.9 + 1^\circ\!.971\,29\,d)\\ \Delta\epsilon &= +\,0^\circ\!.0026 \cos(125^\circ\!.0 - 0^\circ\!.052\,95\,d)\\ &\quad +\,0^\circ\!.0002 \cos(200^\circ\!.9 + 1^\circ\!.971\,29\,d)\end{aligned}$$

where $d = \mathrm{JD} - 245\,1545{\cdot}0$

The corresponding corrections in longitude ($\Delta\lambda$), latitude ($\Delta\beta$), right ascension ($\Delta\alpha$) and declination ($\Delta\delta$) are given by:

$$\begin{aligned}\Delta\lambda &= \Delta\psi\\ \Delta\beta &= 0\\ \Delta\alpha &= (\cos\epsilon + \sin\epsilon\,\sin\alpha\,\tan\delta)\,\Delta\psi - \cos\alpha\,\tan\delta\,\Delta\epsilon\\ \Delta\delta &= \sin\epsilon\,\cos\alpha\,\Delta\psi + \sin\alpha\,\Delta\epsilon\end{aligned}$$

where $\epsilon = 23^\circ\!.44 \qquad \cos\epsilon = 0{\cdot}917 \qquad \sin\epsilon = 0{\cdot}398$

Equatorial rectangular coordinates (x, y, z) referred to the mean equinox of date can be referred to the true equinox of date by the addition of the corrections:

$$\begin{aligned}\Delta x &= -\,y\,\Delta\psi\,\cos\epsilon - z\,\Delta\psi\,\sin\epsilon\\ \Delta y &= +\,x\,\Delta\psi\,\cos\epsilon - z\,\Delta\epsilon\\ \Delta z &= +\,x\,\Delta\psi\,\sin\epsilon + y\,\Delta\epsilon\end{aligned}$$

The elements of the corresponding rotation matrix are:

$$\begin{matrix} 1 & -\,\Delta\psi\,\cos\epsilon & -\,\Delta\psi\,\sin\epsilon\\ +\,\Delta\psi\,\cos\epsilon & 1 & -\,\Delta\epsilon\\ +\,\Delta\psi\,\sin\epsilon & +\,\Delta\epsilon & 1\end{matrix}$$

where in the matrix $\Delta\psi$ and $\Delta\epsilon$ are expressed in radians.

Aberration

The corrections ($\Delta\alpha$, $\Delta\delta$) to the right ascension (α) and declination (δ) of a star, in the sense "apparent place *minus* mean place", due to annual aberration are given by:

$$\begin{aligned}\Delta\alpha &= -\,0^\circ\!.0057\,(\sin\lambda\,\sin\alpha + \cos\lambda\,\cos\epsilon\,\cos\alpha)\,\sec\delta\\ \Delta\delta &= -\,0^\circ\!.0057\,(\sin\lambda\,\cos\alpha\,\sin\delta - \cos\lambda\,(\cos\epsilon\,\sin\alpha\,\sin\delta - \sin\epsilon\,\cos\delta))\end{aligned}$$

where $\lambda = 280^\circ\!.5 + 0^\circ\!.985\,65\,d, \quad \epsilon = 23^\circ\!.44 \quad \cos\epsilon = 0{\cdot}917 \quad \sin\epsilon = 0{\cdot}398{\cdot}$

The total corrections to the geocentric spherical coordinates of a planet due to aberration are given by:

apparent position = geometric position − light-time × rate of change

where the rate of change may be taken to be the first difference divided by the interval of tabulation. The light-time is tabulated with the right ascension and declination; it is the distance (ρ) in au divided by 173·145 au/d.

Interpolation formulae

The following notation is used:

f_n, where n is integral (..., −1, 0, 1, 2, ...), denotes the value of a function $f(t)$ at the tabular argument $(t_0 + nw)$, where w is the interval.

f_p where $0 < p < 1$ unless otherwise specified, denotes the required value of $f(t)$ at a non-tabular value of the argument, $t_0 + pw$.

δ denotes the central-difference operator, which is such that:

$$\delta f_m = f_{m+\frac{1}{2}} - f_{m-\frac{1}{2}}, \qquad \delta^2 f_m = f_{m+1} - 2 f_m + f_{m-1}, \quad \text{etc,}$$

where m is integral for even-order differences and half-integral for odd-order differences.

The f is omitted when there is no ambiguity; for example:

$$\delta_{\frac{1}{2}} = f_1 - f_0, \quad \delta_0^2 = f_1 - 2 f_0 + f_{-1}, \quad \delta_{\frac{1}{2}}^3 = f_2 - 3 f_1 + 3 f_0 - f_{-1}$$

Linear interpolation — maximum error about $0{\cdot}12\ \delta^2$

$$f_p = f_0 + p\,\delta f_{\frac{1}{2}} = (1-p)\,f_0 + p\,f_1$$

Quadratic interpolation — maximum error about $0{\cdot}008\ \delta^3$

$$f_p = f_0 + p\,\delta_{\frac{1}{2}} + B_2\,(\delta_0^2 + \delta_1^2), \quad \text{where}\ B_2 = \tfrac{1}{4}\,p\,(p-1)$$

$$f_{\frac{1}{2}} = (f_0 + f_1)/2 - 0{\cdot}0625\,(\delta_0^2 + \delta_1^2) + \cdots$$

Cubic interpolation — maximum error about $0{\cdot}02\ \delta^4$

$$f_p = f_0 + p\,\delta_{\frac{1}{2}} + G_2\,\delta_0^2 + G_3\,\delta_{\frac{1}{2}}^3, \quad \text{where} \quad G_2 = \tfrac{1}{2}\,p\,(p-1), \quad G_3 = \tfrac{1}{6}\,(p+1)\,p\,(p-1)$$

$$= (1-p)\,f_0 + p\,f_1 + E_2\,\delta_0^2 + F_2\,\delta_1^2, \quad \text{where} \quad E_2 = G_2 - G_3, \quad F_2 = G_3$$

$$= L_{-1}\,f_{-1} + L_0\,f_0 + L_1\,f_1 + L_2\,f_2$$

$$f_{\frac{1}{2}} = (9\,(f_0 + f_1) - f_{-1} - f_{-2})/16 + 0{\cdot}0117\,(\delta_0^4 + \delta_1^4) - \cdots$$

where L_{-1}, L_0, L_1, L_2, are cubic polynomials in p. Their values may be calculated directly from the following formulae or, for certain values of p, taken from the accompanying table.

$$L_{-1} = -\tfrac{1}{6}\,p\,(p-1)\,(p-2) = -((p-3)\,p+2)\,p/6$$
$$L_0 = \tfrac{1}{2}\,(p+1)\,(p-1)\,(p-2) = (((p-2)\,p-1)\,p+2)/2$$
$$L_1 = -\tfrac{1}{2}\,(p+1)\,p\,(p-2) = -((p-1)\,p-2)\,p/2$$
$$L_2 = \tfrac{1}{6}\,(p+1)\,p\,(p-1) = (p^2-1)\,p/6$$

Interpolation coefficients for quarters and tenths

p	B_2	G_2	G_3	E_2	F_2	L_{-1}	L_0	L_1	L_2
0·25	− 0·0469	− 0·0938	− 0·0391	− 0·0547	− 0·0391	− 0·054 69	+ 0·820 31	+ 0·273 44	− 0·039 06
0·50	− 0·0625	− 0·1250	− 0·0625	− 0·0625	− 0·0625	− 0·062 50	+ 0·562 50	+ 0·562 50	− 0·062 50
0·75	− 0·0469	− 0·0938	− 0·0547	− 0·0391	− 0·0547	− 0·039 06	+ 0·273 44	+ 0·820 31	− 0·054 69
0·1	− 0·0225	− 0·0450	− 0·0165	− 0·0285	− 0·0165	− 0·028 50	+ 0·940 50	+ 0·104 50	− 0·016 50
0·2	− 0·0400	− 0·0800	− 0·0320	− 0·0480	− 0·0320	− 0·048 00	+ 0·864 00	+ 0·216 00	− 0·032 00
0·3	− 0·0525	− 0·1050	− 0·0455	− 0·0595	− 0·0455	− 0·059 50	+ 0·773 50	+ 0·331 50	− 0·045 50
0·4	− 0·0600	− 0·1200	− 0·0560	− 0·0640	− 0·0560	− 0·064 00	+ 0·672 00	+ 0·448 00	− 0·056 00
0·5	− 0·0625	− 0·1250	− 0·0625	− 0·0625	− 0·0625	− 0·062 50	+ 0·562 50	+ 0·562 50	− 0·062 50
0·6	− 0·0600	− 0·1200	− 0·0640	− 0·0560	− 0·0640	− 0·056 00	+ 0·448 00	+ 0·672 00	− 0·064 00
0·7	− 0·0525	− 0·1050	− 0·0595	− 0·0455	− 0·0595	− 0·045 50	+ 0·331 50	+ 0·773 50	− 0·059 50
0·8	− 0·0400	− 0·0800	− 0·0480	− 0·0320	− 0·0480	− 0·032 00	+ 0·216 00	+ 0·864 00	− 0·048 00
0·9	− 0·0225	− 0·0450	− 0·0285	− 0·0165	− 0·0285	− 0·016 50	+ 0·104 50	+ 0·940 50	− 0·028 50